Pacific Particle Physics Phenomenology

1st APCTP Workshop
Pacific Particle Physics Phenomenology
Oct. 31 ~ Nov. 2. 1997. SNU. Seoul

Proceedings of the APCTP Workshop

Pacific Particle Physics Phenomenology

Editors

M. Drees
Asia Pacific Center for Theoretical Physics, Korea

C. S. Kim
Yonsei University, Korea

S. K. Kim
Seoul National University, Korea

World Scientific
Singapore • New Jersey • London • Hong Kong

Published by

World Scientific Publishing Co. Pte. Ltd.

P O Box 128, Farrer Road, Singapore 912805

USA office: Suite 1B, 1060 Main Street, River Edge, NJ 07661

UK office: 57 Shelton Street, Covent Garden, London WC2H 9HE

British Library Cataloguing-in-Publication Data
A catalogue record for this book is available from the British Library.

PACIFIC PARTICLE PHYSICS PHENOMENOLOGY
Proceedings of the APCTP Workshop

ISBN 981-02-3623-9

Printed in Singapore.

Editors' Preface

The "Pacific Particle Physics Phenomenology" Workshop, organized by the Asia Pacific Center for Theoretical Physics (APCTP), and held in Seoul, Korea from October 31 to November 2 1997, was attended by almost one hundred physicists from Asia, Europe, and North America. Plenary sessions focussed on overviews of recent progress in particle physics phenomenology in both theory and experiment. The parallel sessions provided opportunities for specialized and lively discussions among experts; very importantly, they also gave many young phenomenologists from the Asia–Pacific region a chance to actively participate in an international meeting.

We thank Prof. Hagiwara of KEK, who originally proposed this meeting, the members of the local organizing committee[a] chaired by Prof. J.B. Choi, and the members of the international advisory committee[b]; they all spent much time and effort to ensure that the scientific program would be of high quality, and of interest to nearly the entire particle physics community in the region. We also thank the APCTP secretaries for their smooth running of the administrative side of the meeting. Last but not least, we thank all the speakers for their comprehensive and illuminating lectures and seminars.

Finally we thank the APCTP for their generous financial support, the JSPS-KOSEF exchange program for their support of many Japanese attendants, and the DFG-KOSEF exchange program as well as the CTP at Seoul National University for their support of a German participant.

Manuel Drees
Choong Sun Kim
Sun Kee Kim

[a] J. Chay, J.B. Choi (Chair), K. Choi, S.Y. Choi, M. Drees, K. Hagiwara, S.D. Hwang, C.S. Kim, S.K. Kim, Y.Y. Keum, P. Ko, W. Namgung

[b] V. Barger, H.Y. Cheng, J. Ellis, T. Huang, K.S. Kang, C.W. Kim, M. Kobayashi, B. McKellar, S. Olsen, Probir Roy, H.S. Song

Contents

THE STATUS OF AND PHYSICS AT BEPC/BES

WEIGUO LI[a]
Institute of High Energy Physics, Chinese Academy of Sciences
19 Yuquan Road, P. O. Box 918, Beijing 100039, China

The status of the Beijing Electron Positron Collider (BEPC) / Beijing Spectrometer (BES) / Beijing Synchrotron Radiation Facility (BSRF) is outlined, and important results obtained at BES so far are summarized. A proposal for promoting High Energy Physics (HEP) in China in the 21th century is introduced.

1 History of BEPC/BES/BSRF

In April of 1983, the project of the Beijing Electron Positron Collider (BEPC) / Beijing Spectrometer (BES) was approved by the Chinese government. BEPC/ BES were in construction from 1984 to 1988. In October of 1988, BEPC, which operates in the c.m. energy region of 3.0-5.6 GeV, and BES became operational, and shortly afterward the first collision at E_{beam} = 1.6 GeV with L = $8 \times 10^{28} cm^{-2}s^{-1}$ was realized and events were observed by BES. In May of 1989, the nominal luminosity value, L=$7 \times 10^{30} cm^{-2}s^{-1}$ at E_{beam} = 2.2 GeV was reached. In 1991 the Beijing Synchrotron Radiation Facility (BSRF) attached to BEPC also became operational and opened to public usage. These achievements were the result of the efforts by the Chinese high energy physics community for many years and the generous support of the Chinese government. It is the first Chinese experimental base for high energy physics, and it has become a research center for τ-charm physics in China.

2 Advantages of BEPC/BES

Because it was working at a series of resonances and particle production thresholds, BEPC/BES had many functional advantages: high event rates, small backgrounds, adequate measurement precision, with a luminosity 4 times higher than that of SPEAR at SLAC in the U.S. . Its main features are:

BEPC has: a higher luminosity: L = $2.6 \times 10^{30} cm^{-2}s^{-1}$ at E_{cm} = 3.1 GeV; a small beam energy spread: δE = 1.4 MeV at E_{cm}=4 GeV; and an energy region covering the charm and τ thresholds.

BES has: adequate momentum and energy measurement precision; a good dE/dx resolution $\sim$ 8%; a large solid angle coverage ($4\pi \times 95\%$ for layer 2 of the main drift chamber (MDC), end cap time of flight (TOF) and shower

[a] liwg@bepc5.ihep.ac.cn

counter (SC)); a good position resolution of SC; a good π/μ identification; and a neutral trigger.

3 Physics Program at BES

Since 1990, BES has accumulated data at different energies for J/ψ, ψ', D_s and τ studies as listed in Table 1. A dedicated energy scan was done in the vicinity of the τ pair production threshold for measuring the τ mass. Based on these data, some important results were obtained.

Table 1, Data Collected at BES

J/ψ	9.0×10^6 events
ψ'	3.6×10^6 events
D_s(4.03 GeV)	22.3 pb^{-1}
τ (near thres.)	5 pb^{-1}
τ (4.03 GeV)	1.0×10^5 events

3.1 Precise measurement of the mass of the τ lepton

By far the most precise measurement of the mass of the τ lepton was made by the threshold scanning method at BES[1,2]. Detailed descriptions of τ scanning are in Ref. 1 and 2. In 1992, the first precise τ mass measurement using only the e/μ channels to identify τ pair was published. In 1996, based on a 5 pb^{-1} τ scanning data sample and more decay channels, a better value was obtained,

$$m_\tau = 1776.96^{+0.18+0.20}_{-0.19-0.16} MeV/c^2$$

The measurement accuracy was improved by a factor of 10 compared with previous measurements, and more importantly, it corrected the central value by 7.1 MeV/c^2. Combining with the most recent measurements[3] of the τ life time of 290.55 ± 1.06 fs and the τ electronic decay branching fraction of 17.786 ± 0.072, lepton universality is found to hold very well.

3.2 Search for glueballs

In the picture of QCD, the force between two quarks is *mediated* by the exchange of gluons, of which there are 8 kinds. The direct gluon-gluon coupling allows the possible existence of glueballs, i.e. bound states of interacting gluons without quarks on the scene at all. Many glueball candidates were reported, but their existence has not yet been firmly established. At BES, two glueball candidate states, $\xi(2230)$ and f(1710) have been carefully studied.

ξ(2230)

It was first observed by MarkIII[4] in the decays

$$J/\psi \to \gamma K^+K^-, \gamma K^0_S K^0_S,$$

but DM2[5] didn't find this signal. BES confirmed the existence of $\xi(2230)$ in $J/\psi \to \gamma K^+K^-, \gamma K^0_S K^0_S$, and found its new decay channels[6] as shown in Fig. 1 and Table 2. From these results, $\xi(2230)$ decays have three features: (1) flavor symmetry; (2) narrow width; and (3) large production rate in J/ψ radiative decays, which strongly favor the glueball interpretation.

Recently BES has studied $\xi(2230)$ in J/ψ radiative decays to $\pi^0\pi^0, \eta\eta, \eta\eta'$, and $\eta'\eta'$. The preliminary results for these decay modes are,

$$Br(J/\psi \to \gamma\xi) \cdot Br(\xi \to \pi^0\pi^0) = (4.8 \pm 2.5 \pm 1.4) \times 10^{-5},$$

with $M_\xi = (2245 \pm 31)$ MeV. Here $\Gamma_\xi = 20$ MeV is fixed in the fitting because the low statistics. The 90% C. L. upper limits on other two decay channels are:

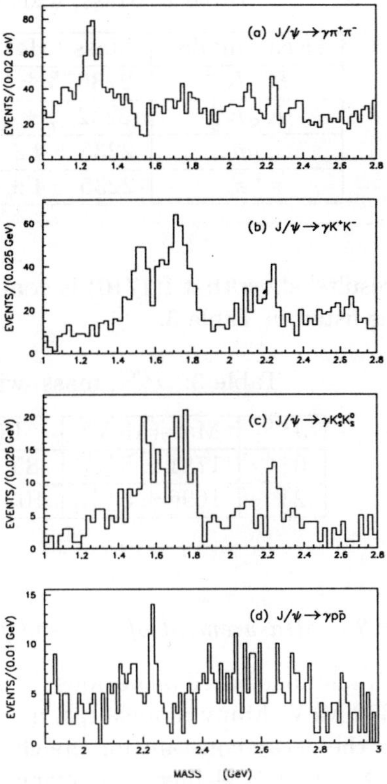

Fig.1: Spectra of J/ψ decay to $\gamma K^+K^-, \gamma K^0_S K^0_S, \gamma\pi^+\pi^-$ and $\gamma p\bar{p}$.

$$Br(J/\psi \to \gamma\xi) \cdot Br(\xi \to \eta\eta) < 5.6 \times 10^{-4};$$
$$Br(J/\psi \to \gamma\xi) \cdot Br(\xi \to \eta\eta') < 2.2 \times 10^{-3}.$$

It is very important to understand the nature of $\xi(2230)$[7]. More J/ψ data will be very valuable to confirm the results and to improve the measurement accuracy.

f(1710)

It was observed by a few experiments, but the reported values of its mass and its spin parity were inconsistent. By using the moment analysis method, BES's

Table 2: Mass, width, and branching ratios of $\xi(2230)$

Decay modes	Mass (MeV)	Γ (MeV)	$\mathrm{Br}(J/\psi \to \gamma X)(10^{-5})$
K^+K^-	$2230^{+6}_{-7} \pm 16$	$20^{+20}_{-15} \pm 17$	$3.3^{+1.6}_{-1.3} \pm 1.2$
$K^0_S K^0_S$	$2232^{+8}_{-7} \pm 15$	$20^{+25}_{-16} \pm 14$	$2.7^{+1.1}_{-0.9} \pm 0.8$
$p\bar{p}$	$2235 \pm 4 \pm 5$	$15^{+12}_{-9} \pm 9$	$1.5^{+0.6}_{-0.5} \pm 0.5$
$\pi^+\pi^-$	$2235 \pm 4 \pm 6$	$19^{+13}_{-11} \pm 12$	$5.6^{+1.8}_{-1.6} \pm 2.0$

results[8] show that f(1710) is composed of two components with different J^{PC}, as listed in Table 3.

Table 3: J^{PC}, mass, width and branching ratios of f(1710)

J^{PC}	Mass(MeV)	Γ(MeV)	$\mathrm{Br}(J/\psi \to \gamma K^+K^-)(10^{-4})$
0^{++}	$1781 \pm 8^{+10}_{-31}$	$85 \pm 24^{+22}_{-19}$	$0.8 \pm 0.1^{+0.3}_{-0.1}$
2^{++}	$1696 \pm 5^{+9}_{-34}$	$103 \pm 18^{+30}_{-11}$	$2.5 \pm 0.4^{+0.9}_{-0.4}$

3.3 *Measurement of f_{D_s} and f_D*

Predictions based on various theoretical models for f_{D_s} and f_D vary from 90 to 350 MeV. Many models can more reliably predict the ratio $f_{D_s} : f_D : f_B$ (f_B is the decay constant for the charged B meson). f_B relates to B−mixing and CKM matrix elements involving b quarks. BES has carried out the world's first direct, model independent measurements of f_{D_s} and f_D by a double tagging method[9]; the results are,

$$f_{D_s} = 430^{+150+40}_{-130-40} \text{ MeV}$$
$$f_D = 300^{+180+80}_{-150-40} \text{ MeV}$$

BES has also measured[10,11],

$$Br(D_s \to \phi\pi) = 3.9^{+5.1+1.8}_{-1.9-1.1}\%;$$
$$Br(D_s \to eX) = 7.7^{+5.7+2.4}_{-4.3-2.1}\%.$$

3.4 *Hadronic decays of $\psi(2s)$*

According to a PQCD prediction, the following ratio holds

$$r \equiv \frac{B(\psi(2s) \to h)}{B(J\psi \to h)} = \frac{B(\psi(2s) \to e^+e^-)}{B(J\psi \to e^+e^-)} = 14.7 \pm 2.3\%$$

Based on $3.6 \times 10^6 \psi(2s)$ events, BES has systematically studied hadronic decays of $\psi(2s)$[12]. 24 decay modes were investigated, of which 14 were studied by BES for the first time. The suppression of $\psi(2s) \to VP$ modes was confirmed and a similar suppression in VT modes, e.g. $\psi(2s) \to \omega f_2, \rho a_2$ was observed for the first time, as shown in Fig. 2.

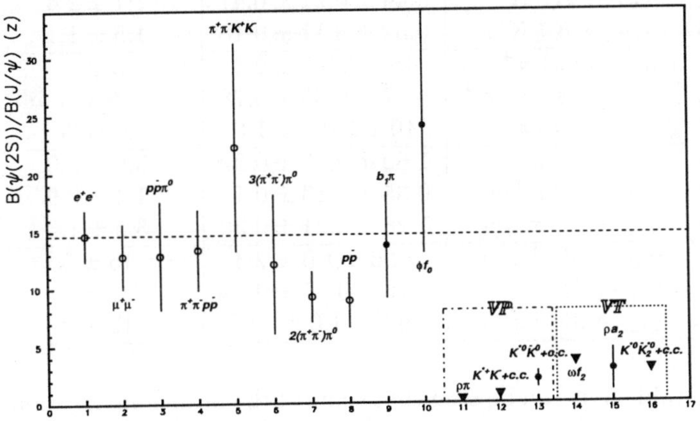

Figure 2: Ratio of decay branching fractions of $\psi(2s)$ to different modes over these of J/ψ.

Systematic studies of χ decay branching fractions were performed using the $\psi(2s)$ data. Table 4 lists branching fractions of some new χ decay modes. And some other measurements are improved in the accuracy compared with current PDG values, as listed in Table 5.

Table 4, Some new, preliminary χ branching fractions.

Decay mode	Branching Fraction
$\chi_{c0} \to p\bar{p}$	$(14.4 \pm 3.7 \pm 3.1) \times 10^{-5}$
$\chi_{c0} \to K_S^0 K_S^0$	$(2.39 \pm 0.14 \pm 0.41) \times 10^{-3}$
$\chi_{c2} \to K_S^0 K_S^0$	$(0.48 \pm 0.17 \pm 0.10) \times 10^{-3}$
$\chi_{c0} \to K^+K^-K^+K^-$	$(2.70 \pm 0.26 \pm 0.46) \times 10^{-3}$
$\chi_{c1} \to K^+K^-K^+K^-$	$(0.38 \pm 0.16 \pm 0.12) \times 10^{-3}$
$\chi_{c2} \to K^+K^-K^+K^-$	$(1.64 \pm 0.25 \pm 0.31) \times 10^{-3}$
$\chi_{c0} \to K_S^0 K^+\pi^- + c.c.$	$< 0.83 \times 10^{-3}$
$\chi_{c1} \to K_S^0 K^+\pi^- + c.c.$	$(2.45 \pm 0.47 \pm 0.52) \times 10^{-3}$
$\chi_{c2} \to K_S^0 K^+\pi^- + c.c.$	$< 1.3 \times 10^{-3}$

Table 5, Some preliminary improved measurement of χ branching fractions, compared with PDG.

Decay mode	BES (in 10^{-3})	PDG (in 10^{-3})
$\chi_{c0} \to \pi^+\pi^-$	$4.27 \pm 0.23 \pm 0.60$	7.5 ± 2.1
$\chi_{c2} \to \pi^+\pi^-$	$1.52 \pm 0.17 \pm 0.29$	1.9 ± 1.0
$\chi_{c0} \to K^+K^-$	$3.44 \pm 0.21 \pm 0.47$	7.1 ± 2.4
$\chi_{c2} \to K^+K^-$	$0.52 \pm 0.11 \pm 0.18$	1.5 ± 1.1
$\chi_{c0} \to \pi^+\pi^-\pi^+\pi^-$	$15.6 \pm 0.5 \pm 2.4$	37 ± 7
$\chi_{c1} \to \pi^+\pi^-\pi^+\pi^-$	$4.77 \pm 0.35 \pm 0.78$	16 ± 5
$\chi_{c2} \to \pi^+\pi^-\pi^+\pi^-$	$10.1 \pm 0.5 \pm 1.6$	22 ± 5
$\chi_{c0} \to \pi^+\pi^-p\bar{p}$	$1.60 \pm 0.23 \pm 0.39$	5.0 ± 2.0
$\chi_{c1} \to \pi^+\pi^-p\bar{p}$	$0.35 \pm 0.13 \pm 0.12$	1.4 ± 0.9
$\chi_{c2} \to \pi^+\pi^-p\bar{p}$	$1.28 \pm 0.21 \pm 0.32$	3.3 ± 1.3
$\chi_{c0} \to 3(\pi^+\pi^-)$	$12.6 \pm 1.0 \pm 2.4$	15 ± 5
$\chi_{c1} \to 3(\pi^+\pi^-)$	$5.76 \pm 0.67 \pm 0.94$	22 ± 8
$\chi_{c2} \to 3(\pi^+\pi^-)$	$9.1 \pm 1.0 \pm 1.6$	12 ± 8

From Table 5, it can be seen that some of the branching fractions are much different from PDG values; these discrepancies should be resolved.

4 BSRF

BSRF is the only available light source with light energy ranging from UV/VUV to hard X-rays in China. At present, there are 11 experimental stations at 7 beamlines: topography, EXAFS, diffuse scattering, PES, diffraction, small angle scattering, fluorescence, high pressure diffraction, soft X-ray, lithography and biological spectroscopy. A few hundred users from about 80 institutes have used this facility since 1991, with a total dedicated beam time of 17900 hours. During these years, about 290 results have been published. The research topics are related to material science, condensed matter, atomic and molecular physics, biology, medicine, chemistry, enviromental science, micro-electronics, etc.

5 Upgrade and the future HEP in China

From 1993 to 1997, BEPC/BES/BSRF has been upgraded. The following achievements were obtained.

BEPC has: increased luminosity by a factor of 1.5 so far; significantly reduced beam associated backgrounds; realized the full energy injection from the linac at J/ψ energy; and improved the control and monitor system. BES II has: replaced the aluminum beam pipe by a beryllium one; replaced the central drift chamber (CDC) by a vertex chamber (VC) with 12 radical readout planes and a spatial resolution of about 80 μm; built a new MDC, new barrel TOF counters with an average time resolution of about 200 ps and new luminosity monitor counters; established a new on-line data acquisition system, reducing the average dead time for reading out an event from 23 to 10 ms; and modified barrel shower counter (BSC) HV distribution, with energy resolution of about $20\%/\sqrt{E}$ (GeV).

A serious problem prevents BES II from taking data at present. The feedthroughs of the new MDC, which were made in U.S., drew too much dark current under high voltage and a small fraction of them spark. The problem was traced to inappropriate temperature control during the molding process of the feedthroughs. A split HV scheme is in preparation to enable BES II to take data at the end of 1997. A new MDC will also have to be made.

BSRF has: upgraded the existing 7 beamlines and relevant 11 experimental stations; constructed two new beamlines with 3W1 multi-period wiggler magnets for soft and hard X-rays, which increased light intensity by a factor of 10.

To further develop HEP in China, a proposal to build the Beijing Tau–Charm Factory (BTCF), a high precision collider with $L_{peak} = 1\times10^{33}cm^{-2}s^{-1}$ and a detector with superior performance, was put forward in 1994. Since then, Chinese physicists have been actively involved in feasilibity studies of the BTCF. A series of international workshops on τcF concluded that a τcF is still a very valuable facility after the B factories at SLAC and KEK are built. In general, B factories and τcF are complementary. A τcF possesses many advantages for certain studies, and it is unique for some subjects in the field of τ-charm physics, such as the search for glueballs, and searches for mixing and CP violation in τ and charm decays. It was strongly suggested by the international HEP community that China should hold a leading position in the construction of a τcF, for several reasons:

- Because of the successful BEPC/BES project, a team of experienced physicists and engineers is in place. A τcF is a natural extension of the BEPC/BES.

- Some of the existing facilities at BEPC can be used, such as the linac and infrastructure.

- Since a τcF costs much less than a machine at higher energy, it is more suitable for China as a developing country with limited resources.

From 1995 to 1996, a BTCF feasibility study was carried out[13]. Based on the study of previous designs and developments of similar machines and detectors, on theoretical calculations and simulations, and on some small scale tests, the following has been achieved:

- The physics objective of BTCF has been understood more deeply and extensively, its expected reach has been studied and the requirements for the machine and detector defined. The key technologies to meet the above-mentioned specifications have been recognized and as a result, various schemes for dealing with specific problems have been worked out.
- A reference design of the machine and detector has been formed and R&D items for completing the conceptual design specified.
- A framework for domestic collaboration has been formed, and an international collaboration in some key selected fields has been started.

Another possibility of future improvement of BEPC/BES is to further upgrade BEPC to a luminosity of $10^{32} cm^{-2} s^{-1}$ and corresponding BES subsystems to improve its performances. Now we are putting forward an application for an R&D project, which involves both BTCF and a further upgrade of BEPC/BES, to the Chinese government.

6 The measurement of R in 2-5 GeV with BES II

The QED running coupling constant $\alpha(s)$ and the anomalous magnetic moment of the muon a_μ are two fundamental quantities for precision tests of the Standard Model. The current uncertainties of $\alpha(s)$ and a_μ are dominated by the contribution from poor R value measurements, with uncertainties in the range of 15-20% in the energy region of 1-5 GeV. Our plan is to measure R at c.m. energies between 2 to 5 GeV to a precision below 10%, and hopefully to reduce the measurement error to 5% with BES II. To reach such a precision, good performance of machine and detector are required. Recently a pre-study for an R measurement has being carried out by the R study group at BES, such as hadronic event selection, M.C. simulation, machine study at BEPC to find the achievable luminosity, especially in the energy region of 2-2.5 GeV, luminosity measurement with large angle Bhabha events, etc.

7 Summary

Only with the painstaking efforts of Chinese HEP physicists and engineers in the past decades, especially through the construction and operation of

BEPC/BES, China has acquired an existing experimental base, obtained some notable physics results, developed an experienced team of physicists and engineers, including the younger generation, and had strong support from the international and domestic HEP communities. In the coming century, challenge and hope co-exist for HEP in China. We will continue to win over the next step for developing HEP, BTCF or a further upgrade of BEPC/BES.

References

1. BES Collaboration, Phys. Rev. Lett. **69**(1992)3021
2. BES Collaboration, Phys. Rev. D**53**(1996)20
3. Weiguo Li, *The Tau Physics*, talk given at 18th Lepton-Photons Conference in HAmburg, July, 1997.
4. R.M. Baltrusait et al., Phys. Rev. Lett. **56**(1986)107
5. J. Augustin et al., Phys. Rev. Lett. **60**(1988)2238
6. BES Collaboration, Phys. Rev. Lett. **76**(1996)3502
7. J. Li, in Proceedings of International Conference on High Energy Physics, eds. Z. Ajduk and A.K.Wroblewski, World Scientific, 1996, pp.504-506
8. BES Collaboration, Phys. Rev. Lett. **77**(1996)3959
9. BES Collaboration, Phys. Rev. Lett. **74**(1995)4599
10. BES Collaboration, Phys. Rev. D**52**(1995)3781
11. BES Collaboration, Phys. Rev. D**56**(1997)3779
12. Y.S. Zhu, in Proceedings of International Conference on High Energy Physics, eds. Z. Ajduk and A.K.Wroblewski, World Scientific, 1996, pp.507-510
13. Feasibility Study Report on Beijing Tau-Charm Factory, IHEP-BTCF Report-03, October, 1996

SUPERSYMMETRIC THEORIES WITH EXPLICIT R–PARITY VIOLATION

PROBIR ROY
Tata Institute of Fundamental Research, Mumbai, India

1 R_p and motivation for explicit $\not{R}_p$

The superfield spectrum of the Minimal Supersymmetric Standard Model MSSM[1] can be written in usual notation as

$$Q_i = \begin{pmatrix} U_i \\ D_i \end{pmatrix}, \; L_i = \begin{pmatrix} N_i \\ E_i \end{pmatrix}, \; H_u = \begin{pmatrix} H_u^+ \\ H_u^0 \end{pmatrix}, \; H_d = \begin{pmatrix} H_d^0 \\ H_d^- \end{pmatrix}, \; \bar{U}_i, \bar{D}_i, \bar{E}_i$$

and its superpotential as

$$W_{MSSM} = h_{ij}^u Q_i \cdot H_i U_j + h_{ij}^d Q_i \cdot H_d \bar{D}_j + h_{ij}^e L_i \cdot H_d \bar{E}_j + \mu H_u \cdot H_d. \qquad (1)$$

(1) has been constructed so as to conserve R-parity (even for particles, odd for sparticles)

$$R_p \equiv (-1)^{3B+L+2s},$$

B, L and s being baryon no., lepton no. and spin. This is equivalent to the conservation of matter parity which transforms the superfields as

$$(L_i, \bar{E}_i, Q_i, \bar{U}_i, \bar{D}_i) \rightarrow -(L_i, \bar{E}_i, Q_i, \bar{U}_i, \bar{D}_i),$$

$$(H_u, H_d, V_\gamma, V_Z, V_W^+, V_W^-) \rightarrow (H_u, H_d, V_\gamma, V_Z, V_W^+, V_W^-).$$

Three spectacular consequences ensue from R_p-conservation. First, each vertex involving sparticles has them in a pair. Second, the lightest supersymmetric particle (LSP $\tilde{\chi}_1^0$: a weakly interacting neutralino, escaping through detectors, being neutral on cosmological grounds) is an excellent candidate for cold dark matter. Finally, each sparticle, pair-produced in a collider, would decay within the detector into particles and $\tilde{\chi}_1^0$, the latter being characterized by a hard ($\not{E}_T$) signature.

R_p–conservation in supersymmetric theories was phenomenologically postulated [2] by the need to avoid catastrophic proton decays induced by $\not{R}_p$ vertices. However, such decays need to violate both B and L. So, in place of R_p, it suffices to have another discrete symmetry such as B-parity transforming the superfields as

$$(Q_i, \bar{U}_i, \bar{D}_i) \rightarrow -(Q_i, \bar{U}_i, \bar{D}_i),$$

$$(L_i, \bar{E}_i, H_u, H_d, V_\gamma, V_Z, V_W^+, V_W^-) \to (L_i, \bar{E}_i, H_u, H_d, V_\gamma, V_Z, V_W^+, V_W^-).$$

Analogously, one can have an L-parity instead. Since our presently scanty knowledge of high scale theories is unable [3] to discriminate between various possible residual discrete symmetries at low energies, it is quite reasonable to consider [4] $\not{R}_p$ (but L- or B-conserving) interactions near weak-scale energies. In fact, supersymmetric grand unified theories have been constructed [5] which have $\not{R}_p$ interactions present at low energies. There are even stringy scenarios [6] which predict them. On the other hand, a spontaneous violation of R_p requires [7] an extra SM-singlet and hence a nonminimal spectrum; so we will restrict ourselves to explicit R_p-violation. The LSP, in this scenario, is neither stable nor necessarily neutral.

2 MSSM → $\not{R}_p$ MSSM

All renormalizable supersymmetric $\not{R}_p$ interactions are incorporated by the addition of new terms to the MSSM superpotential

$$W_{\not{R}_p} = \frac{1}{2}\lambda_{[ij]k} L_i \cdot L_j \bar{E}_k + \lambda'_{ijk} L_i \cdot Q_j \bar{D}_k + \frac{1}{2}\lambda''_{i[jk]} \bar{U}_i \bar{D}_j \bar{D}_k + \epsilon_i L_i \cdot H_u, \quad (2)$$

i, j, k being flavor indices and square brackets on them testifying to antisymmetry in the bracketted indices. The last RHS term in (2) can be "absorbed" in the μ-term of (1) by the transformation $H_d \to H'_d = (\mu^2 + \epsilon_j\epsilon_j)^{-1/2}(\mu H_d - \epsilon_i L_i)$. This will, in general, lead to $\not{L}$ terms in the scalar potential and, unless one is careful, can lead to nonzero sneutrino VEVs $\langle\tilde{\nu}_i\rangle \neq 0$. Moreover, the abovementioned "absorption" works at a certain energy scale. At a different energy the μ and ϵ_i parameters will be different, on account of separate renormalization group evolutions, leading to a "reappearance" of the $\epsilon_i L_i \cdot H_u$ terms in W. However, the effects of such terms are smaller [8] than the bounds on the λ-, λ'-, λ''-couplings that we discuss below and so henceforth we will ignore the $\epsilon_i L_i \cdot H_u$ piece in (2b).

We still have 9 λ-type, 27 λ'-type and 9 λ''-type, i.e. 45 possible, independent extra, couplings. The $\not{R}_p$ Lagrangian density reads

$$\begin{aligned}\mathcal{L}_{\not{R}_p} \;=\;& \lambda_{[ij]k}\Big[\tilde{\nu}_{iL}\bar{e}_{kR}e_{jL} + \tilde{e}_{jL}\bar{e}_{kR}\nu_{iL} + \tilde{e}^\star_{kR}\overline{(\nu_{iL})^C}e_{jL} - \tilde{\nu}_{jL}\bar{e}_{kR}e_{iL} \\ & -\tilde{e}_{iL}\bar{e}_{kR}\nu_{jL} + \tilde{e}^\star_{kR}\overline{(\nu_{jL})^C}e_{iL}\Big] + \lambda'_{ijk}\Big[\tilde{\nu}_{iL}\bar{d}_{kR}d_{jL} + \tilde{d}_{jL}\bar{d}_{kR}\nu_{iL} \\ & +\tilde{d}^\star_{kR}\overline{(\nu_{iL})^C}d_{jL} - \tilde{e}_{iL}\bar{d}_{kR}u_{jL} - \tilde{u}_{jL}\bar{d}_{kR}e_{jL} - \tilde{d}^\star_{kR}\overline{(e_{iL})^C}u_{jL}\Big] + \lambda''_{i[jk]}\epsilon_{\alpha\beta\gamma}\end{aligned}$$

$$\left[\tilde{u}^{\star}_{iR\alpha}\bar{d}_{kR\beta}d^{C}_{jR\gamma} + \tilde{d}_{jR\beta}\bar{d}_{kR\gamma}u^{C}_{iR\alpha} + \tilde{d}^{\star}_{kR\gamma}\overline{(u_{iR\alpha})^{C}}d_{jR\beta}\right] + h.c. \qquad (3)$$

Generally, except in a specific case, we take the couplings to be real. Moreover, in analogy with the Yukawa couplings of MSSM, we anticipate a generation hierarchy in their strengths with the third generation ones expected to be the largest. Furthermore, the simultaneous presence of λ or λ' *and* λ'' type of couplings is extremely unlikely given the lack of observation of either baryon nonconserving nucleon decay or double-nucleon annihilation in nuclei. Indeed, these constraints imply[8] $|\lambda'_{11k}\lambda''_{11k}| \lesssim 2 \times 10^{-27}\left(m_{\tilde{k}}/100 \text{ GeV}\right)^2$, $|\lambda_{233}\lambda''_{112}| \lesssim 10^{-14}\left(m_{\tilde{k}}/100 \text{ GeV}\right)^2$ and $|(\lambda_{...} \text{ or } \lambda'_{...})\lambda''_{...}| < 10^{-10}\left(m_{\tilde{k}}/100 \text{ GeV}\right)^2$, where $m_{\tilde{k}}$ is the mass of a squark of flavor k (generic in the case of 6c) and $\cdots$ stand for any any set of three flavor indices. Because (6) is so stringent, we will henceforth consider either L-violating and B-conserving λ, λ'-couplings *or* B-violating, L-conserving λ''-ones.

3 Bounds on $\not{R}_p$ couplings

In deriving bounds on the absolute values of couplings, we will assume that one $\not{R}_p$ operator acts at a time. Known upper bounds on various λ- and λ'' type of couplings, along with the physical process which is the source of each bound, are listed in Table 1. Specific references to the original derivation of these bounds may be found in the review by Bhattacharyya[10]. A similar listing has also been done, for whatever bounds that exist on the λ' couplings.

An interesting recent development has been the derivation[16] of bounds on products of different $\not{R}_p$ couplings that are more stringent than the products of the individual bounds available. These follow if, instead of assuming the action of only one operator, we allow the possibility of two different operators acting at two different vertices in a Feynman diagram. Some of these bounds are tabulated below with the corresponding processes cited. Additional results may be found in the recent papers of Jang et al[16].

4 Collider-specific phenomenology

Signatures for $\not{R}_p$ interactions in collider experiments involve two aspects of the concerned processes: first, production of some sparticles and second their specific decay modes. If sparticle pair-production is utilized, as is usually the case, only MSSM vertices are involved in the production process, whereas $\not{R}_p$ shows up through the sparticle decays that follow. Single sparticle production is, however, possible with $\not{R}_p$ vertices and can be exploited sometimes.

Table 1: Upper bounds on R_p couplings for $\tilde{m} = 100$ GeV. The numbers with ($\star$) correspond to 2σ limits and those with (‡) are basis-dependent limits.

PART A

Coupling	Bound	Source	Coupling	Bound	Source
$\lvert\lambda_{121}\rvert$	$0.05^\star$	CC univ.	$\lvert\lambda''_{112}\rvert$	10^{-6}	$NN \to K$'s
$\lvert\lambda_{122}\rvert$	$0.05^\star$	CC univ.	$\lvert\lambda''_{113}\rvert$	10^{-5}	$n\bar{n}$ osc.
$\lvert\lambda_{123}\rvert$	$0.05^\star$	CC univ.	$\lvert\lambda''_{123}\rvert$	1.25	Pert. unit.
$\lvert\lambda_{131}\rvert$	0.06	$\Gamma(\tau \to e\nu\bar{\nu})/\Gamma(\tau \to \mu\nu\bar{\nu})$	$\lvert\lambda''_{212}\rvert$	1.25	Pert. unit.
$\lvert\lambda_{132}\rvert$	0.06	$\Gamma(\tau \to e\nu\bar{\nu})/\Gamma(\tau \to \mu\nu\bar{\nu})$	$\lvert\lambda''_{213}\rvert$	125	Pert. unit.
$\lvert\lambda_{133}\rvert$	0.003	ν_e-mass	$\lvert\lambda''_{223}\rvert$	1.25	Pert. unit.
$\lvert\lambda_{231}\rvert$	0.06	$\Gamma(\tau \to e\nu\bar{\nu})/\Gamma(\tau \to \mu\nu\bar{\nu})$	$\lvert\lambda''_{312}\rvert$	10^{-5}	$n\bar{n}$ osc.
$\lvert\lambda_{232}\rvert$	0.06	$\Gamma(\tau \to e\nu\bar{\nu})/\Gamma(\tau \to \mu\nu\bar{\nu})$	$\lvert\lambda''_{313}\rvert$	10^{-5}	$n\bar{n}$ osc.
$\lvert\lambda_{233}\rvert$	0.06	$\Gamma(\tau \to e\nu\bar{\nu})/\Gamma(\tau \to \mu\nu\bar{\nu})$	$\lvert\lambda''_{323}\rvert$	0.50	Γ^Z_ℓ/Γ^Z_h

PART B

Coupling	Bound	Source	Coupling	Bound	Source	Coupling	Bound	Source
$\lvert\lambda'_{111}\rvert$	0.00035	$(\beta\beta)_{0\nu}$	$\lvert\lambda'_{211}\rvert$	0.09	R_π	$\lvert\lambda'_{311}\rvert$	0.10	$\tau^- \to \pi^-\nu_\tau$
$\lvert\lambda'_{112}\rvert$	$0.02^\star$	CC univ.	$\lvert\lambda'_{212}\rvert$	0.09	R_π	$\lvert\lambda'_{312}\rvert$	0.10	$\tau^- \to \pi^-\nu_\tau$
$\lvert\lambda'_{113}\rvert$	$0.02^\star$	CC univ.	$\lvert\lambda'_{213}\rvert$	0.09	R_π	$\lvert\lambda'_{313}\rvert$	0.10	$\tau^- \to \pi^-\nu_\tau$
$\lvert\lambda'_{121}\rvert$	$0.035^\star$	APV	$\lvert\lambda'_{221}\rvert$	0.18	D-decay	$\lvert\lambda'_{321}\rvert$	$0.20^\ddagger$	$D^0\bar{D}^0$ mix.
$\lvert\lambda'_{122}\rvert$	0.02	ν_e-mass	$\lvert\lambda'_{222}\rvert$	0.18	D-decay	$\lvert\lambda'_{322}\rvert$	$0.20^\ddagger$	$D^0\bar{D}^0$ mix.
$\lvert\lambda'_{123}\rvert$	$0.20^\ddagger$	$D^0 \leftrightarrow \bar{D}^0$	$\lvert\lambda'_{223}\rvert$	0.18	D-decay	$\lvert\lambda'_{323}\rvert$	$0.20^\ddagger$	$D^0\bar{D}^0$ mix.
$\lvert\lambda'_{131}\rvert$	$0.035^\star$	APV	$\lvert\lambda'_{231}\rvert$	$0.22^\star$	ν_μ d.i.	$\lvert\lambda'_{331}\rvert$	0.48	R_τ (LEP)
$\lvert\lambda'_{132}\rvert$	0.34	R_e (LEP)	$\lvert\lambda'_{232}\rvert$	0.36	R_μ	$\lvert\lambda'_{332}\rvert$	0.48	R_τ (LEP)
$\lvert\lambda'_{100}\rvert$	0.0007	ν_e-mass	$\lvert\lambda'_{233}\rvert$	0.36	R_μ	$\lvert\lambda'_{333}\rvert$	0.48	R_τ (LEP)

4.1 Tevatron → LHC

In such hadron colliders the possible production and decay-chains are

$$q\bar{q} \text{ or } gg \to \tilde{q}\tilde{q}^\star \to q\bar{q}\tilde{\chi}^0_1\tilde{\chi}^0_1, \quad \tilde{\chi}^0_1 \to \ell\bar{\ell}\nu \text{ or } q'\bar{q}\ell. \tag{4}$$

Whether $\tilde{\chi}^0_1$ decays largely into $\ell^+\ell^-\nu$ or two jets and a lepton depends on which of the operators $L_i \cdot L_j\bar{E}_k$ and $L_i \cdot Q_j\bar{D}_k$ in (2b) is dominant. These $\ell\ell\bar{\ell}\bar{\ell}$ $\not{E}_T$ or $jjjj\ell\ell$ signals consistute very characteristic signatures[17] of R_p. Another process involving R_p interactions in both production and decay is

$$g + d \to \tilde{u}_j + \ell^-, \quad \tilde{u}_j \to d_k + \bar{\ell}, \tag{5}$$

Table 2: Upper bounds on some important product couplings for $\tilde{m} = 100$ GeV.

Combination	Bound	Source	Combination	Bound	Source
$\lvert\lambda'_{11k}\lambda''_{11k}\rvert$	10^{-27}	Proton decay	$\lvert\lambda'_{ijk}\lambda''_{lmn}\rvert$	10^{-10}	Proton decay
$\lvert\lambda_{1j1}\lambda_{1j2}\rvert$	$7 \cdot 10^{-7}$	$\mu \to 3e$	$\lvert\lambda_{231}\lambda_{131}\rvert$	$7 \cdot 10^{-7}$	$\mu \to 3e$
$\lvert\text{Im } \lambda'_{i12}\lambda'^{*}_{i21}\rvert$	$8 \cdot 10^{-12}$	ϵ_K	$\lvert\lambda'_{i12}\lambda'_{i21}\rvert$	$1 \cdot 10^{-9}$	Δm_K
$\lvert\lambda'_{i13}\lambda'_{i31}\rvert$	$8 \cdot 10^{-8}$	Δm_B	$\lvert\lambda'_{1k1}\lambda'_{2k2}\rvert$	$8 \cdot 10^{-7}$	$K_L \to \mu e$
$\lvert\lambda'_{1k1}\lambda'_{2k1}\rvert$	$5 \cdot 10^{-8}$	$\mu\text{Ti} \to e\text{Ti}$	$\lvert\lambda'_{11j}\lambda'_{21j}\rvert$	$5 \cdot 10^{-8}$	$\mu\text{Ti} \to e\text{Ti}$
$\lvert\lambda''_{i32}\lambda''_{i21}\rvert$	0.008	$B^{\pm} \to K^0 K^{\pm}$	$\lvert\lambda''_{i31}\lambda''_{i21}\rvert$	0.006	$B^{\pm} \to K^0 \pi^{\pm}$

leading to a $\ell\bar{\ell}j$ final state. The lower limits of $O(250)$ GeV, derived from Tevatron data on squark and gluino masses in $MSSM$, get diluted to about $O(100-130)$ GeV in $R\!\!\!/_p$ $MSSM$. Another important point is that λ'-couplings make additional contributions to top semileptonic decay $t \to b\ell^+\nu$, leading to a violation of $e - \mu$ universality.

4.2 HERA

Resonant squark production in deep inelastic ep scattering is possible via $R\!\!\!/_p$ interactions, e.g.

$$\begin{aligned}
e^+ d &\to \tilde{u} \to (e^+ d_j, \bar{\nu} u_j, \tilde{\chi}^0_1, u_j, \tilde{\chi}^+_1 d_j), \\
e^- u &\to \tilde{d} \to (\bar{e} u_j, \bar{\nu}_e d_j, \tilde{\chi}^0_1 d_j, \tilde{\chi}^-_1 u_j), \\
\tilde{\chi}^0_1 &\to (\ell^{\pm} \text{ or } \nu) jj \text{ or } jj\ell, \\
\tilde{\chi}^{\pm}_1 &\to (\ell^{\pm} \text{ or } \nu) jj \text{ or } jj\ell,
\end{aligned}$$

leading to a variety of final states containing multileptons + $E\!\!\!/_T$ or multileptons + jets. For these signals to be observable, assuming that the squark masses are within the energy reach, one will need the relevant λ' couplings to be between[18]

$O(10^{-2})$ and $O(10^{-1})$ in magnitude. A point to note is that these are precisely the couplings that are relevant to atomic parity violation (cf. Table 1).

4.3 LEP 2 and NLC

In e^+e^- colliders one can look for LSP pair-production followed by $\not{R}_p$ decays:

$$e^+e^- \to \tilde{\chi}_1^0\tilde{\chi}_1^0, \quad \tilde{\chi}_1^0 \to \ell^+\ell^-\nu \text{ or } jj\ell. \tag{6}$$

Thus one can again have quadrilepton + $\not{E}_T$ or dilepton + multijet final states. Particularly interesting[14] are final states containing an isolated likesign ditau with multijets. These signals can be visible even if the concerned λ or λ' couplings are as low as $O(10^{-4})$.

5 Cosmological implications

5.1 Baryogenesis

Any baryon asymmetry, generated at temperatures pertaining to the GUT scale, is in danger of being washed out by sphaleron-induced processes if there are additional $\not{B} - \not{L}$ interactions at the electroweak scale. The requirement of the preservation of primordial baryon asymmetry was alleged to impose strong ($< 10^{-7}$) constraints on the strengths of $\not{R}_p$ interactions which violate $\not{B} - \not{L}$. However, the conservation of $\frac{1}{3}B - L_i$, where i is any $\not{B} - \not{L}$ *one* generation, has been shown[19] to suffice for this purpose – thereby leaving most of the $\not{R}_p$ couplings cosmologically unconstrained.

5.2 LSP decay

The simplest possible decay mode of the lightest neutralino is $\tilde{\chi}_1^0 \to \bar{e}u\bar{d}$ with the rate given by Dawson's formula[20] $\tau_{\tilde{\chi}_0}^{-1} = (128\pi^2)^{-1}3\alpha|\lambda'_{121}|^2 M_{\tilde{\chi}_1^0}^5 m_{\tilde{e}}^{-4}$. The requirement of the LSP decaying within a detector can be quantified as $c\gamma_L\tau_{\tilde{\chi}_0} < 1m$ and leads to the constraint

$$|\lambda'_{121}| > 1.4 \times 10^{-6}\sqrt{\gamma_L}\,(m_{\tilde{e}}/200 \text{ GeV})^2\,(100 \text{ GeV}/M_{\tilde{\chi}})^{5/2}\,, \tag{7}$$

where γ_L is the Lorentz boost factor. On the other hand, from the experimental absence of a long-lived relic LSP whose decay would lead to detectable upward going muons, one can infer[21] that $1 \text{ s} < \tau_{\tilde{\chi}} < 10^{17}$ yrs leading to the forbidden interval $10^{-10} \not< |\lambda, \lambda', \lambda''| \not< 10^{-22}$. Note that $\tau_{\tilde{\chi}} > 10^{17}$ yrs would make $\not{R}_p$ $MSSM$ indistinguishable from $MSSM$. Furthermore, the interval $10^{-8}s < \tau_{\tilde{\chi}_1} < 1s$ is practically out of observational reach. Of course, with an unstable LSP, one needs another candidate particle (axion?) for cold dark matter.

Summary and outlook

We conclude by emphasizing four main points. (1) There is presently no credible theoretical objection against the presence of $\not{R}_p$ interactions near the weak scale. (2) Though many of the $\not{R}_p$ couplings have been constrained in Tables 1-3, several (e.g. $\lambda'_{322}, \lambda'_{323}$ etc.) are totally unconstrained. (3) Vigorous seaches are in progress and detection strategies are under formulation for collider signatures of $\not{R}_p$ in terms of excess lepton-jet combinations at HERA, violation of $e - \mu$ universality in top semileptonic decay at the Tevatron and isolated ditaus and jets in LEP 2 and NLC. (4) It is highly desirable to find some way of discriminating between different CDM candidates such as a neutralino and an axion.

I compliment the P^4 workshop organizers for a job well-done.

References

1. H.E. Haber and G.L. Kane, *Phys. Rep.* **117**, 75 (1975).
2. G. Farrar and P. Fayet, *Phys. Lett.* B **76**, 575 (1978).
3. L.E. Ibanez and G.G. Ross, *Phys. Lett.* B **260**, 1991 (;) *Nucl. Phys.* B **368**, 3 (1992). T. Banks and M. Dine, *Phys. Rev.* D **45**, 1424 (1992).
4. L.J. Hall and M. Suzuki, *Nucl. Phys.* B **231**, 419 (1984).
5. D. Brahm and L.J. Hall, *Phys. Rev.* D **40**, 2449 (1989). K. Tamvakis, *Phys. Lett.* B **383**, 207 (1996). G.F. Giudice and R. Rattazzi, hep-ph/9704339.
6. M.C. Bento, L.J. Hall and G.G. Ross, *Nucl. Phys.* B **292**, 400 (1987).
7. C.S. Aulakh and R.N. Mohapatra, *Phys. Lett.* B **119**, 136 (1982). A. Santamaria and J.W.F. Valle, *Phys. Lett.* B **195**, 423 (1987).
8. R. Hempfling, *Nucl. Phys.* B **478**, 3 (1996). B. Mukhopadhay and S. Roy, *Phys. Rev.* D **55**, 7020 (1997).
9. J.L. Goity and M. Sher, *Phys. Lett.* B **346**, 69 (1995). A.Yu. Smirnov and F. Vissani, *Phys. Lett.* B **380**, 317 (1996).
10. V. Barger, G.F. Giudice and T. Han, *Phys. Rev.* D **40**, 2987 (1989). G. Bhattacharyya, hep-ph/9709396. H. Dreiner, hep-ph/9707435.
11. B. Brahmachari and P. Roy, *Phys. Rev.* D **50**, 39 (1994), err. *ibid.* **D51**, 193 (1995).
12. D. Chang and W-K. Keung, *Phys. Lett.* B **389**, 294 (1996).
13. G. Bhattacharyya, D. Choudhury and K. Sridhar, *Phys. Lett.* B **355**, 193 (1995).
14. R.M. Godbole, P. Roy and X. Tata, *Nucl. Phys.* B **401**, 67 (1993).
15. M. Hirsch, H.V. Klapdor-Kleingrothaus and S.G. Kovalenko, *Phys. Rev.*

Lett. **75**, 2276 (1995). K.S. Babu and R.N. Mohapatra, *Phys. Rev. Lett.* **75**, 2276 (1995). D.K. Gosh, S. Raychaudhuri and K. Sridhar, *Phys. Lett.* B **396**, 117 (1997).
16. D. Choudhury and P. Roy, *Phys. Lett.* B **378**, 153 (1996). J-H. Jang, J.K. Kim and J.S. Lee, *Phys. Rev.* D **55**, 7296 (1997).
17. H. Dreiner and G.G. Ross, *Nucl. Phys.* B **365**, 597 (1991). D.P. Roy, *Phys. Lett.* B **128**, 270 (1992).
18. D. Choudhury and S. Raychaudhuri, *Phys. Lett.* B **401**, 54 (1997).
19. H. Dreiner and G.G. Ross, *Nucl. Phys.* B **410**, 183 (1993).
20. S. Dawson, *Nucl. Phys.* B **261**, 297 (1985).
21. E.A. Baltz and P. Gondolo, hep-ph/9704411.

FLAVOUR CHANGING NEUTRAL CURRENT PROCESSES AND CKM PHENOMENOLOGY

A. ALI

Deutsches Elektronen-Synchrotron DESY, Notkestrasse 85, D-22603 Hamburg, FRG

Some selected topics in flavour changing neutral current processes in B decays are discussed in the context of the Standard Model. The emphasis here is on the determination of the CKM matrix elements from ongoing and forthcoming experiments. I discuss: (i) CKM-allowed radiative decays $B \to X_s + \gamma$, (ii) CKM-suppressed radiative decays $B \to X_d + \gamma$, $B^\pm \to \rho^\pm + \gamma$ and $B^0 \to (\rho^0, \omega) + \gamma$, (iii) B^0 - $\overline{B^0}$-mixings, and (iv) the Fleischer-Mannel bound in $B \to K\pi$ decays.

1 Determination of $|V_{ts}|$ from $\mathcal{B}(B \to X_s + \gamma)$

1.1 *Experimental status*

Electromagnetic penguins in B decays were first measured in 1993 by the CLEO collaboration through the exclusive decay $B \to K^\star + \gamma$ [1], followed by the measurement of the inclusive decay $B \to X_s + \gamma$ in 1994 [2]. The present CLEO measurements can be summarized as [3]:

$$\begin{aligned} \mathcal{B}(B \to X_s + \gamma) &= (2.32 \pm 0.57 \pm 0.35) \times 10^{-4}, \\ \mathcal{B}(B \to K^\star + \gamma) &= (4.2 \pm 0.8 \pm 0.6) \times 10^{-5}, \end{aligned} \tag{1}$$

which yield an exclusive-to-inclusive ratio: $R_{K^*} \equiv \Gamma(B \to K^\star + \gamma)/\Gamma(B \to X_s + \gamma) = (18.1 \pm 6.8)\%$. Very recently, the inclusive radiative decay has also been reported by the ALEPH collaboration with a (preliminary) branching ratio [4]:

$$\mathcal{B}(H_b \to X_s + \gamma) = (3.29 \pm 0.71 \pm 0.68) \times 10^{-4}. \tag{2}$$

The branching ratio in (2) involves a different weighted average of the various B-mesons and Λ_b baryons produced in Z^0 decays (hence the symbol H_b) than the corresponding one given in (1), which has been measured in the decay $\Upsilon(4S) \to B^+B^-, B^0\overline{B^0}$.

In the context of the SM, the principal interest in the inclusive branching ratio $B \to X_s + \gamma$ lies in that it determines the ratio of the Cabibbo-Kobayashi-Maskawa (CKM) matrix elements $|V_{ts}^* V_{tb}/V_{cb}|$. Since $|V_{cb}|$ and $|V_{tb}|$ have been directly measured, the measurements (1) and (2) yield $|V_{ts}|$. The ratio R_{K^*} constrains the form factor in $B \to K^* + \gamma$. Estimates of this ratio in the QCD sum rule and lattice QCD [5] are in agreement with the present measurements.

1.2 SM estimates of $\mathcal{B}(B \to X_s + \gamma)$ *and* $\mathcal{B}(H_b \to X_s + \gamma)$

The leading contribution to the decay $b \to s + \gamma$ arises at one-loop from the so-called penguin diagrams. With the help of the unitarity of the CKM matrix, the decay matrix element in the lowest order can be written as:

$$\mathcal{M}(b \to s + \gamma) = \frac{G_F}{\sqrt{2}} \frac{e}{2\pi^2} \lambda_t (F_2(x_t) - F_2(x_c)) q^\mu \epsilon^\nu \bar{s} \sigma_{\mu\nu} (m_b R + m_s L) b \, . \quad (3)$$

where G_F is the Fermi coupling constant, $e = \sqrt{4\pi\alpha_{\text{em}}}$, $x_i = m_i^2/m_W^2$; $i = u, c, t$ are the scaled quark mass ratios, and q_μ and ϵ_μ are, respectively, the photon four-momentum and polarization vector. Here, $F_2(x_i)$ is the Inami-Lim function derived from the (1-loop) penguin diagrams [6] and the CKM-matrix element dependence is factorized in $\lambda_t \equiv V_{tb} V_{ts}^*$. As the inclusive decay widths of the B hadrons are proportional to $|V_{cb}|^2$, the measurement of $\mathcal{B}(B \to X_s + \gamma)$ can be readily interpreted in terms of the CKM-matrix element ratio $\lambda_t/|V_{cb}|$. The effective Hamiltonian in the SM can be written as (keeping operators up to dimension 6) [7],

$$\mathcal{H}_{eff}(b \to s + \gamma) = -\frac{4G_F}{\sqrt{2}} \lambda_t \sum_{i=1}^{8} C_i(\mu) \mathcal{O}_i(\mu). \quad (4)$$

Expressing the branching ratio $\mathcal{B}(B \to X_s + \gamma)$ in terms of the semileptonic decay branching ratio $\mathcal{B}(B \to X \ell \nu_\ell)$,

$$\mathcal{B}(B \to X_s \gamma) = [\frac{\Gamma(B \to \gamma + X_s)}{\Gamma_{SL}}]^{th} \, \mathcal{B}(B \to X \ell \nu_\ell), \quad (5)$$

the theoretical part denoted by $[...]^{th}$ has been calculated in the NLO accuracy [8,9,10]. The power corrections have been obtained assuming that the decay $B \to X_s + \gamma$ is dominated by the magnetic moment operator O_7. The calculated $1/m_b^2$ correction is found to be innocuous (contributing at about 1%) in $\mathcal{B}(B \to X_s + \gamma)$. The correction proportional to $1/m_c^2$, resulting from the interference of the operators O_2 and O_7 in $H_{eff}(b \to s\gamma)$, has also been worked out [11,12,13]. Expressing this symbolically as

$$\frac{\Gamma(B \to X_s + \gamma)}{\Gamma^0(B \to X_s + \gamma)} = 1 + \frac{\delta_c}{m_c^2}, \quad (6)$$

one finds [13] $\delta_c/m_c^2 \simeq +0.03$.

Using $|V_{ts}^* V_{tb}/V_{cb}| = 0.976 \pm 0.010$, obtained from the CKM-unitarity [14], and the current measurements [15]: $\mathcal{B}(B \to X \ell \nu_\ell) = (10.49 \pm 0.46)\%$ (at $\Upsilon(4S)$)

and $\mathcal{B}(H_b \to X \ell \nu_\ell) = (11.16 \pm 0.20)\%$ (at Z^0), yields

$$\begin{aligned} \mathcal{B}(B \to X_s + \gamma) &= (3.51 \pm 0.32) \times 10^{-4} , \\ \mathcal{B}(H_b \to X_s + \gamma) &= (3.76 \pm 0.30) \times 10^{-4} , \end{aligned} \tag{7}$$

to be compared with the CLEO measurement $\mathcal{B}(B \to X_s + \gamma) = (2.32 \pm 0.67) \times 10^{-4}$ and the ALEPH measurement $\mathcal{B}(H_b \to X_s + \gamma) = (3.29 \pm 0.98) \times 10^{-4}$, respectively.

Treating the CKM factor $|V_{ts}^* V_{tb}/V_{cb}|$ as a free parameter, to be fitted by comparing NLO SM-based theory and experiments, one has [5]

$$|\frac{V_{ts}^* V_{tb}}{V_{cb}}| = 0.84 \pm 0.10 \ . \tag{8}$$

CKM unitarity gives $|\frac{V_{ts}^* V_{tb}}{V_{cb}}| \simeq |V_{cs}|$; this equality holds numerically (within present precision) if one compares eq. (8) with $|V_{cs}| = 1.01 \pm 0.18$, determined from charmed hadron decays [14]. Using the value of $|V_{tb}|$ measured by the CDF collaboration [16], $|V_{tb}| = 0.99 \pm 0.15$, and noting that $|V_{cb}| = 0.0393 \pm 0.0028$ [17], finally yields $|V_{ts}| = 0.033 \pm 0.007$. This is probably as direct a determination of $|V_{ts}|$ as we will ever see, as the decay $t \to W + s$ is too daunting to measure due to the low tagging efficiency of the s-quark jet.

2 The decay $B \to X_d + \gamma$ and constraints on the CKM parameters

In close analogy with the $B \to X_s + \gamma$ case, the complete set of dimension-6 operators relevant for the processes $b \to d\gamma$ and $b \to d\gamma g$ can be written as:

$$\mathcal{H}_{eff}(b \to d) = -\frac{4G_F}{\sqrt{2}} \, \xi_t \sum_{j=1}^{8} C_j(\mu) \, \hat{O}_j(\mu), \tag{9}$$

where $\xi_j = V_{jb} \, V_{jd}^*$ with $j = u, c, t$. The current-current operators $\hat{O}_j$, $j = 1, 2$, have implicit in them CKM factors ξ_c and ξ_u. We shall use the Wolfenstein parameterization [18], in which case the matrix is determined in terms of the four parameters $A, \lambda = \sin\theta_C$, ρ and η, and one can express the above factors as :

$$\xi_u = A \, \lambda^3 \, (\rho - i\eta), \quad \xi_c = -A \, \lambda^3, \quad \xi_t = -\xi_u - \xi_c. \tag{10}$$

We note that all three CKM-angle-dependent quantities ξ_j are of the same order of magnitude, $O(\lambda^3)$. The branching ratio $\mathcal{B}(B \to X_d + \gamma)$ in the SM can be quite generally written as:

$$\begin{aligned} \mathcal{B}(B \to X_d + \gamma) = D_1 \lambda^2 \\ \{(1-\rho)^2 + \eta^2 - (1-\rho) D_2 - \eta D_3 + D_4\}, \end{aligned} \tag{11}$$

where the functions D_i depend on various parameters such as m_t, m_b, m_c, μ, and α_s. These functions were calculated in the LL approximation some time ago [19]; since then their estimates have been improved [20], transcribing the NLO calculations done for $B \to X_s + \gamma$, thereby reducing the scale dependence in $\mathcal{B}(B \to X_d + \gamma)$ which is marked in the LL result.

To get an estimate of $\mathcal{B}(B \to X_d + \gamma)$, the CKM parameters ρ and η have to be constrained from the unitarity fits [21]. Allowing these parameters to vary over the entire allowed domain, one gets (at 95% C.L.) [20]:

$$6.0 \times 10^{-6} \leq \mathcal{B}(B \to X_d + \gamma) \leq 3.0 \times 10^{-5}. \tag{12}$$

The present theoretical uncertainty in this rate is a factor 5, almost all of which is due to the CKM parameters, which shows that even a modest (say, $\pm 30\%$) measurement of $\mathcal{B}(B \to X_d + \gamma)$ will have a very significant impact on CKM phenomenology. The decay $B \to X_d + \gamma$ is of interest from the point of view of direct CP violation as well. The CP asymmetry in this mode is estimated as $O(10\%)$. This should also reflect itself in the rate asymmetry in some exclusive decays such as $B^\pm \to \rho^\pm + \gamma$.

3 CKM-suppressed exclusive radiative decays $B \to V + \gamma$

Exclusive radiative B decays $B \to V + \gamma$, with $V = K^*, \rho, \omega$, are also potentially very interesting for CKM phenomenology [22]. Extraction of CKM parameters from these decays would, however, require a trustworthy estimate of the (short distance) SD- and (long distance) LD-contributions to the decay amplitudes. Present estimates based on QCD sum rules [23,24] and soft-scattering models [25] yield small ($O(5\%)$) to moderate ($O(20\%)$) LD-contributions in these decays depending on the decay modes. More importantly, data on neutral and charged B mesons can be used to disentangle the LD-effects directly.

The SD-contribution in the exclusive decays $(B^\pm, B^0) \to (K^{*\pm}, K^{*0}) + \gamma$, $(B^\pm, B^0) \to (\rho^\pm, \rho^0) + \gamma$, $B^0 \to \omega + \gamma$ and the corresponding B_s decays, $B_s \to \phi + \gamma$, and $B_s \to K^{*0} + \gamma$, involve the magnetic moment operator $\mathcal{O}_7$ and the related one obtained by the change $s \to d$, $\hat{O}_7$. Keeping only the SD-contribution leads to obvious relations among the exclusive decays, exemplified here by the decay rates for $(B^\pm, B^0) \to (\rho^\pm, \rho^0) + \gamma$ and $(B^\pm, B^0) \to (K^{*\pm}, K^{*0}) + \gamma$:

$$\frac{\Gamma((B^\pm, B^0) \to (\rho^\pm, \rho^0) + \gamma)}{\Gamma((B^\pm, B^0) \to (K^{*\pm}, K^{*0}) + \gamma)} \simeq \kappa_{u,d} \left[\frac{|V_{td}|}{|V_{ts}|}\right]^2 , \tag{13}$$

where $\kappa_i \equiv [F_S(B_i \to \rho\gamma)/F_S(B_i \to K^*\gamma)]^2$ involves the ratio of the transition form factors (apart from the small phase-space dependent correction) in the indicated radiative B decays; this ratio is unity in the $SU(3)$ limit. Likewise, assuming dominance of SD physics gives relations among various decay rates

$$\Gamma(B^{\pm} \to \rho^{\pm}\gamma) = 2\,\Gamma(B^0 \to \rho^0\gamma) = 2\,\Gamma(B^0 \to \omega\gamma)\ , \tag{14}$$

where the first equality holds due to the isospin invariance, and in the second $SU(3)$ symmetry has been assumed.

The LD-amplitudes in radiative B decays from the light quark intermediate states necessarily involve CKM matrix elements other than V_{td} or V_{ts}. In the CKM-suppressed decays $B \to V + \gamma$ the LD-contributions are dominantly induced by the matrix elements of the four-Fermion operators $\hat{O}_1$ and $\hat{O}_2$. Using factorization, the LD-amplitude in the decay $B^{\pm} \to \rho^{\pm}+\gamma$ can be written in terms of two form factors F_1^L and F_2^L. These are found to be numerically close to each other in the QCD sum rule approach [23,24], $F_1^L \simeq F_2^L \equiv F_L$, hence the ratio of the LD- and the SD- contributions reduces to a number

$$\mathcal{A}_{long}/\mathcal{A}_{short} = R_{L/S}^{B^{\pm}\to\rho^{\pm}\gamma} \cdot \frac{V_{ub}V_{ud}^*}{V_{tb}V_{td}^*}\ . \tag{15}$$

where $R_{L/S}^{B^{\pm}\to\rho^{\pm}\gamma}$ is estimated as [24]:

$$R_{L/S}^{B^{\pm}\to\rho^{\pm}\gamma} = -0.30 \pm 0.07\ . \tag{16}$$

The CKM ratio in Eq. (15) is constrained by unitarity to lie in the range [21] $0.22 \leq V_{ub}V_{ud}^*/V_{tb}V_{td}^* \leq 0.71$, with the central value around 0.33. This would put the LD-corrections at $O(10-20)\%$. The analogous LD-contributions to the neutral B decays $B^0 \to \rho\gamma$ and $B^0 \to \omega\gamma$ are expected to be much smaller due to the electric charge and QCD colour factors, and are estimated to be [24,25] $R_{L/S}^{B^0\to\rho\gamma} \simeq R_{L/S}^{B^0\to\omega\gamma} = 0.05$.

The SD relations (14), get modified by including the LD-contributions. Likewise, the ratios of the CKM-suppressed and CKM-allowed decay rates given in eq. (13) get modified due to the LD contributions. The effect of the LD-contributions is estimated to be modest but not negligible in charged B decays, introducing an uncertainty of $O(15\%)$, comparable to the uncertainty in the overall normalization due to the $SU(3)$-breaking effects in the quantity κ_u. Neutral B-meson radiative decays are less-prone to the LD-effects, and hence one expects that to a good approximation (say, better than 10%) the ratio of the decay rates for neutral B meson obtained in the approximation of SD-dominance remains valid [22]:

$$\frac{\Gamma(B^0 \to \rho\gamma, \omega\gamma)}{\Gamma(B \to K^*\gamma)} = \kappa_d \lambda^2[(1-\rho)^2 + \eta^2]\ , \tag{17}$$

where this relation holds for each of the two decay modes separately.

Finally, combining the estimates for the LD- and SD-form factors [24,22], and restricting the Wolfenstein parameters in the allowed range given earlier, yields: $\mathcal{B}(B^\pm \to \rho^\pm\gamma) = (1.5 \pm 1.1) \times 10^{-6}$,$\mathcal{B}(B^0 \to \rho\gamma) \simeq \mathcal{B}(B^0 \to \omega\gamma) = (0.65 \pm 0.35) \times 10^{-6}$. The large range reflects to a large extent the poor knowledge of the CKM matrix elements and hence experimental measurements of these branching ratios will contribute greatly to determine the Wolfenstein parameter ρ and η. Present experimental limits (at 90% C.L.) are [3]: $\mathcal{B}(B^\pm \to \rho^\pm\gamma) < 1.1 \times 10^{-5}$, $\mathcal{B}(B^0 \to \rho\gamma) < 3.9 \times 10^{-5}$ and $\mathcal{B}(B^0 \to \omega\gamma) < 1.3 \times 10^{-5}$. The constraints on the parameters (ρ, η) following from them are, however, not yet competitive to the ones which follow from CKM-unitarity and the lower bound on the mass difference ratio $\Delta M_s/\Delta M_d$ in the B_s^0 - $\overline{B_s^0}$ and B_d^0 - $\overline{B_d^0}$ sectors, which we discuss next.

4 ΔM_s (and x_s), Unitarity Triangle and CP Phases

Mass differences in the B_s^0-$\overline{B_s^0}$ and B_d^0-$\overline{B_d^0}$ systems are dominated in the standard model by t-quark exchange. Of the two, ΔM_d is well-measured [15], $\Delta M_d = (0.464 \pm 0.018)(\text{ps})^{-1}$, which can be translated into a determination of the matrix element $|V_{td}|$, yielding [21] $|V_{td}| = (9.0 \pm 2.6) \times 10^{-3}$, where the error reflects dominantly theoretical uncertainty on the hadronic quantity, for which $f_{B_d}\sqrt{\hat{B}_{B_d}} = (200 \pm 40)$ MeV has been used. The mass difference ΔM_s depends quadratically on the CKM factor $|V_{ts}^* V_{tb}|$, which has been measured from the branching ratio $B \to X_s + \gamma$, yielding $|V_{ts}^* V_{tb}| = 0.033 \pm 0.007$. However, as the error on this quantity is more than a factor 2 larger than what one gets from the CKM unitarity, we use here the value obtained by unitarity, $|V_{ts}^* V_{tb}| = 0.038 \pm 0.0027$. This yields:

$$
\begin{aligned}
\Delta M_s &= (12.0 \pm 2.0) \, \frac{f_{B_s}^2 B_{B_s}}{(230 \text{ MeV})^2} \, (\text{ps})^{-1} \; , \\
x_s &= (18.2 \pm 3.0) \, \frac{f_{B_s}^2 B_{B_s}}{(230 \text{ MeV})^2} \; ,
\end{aligned}
\tag{18}
$$

where we have used $\tau(B_s) = (1.52 \pm 0.07)$ ps to calculate x_s. The choice $f_{B_s}\sqrt{\hat{B}_{B_s}} = 230$ MeV corresponds to the central value given by lattice-QCD estimates. Allowing the coefficient to vary by $\pm 2\sigma$, and taking the central value for $f_{B_s}\sqrt{\hat{B}_{B_s}}$, gives

$$
8.0 \, (\text{ps})^{-1} \;\leq\; \Delta M_s \leq 16.0 \, (\text{ps})^{-1} \; ,
$$

$$12.2 \leq x_s \leq 24.2 \,. \tag{19}$$

It is difficult to ascribe a confidence level to this range due to the dependence on the unknown coupling constant factor. The present lower bound (from LEP) on ΔM_s is: $\Delta M_s > 10.0\ (\mathrm{ps})^{-1}$ (95% C.L.)[26]. A useful quantity for the CKM phenomenology is the ratio

$$\frac{\Delta M_s}{\Delta M_d} = \frac{\hat{\eta}_{B_s} M_{B_s} \left(f_{B_s}^2 B_{B_s}\right)}{\hat{\eta}_{B_d} M_{B_d} \left(f_{B_d}^2 B_{B_d}\right)} \left|\frac{V_{ts}}{V_{td}}\right|^2 , \tag{20}$$

in which all dependence on the t-quark mass drops out, leaving the square of the ratio of CKM matrix elements, multiplied by a factor which reflects $SU(3)_{\text{flavour}}$-breaking effects.

The present lower bound on ΔM_s from LEP and the world average of ΔM_d can be used to put a bound on the ratio $\Delta M_s/\Delta M_d$, yielding $\Delta M_s/\Delta M_d \geq 20.4$ at 95% C.L. This is significantly better than the lower bound on this quantity from the CKM fits[21]: $\Delta M_s/\Delta M_d \geq 11.4\ (\xi_s/1.15)^2$, where $\xi_s \equiv (f_{B_s}\sqrt{\hat{B}_{B_s}})/(f_{B_d}\sqrt{\hat{B}_{B_d}})$ is theoretically estimated to be $\xi_s = (1.15 \pm 0.05)$. The 95% confidence limit on $\Delta M_s/\Delta M_d$ can be turned into a bound on the CKM parameter space (ρ, η) by choosing a value for the SU(3)-breaking parameter ξ_s. We assume three representative values: $\xi_s^2 = 1.21$, 1.32 and 1.44, and display the resulting constraints in Fig. 1. From this graph we see that the LEP bound now restricts the allowed ρ-η region for all three values of ξ_s^2. We note that the bound on $\Delta M_s/\Delta M_d$ removes a large (otherwise allowed) negative-ρ region. The resulting fits of the CKM triangle taking into account this bound are necessarily asymmetric around $\rho = 0$. The best fit solutions are around $\rho = 0.11, \eta = 0.33$, as also noted by Paganini et al.[27].

The CP-violating asymmetries in B decays can be expressed straightforwardly in terms of the CKM parameters ρ and η. The 95% C.L. constraints on ρ and η just discussed can be used to predict the correlated ranges of the angles of the CKM unitarity triangle α, β and γ in the standard model. These correlations were worked out earlier without the $\Delta M_s/\Delta M_d$ bound[21,7] and are being updated here. It is found that the effect of the LEP bound on $\Delta M_s/\Delta M_d$ is still marginal on the $\sin 2\beta$ - $\sin 2\alpha$ correlation; a small region in $\sin 2\beta$ below $\sin 2\beta = 0.3$ is now removed, with the (95% C.L.) range for $\sin 2\alpha$ remaining practically unconstrained with or without the LEP bound. The correlation in the angles α - γ is more susceptible to the aforementioned bound. The allowed range for this correlation also depends more sensitively on the value of the quantity B_K than is the case for the $\sin 2\beta$ - $\sin 2\alpha$ correlation. The previously allowed range[21,7] (at 95% C.L.): $\gamma = (90 \pm 50)^\circ$ was based on using the bag

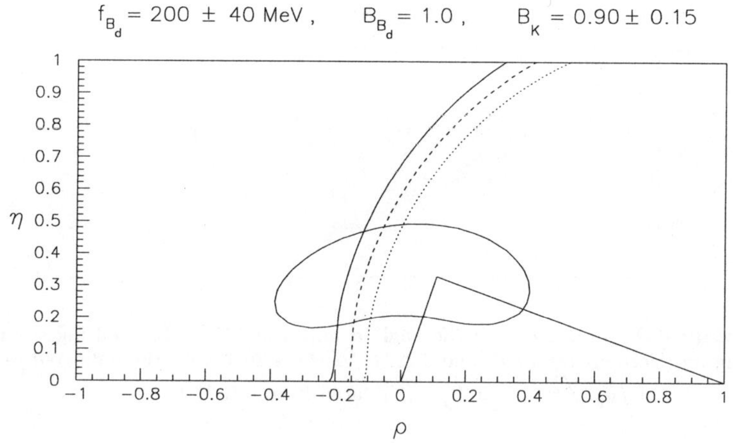

Figure 1: Constraints in ρ-η space from the LEP bound $\Delta M_s/\Delta M_d > 20.4$. The bounds are presented for 3 choices of the SU(3)-breaking parameter: $\xi_s^2 = 1.21$ (dotted line), 1.32 (dashed line) and 1.44 (solid line). In all cases, the region to the left of the curve is ruled out. (Updated from ref. 21.)

parameter to have a value in the range $B_K = 0.9 \pm 0.1$. Taking a somewhat larger error on the advice of our lattice colleagues, $B_K = 0.9 \pm 0.15$, this range becomes larger. One now gets $\gamma = (90 \pm 58)^\circ$ without the LEP bound. The corresponding (95 % C.L.) range with the LEP bound is: $32^\circ \leq \gamma \leq 120^\circ$. Thus, values of the angle γ in excess of 120° are now excluded by the LEP bound on $\Delta M_s/\Delta M_d$. Moreover, the allowed range in γ is no longer symmetric around $\gamma = 90^\circ$. This is illustrated in Fig. 2, which shows the region in α-γ space allowed by the present data.

The ranges for the CP-violating rate asymmetries parametrized by $\sin 2\alpha$, $\sin 2\beta$ and and $\sin^2\gamma$ are determined at 95% C.L. to be:

$$\begin{aligned} -1.0 &\leq \sin 2\alpha \leq 1.0 \ , \\ 0.30 &\leq \sin 2\beta \leq 0.88 \ , \\ 0.27 &\leq \sin^2\gamma \leq 1.0 \ . \end{aligned} \tag{21}$$

The CP asymmetries just discussed will be measured through rate asymmetries in non-leptonic B decays, such as $\overset{(-)}{B_d} \to J/\psi K_S$ ($\sin 2\beta$), $\overset{(-)}{B_d} \to \pi^+\pi^-$ ($\sin 2\alpha$) and $\overset{(-)}{B_s} \to D_s^\pm K^\mp$ (or $B^\pm \to \overset{(-)}{D} K^\pm$) ($\sin^2\gamma$).

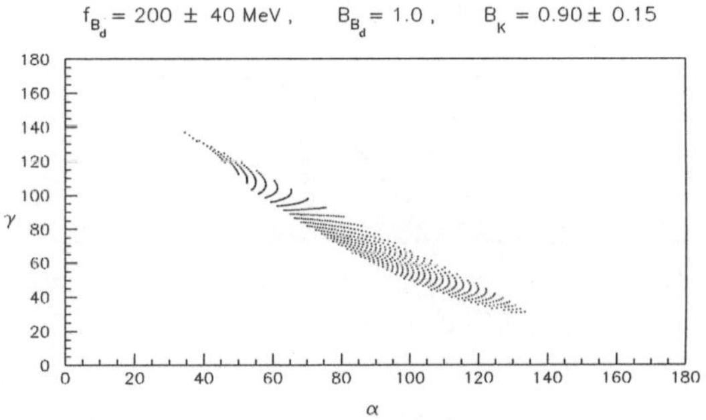

Figure 2: Allowed values (in degrees) of the angle α and γ at 95% C.L. resulting from the CKM fits taking into account the LEP bound $\Delta M_s/\Delta M_d > 20.4$, with the indicated ranges of $f_{B_d}\sqrt{\hat{B}_{B_d}}$ and B_K. (Updated from ref. 21.)

5 The Fleischer-Mannel bound in $B \to K\pi$ decays

Recently, the role of the decays $B^0 \to \pi^\mp K^\pm$ and $B^\pm \to \pi^\pm K^0$ in determining the angle γ has been emphasized by Fleischer and Mannel [28]. Assuming that the so-called Tree (current-current) and QCD penguins are the dominant contributions in these decays, one can write the amplitudes as

$$\begin{aligned} \Gamma(B^0 \to \pi^\mp K^\pm) &\propto |A_P|^2(1 - 2r\cos\gamma\cos\delta + r^2) \ , \\ \Gamma(B^\pm \to \pi^\pm K^0) &\propto |A_P|^2 \ , \end{aligned}$$

where $r = |A_T|/|A_P|$ is the ratio of the Tree and Penguin amplitudes, and δ is the strong phase difference involving these amplitudes. This leads to the Fleischer-Mannel relation (an average over CP-conjugate modes is implied),

$$R \equiv \frac{\Gamma(B^0 \to \pi^\mp K^\pm)}{\Gamma(B^\pm \to \pi^\pm K^0)} = 1 - 2\,r\cos\gamma\cos\delta + r^2 \geq \sin^2\gamma \ . \tag{22}$$

From this, constraints on γ of the form

$$0^\circ \leq \gamma \leq \gamma_0 \ \wedge \ 180^\circ - \gamma_0 \leq \gamma \leq 180^\circ \tag{23}$$

follow, where γ_0 is the maximum value of γ, which are complementary to the ones from the CKM unitarity fits discussed above. For $R < 1$, a useful bound on γ emerges, $\gamma_o^{\max} = \cos^{-1}(\sqrt{1-R})$. At present R has the value [29]: $R = 0.65 \pm 0.40$. For the central value of R, for example, this yields $\gamma_o^{\max} = 54^\circ$, implying $0^\circ \leq \gamma \leq 54^\circ \ \wedge \ 126^\circ \leq \gamma \leq 180^\circ$. One then finds that there is no overlap in the range $\gamma > 90^0$ (corresponding to $\rho < 0$ in Fig. 1) with the allowed range from the CKM fits, but there is significant overlap in the range $\gamma < 90°$ (or $\rho > 0$ in Fig. 1), yielding $32° \leq \gamma \leq 54°$ for the assumed value of R. However, as R increases, the overlap with the parameter space allowed by the CKM-unitarity increases. Hence, with the present experimental error on R, no useful bound on γ follows: For the Fleischer-Mannel bound to be useful the ratio R must be significantly less than 1. With model-dependent estimates of r and δ, such as those obtained in the context of factorization models, sharper constraints follow on the CKM parameters. For example, r is estimated to lie in the range $0.15 \leq r \leq 0.30$ in such models, one then finds [30] that $R = 0.65 \pm 0.40$ implies $\rho \geq 0$ (at $\pm 1\sigma$). Possible dilution of the Fleischer-Mannel bounds from neglected contributions (such as annihilation, electroweak penguins and soft rescattering) has been anticipated [31]. Some of these contributions may introduce significant uncertainties in the extraction of γ from this method. These issues will surely be studied further in future with more data available. It is too early to say if the ratio R (as well as other related ones) will make a big impact on the CKM phenomenology, but at any rate it remains a potentially interesting method to measure the otherwise daunting angle γ.

Acknowledgments

I would like to acknowledge collaboration with Hrachia Asatrian, Christoph Greub and David London on many aspects of B physics discussed here. In particular, David London provided the two updated figures shown in this report. Thanks are due to Manuel Drees and the organizing committee of the APCTP workshop in Seoul for holding an exciting meeting. The hospitality of Prof. H.S. Song at Seoul National University is gratefully acknowledged. This work is partially supported by the German-Korean scientific exchange programme DFG-446-KOR-113/72/0.

References

1. R. Ammar et al. (CLEO Collaboration), *Phys. Rev. Lett.* **71**, 674 (1993).
2. M.S. Alam et al. (CLEO Collaboration), *Phys. Rev. Lett.* **74**, 2885 (1995).
3. T. Skwarnicki, preprint HEPSY 97-03, hep-ph/9712253.
4. P.G. Colrain and M.I. Williams (ALEPH Collaboration), contributed paper # 587 , EPS Conference, Jerusalem, August 1997.

5. A. Ali, DESY 97-192, hep-ph/9709507, to be published in Proc. of the Seventh Int. Symp. on Heavy Flavors, Santa Barbara, Calif., 1997.
6. T. Inami and C.S. Lim, *Prog. Theor. Phys.* **65**, 297 (1981).
7. A. Ali, *Acta Physica Polonica* B **27**, 3529 (1996).
8. K. Chetyrkin, M. Misiak and M. Münz, *Phys. Lett.* B **400**, 206 (1997).
9. C. Greub and T. Hurth, hep-ph/9708214.
10. A.J. Buras, A. Kwiatkowski and N. Pott, hep-ph/9707482.
11. M.B. Voloshin, *Phys. Lett.* B **397**, 275 (1997).
12. A. Khodjamirian et al., *Phys. Lett.* B **402**, 167 (1997); Z. Ligeti, L. Randall and M.B. Wise, *Phys. Lett.* B **402**, 178 (1997); A.K. Grant et al., *Phys. Rev.* D **56**, 3151 (1997).
13. G. Buchalla, G. Isidori and S.J. Rey, hep-ph/9705253.
14. R.M. Barnett et al. (Particle Data Group), *Phys. Rev.* D **54**, 1 (1996).
15. M. Feindt, plenary talk, EPS Conference, Jerusalem, August 1997.
16. M. Narain, in Proc. of the Seventh Int. Symp. on Heavy Flavors, Santa Barbara, Calif., 1997.
17. L. Gibbons, in Proc. of the Seventh Int. Symp. on Heavy Flavors, Santa Barbara, Calif., 1997.
18. L. Wolfenstein, *Phys. Rev. Lett.* **51**, 1845 (1983)
19. A. Ali and C. Greub , *Phys. Lett.* B **287**, 191 (1992).
20. A. Ali, H.M. Asatrian, and C. Greub, DESY 97-255 (1997).
21. A. Ali and D. London, *Nucl. Phys.* B(Proc. Suppl.) **54A**, 297 (1997).
22. A. Ali, V.M. Braun and H. Simma, *Z. Phys.* C **63**, 437 (1994).
23. A. Khodzhamirian, G. Stoll, and D. Wyler, *Phys. Lett.* B **358**, 129 (1995).
24. A. Ali and V.M. Braun, *Phys. Lett.* B **359**, 223 (1995).
25. J.F. Donoghue, E. Golowich and A.A. Petrov, *Phys. Rev.* D **55**, 2657 (1997).
26. H.G. Moser, talk at the Heavy Flavor Workshop, Rostock, 1997.
27. P. Paganini, F. Parodi, R. Roudeau and A. Stochi, hep-ph/9711261.
28. R. Fleischer and T. Mannel, preprint TTP97-17 (hep-ph/9704423).
29. R. Godang et al. (CLEO Collaboration), preprint CLNS 97-1522, CLEO 97-27 (1997).
30. A. Ali and C. Greub, hep-ph/9707251, to appear in Phys. Rev. D.
31. J.M. Gerard and J. Weyers, preprint UCL-IPT-97-18 (hep-ph/9711469); M. Neubert, preprint CERN-TH/97-342 (hep-ph/9712224); A.F. Falk et al., preprint JHU-TIPAC-97018 (hep-ph/9712225).

Recent Progress in Exclusive Charmless Hadronic B Decays: Status of $B \to \eta' K$ Decay

Hai-Yang Cheng
Institute of Physics, Academia Sinica, Taipei, Taiwan 115, Republic of China

Recent progress in exclusive charmless hadronic decays of the B meson is discussed.

1 Introduction

Recently there has been a remarkable progress in the study of exclusive charmless B decays, both experimentally and theoretically. On the experimental side, many new two-body decay modes were discovered by CLEO [1]:

$$B \to \eta' K^+,\ \eta' K_S^0,\ \pi^+ K_S^0,\ \pi^0 K^+,\ \omega K^+,\ \omega K_S^0,\ \omega\pi^+,\ \phi K^*. \tag{1}$$

Moreover, CLEO has improved upper limits for many other channels. Therefore, it is a field whose time has finally arrived. On the theoretical aspect, there are two important issues to be addressed: (i) the renormalization scheme and scale dependence of hadronic matrix elements, and (ii) nonfactorizable effects in charmless B decays. A fascinating progress in dealing with the abovementioned theoretical issues has been made over the last few years. In this talk I'll first discuss the theoretical progress and then proceed to elaborate the decay $B \to \eta' K$ which has received a lot of attention recently.

2 Renormalization scale and scheme dependence of hadronic matrix element

The relevant effective weak Hamiltonian for hadronic weak B decay is of the form

$$\mathcal{H}_{\rm eff} = \frac{G_F}{\sqrt{2}}\left[\lambda_u(c_1O_1^u + c_2O_2^u) + \lambda_c(c_1O_1^c + c_2O_2^c) - \lambda_t\sum_{i=3}^{10} c_iO_i\right], \tag{2}$$

where $\lambda_u = V_{ub}V_{uq}^*$. The Wilson coefficients $c_i(\mu)$ in Eq. (2) have been evaluated at the renormalization scale $\mu \sim m_b$ to the next-to-leading order. Beyond the leading logarithmic approximation, they depend on the choice of the renormalization scheme. The mesonic matrix elements are customarily evaluated under the factorization hypothesis. In the naive factorization approach, the relevant Wilson coefficient functions for color-allowed external W-emission (or so-called "class-I") and color-suppressed (class-II) internal W-emission amplitudes are given by $a_1 = c_1 + c_2/N_c$, $a_2 = c_2 + c_1/N_c$, respectively, with N_c the number of colors. Inspite of its tremendous simplicity, naive factorization encounters two major difficulties. First, it never works for the decay rate of class-II decay modes, though it usually operates for class-I transition. Second, the hadronic matrix element under factorization is renormalization scale μ independent as the vector or axial-vector current is partially conserved. Consequently, the amplitude $c_i(\mu)\langle O\rangle_{\rm fact}$ is not truly physical as the scale dependence of Wilson coefficients does not get compensation from the matrix elements. The first difficulty indicates that it is inevitable and mandatory to take into account nonfactorizable contributions, especially for class-II decays, to render the color suppression of internal W emission ineffective. The second difficulty also should not occur since the matrix elements of four-quark operators ought to be evaluated in the same renormalization scheme as that for Wilson coefficients and renormalized at the same scale μ.

To circumvent the aforementioned second problem, one should evaluate perturbative QCD and electroweak corrections to the hadronic weak matrix elements parametrized by the matrices $\hat{m}_s$ and $\hat{m}_e$, respectively so that [2]

$$c_i(\mu)\langle O_i(\mu)\rangle = c_i(\mu)\left[\mathrm{I} + \frac{\alpha_s(\mu)}{4\pi}\hat{m}_s(\mu) + \frac{\alpha}{4\pi}\hat{m}_e(\mu)\right]_{ij}\langle O_j^{\rm tree}\rangle \equiv c_j^{\rm eff}\langle O_j^{\rm tree}\rangle. \quad (3)$$

Then factorization is applied to the matrix elements of tree operators; that is, before employing factorization to the four-quark operators, it is necessary to absorb all corrections to $O^{\rm tree}$ into the effective coefficients $c_i^{\rm eff}$. One-loop penguin corrections and vertex corrections to the operators O_{1-6} have been calculated in [2,3,4]. One can explicitly check that the effective Wilson coefficients $c_i^{\rm eff}$ are indeed renormalization scheme independent and approximately renormalization scale independent [5]. It should be stressed that the effective penguin coefficients $c_i^{\rm eff}$ ($i = 3, \cdots, 10$) take into account the effect of penguin diagrams not only with top quark exchange characterized by the Wilson coefficients c_i but also with internal u and c quarks induced by the tree operators O_1^u and O_1^c, respectively.

3 Nonfactorizable effects in charmless hadronic B decay

As stressed in the last section, it is mandatory to take into account the nonfactorizable effects, especially for class-II modes. For $B \to PP$ or VP decays, nonfactorizable contributions can be lumped into the effective parameters a_1

and a_2 [6]:

$$a_1^{\rm eff} = c_1 + c_2\left(\frac{1}{N_c} + \chi_1\right), \qquad a_2^{\rm eff} = c_2 + c_1\left(\frac{1}{N_c} + \chi_2\right), \tag{4}$$

where χ_i are nonfactorizable terms and receive main contributions from the color-octet current operators. Phenomenological analyses of two-body decay data of D and B mesons indicate that while the generalized factorization hypothesis in general works reasonably well, the effective parameters $a_{1,2}^{\rm eff}$ do show some variation from channel to channel, especially for the weak decays of charmed mesons [6,7]. An eminent feature emerged from the data analysis is that $a_2^{\rm eff}$ is negative in charm decay, whereas it becomes positive in bottom decay [6,8,9]:

$$a_2^{\rm eff}(D \to \bar{K}\pi) \sim -0.50\,, \quad a_2^{\rm eff}(B \to D\pi) \sim 0.26\,, \tag{5}$$

which in turn imply

$$\chi_2(\mu \sim m_c;\ D \to \bar{K}\pi) \sim -0.36\,, \quad \chi_2(\mu \sim m_b;\ B \to D\pi) \sim 0.11\,. \tag{6}$$

The observation $|\chi_2(B)| \ll |\chi_2(D)|$ is consistent with the intuitive picture that soft gluon effects become stronger when the final-state particles move slower, allowing more time for significant final-state interactions after hadronization [6]. Phenomenologically, it is often to treat the number of colors N_c as a free parameter and fit it to the data. Theoretically, this amounts to defining an effective number of colors by $1/N_c^{\rm eff} \equiv (1/N_c) + \chi$. It is clear from Eq. (6) that

$$N_c^{\rm eff}(D \to \bar{K}\pi) \gg 3, \quad N_c^{\rm eff}(B \to D\pi) \sim 2. \tag{7}$$

It is natural to ask that does the naive factorization approach also fail in charmless B decays ? If so, how large is the nonfactorizable effect ? Since the energy release in charmless two-body decays of the B meson is generally slightly larger than that in $B \to D^{(*)}\pi$, it is expected that $N_c^{\rm eff}$ for the B decay into two light mesons is close to $N_c^{\rm eff}(B \to D\pi) \approx 2$. It is pointed out in [10] that the parameters a_2, a_3 and a_5 are strongly dependent on $N_c^{\rm eff}$ and the rates dominated by these coefficients can have large variation. For example, the decay widths of $B^- \to \omega K^{(*)-}$, $B^0 \to \omega K^0$, ρK^{*0}, $B_s \to \eta\omega$, $\eta\phi$, $\omega\phi, \cdots$, etc. have strong N_c dependence [10]. We have shown recently in [11] that the branching ratio of $B^- \to \omega K^-$ has its lowest value of order 1×10^{-6} near $N_c^{\rm eff} \sim 3 - 4$ and hence the naive factorization with $N_c^{\rm eff} = 3$ is ruled out by experiment [1], $\mathcal{B}(B^\pm \to \omega K^\pm) = \left(1.5^{+0.7}_{-0.6} \pm 0.3\right) \times 10^{-5}$.

However, it is not easy to discern between $N_c^{\rm eff} = \infty$ and $N_c^{\rm eff} = 2$ in $B \to \omega K$ decays, and it becomes important to have a more decisive test on

$N_c^{\rm eff}$. For this purpose, we shall focus on the decay modes dominated by the tree diagrams and sensitive to the interference between external and internal W-emission amplitudes. The fact that $N_c^{\rm eff}=2$ ($N_c^{\rm eff}=\infty$) implies constructive (destructive) interference will enable us to differentiate between them. Good examples are the class-III modes: $B^\pm\to\pi^0\pi^\pm,\ \eta\pi^\pm,\ \pi^0\rho^\pm,\ \omega\pi^\pm,\cdots$. We found that [11] the averaged branching ratio of $B^\pm\to\omega\pi^\pm$ has its lowest value of order 2×10^{-6} at $N_c^{\rm eff}=\infty$ and then increases with $1/N_c^{\rm eff}$. Since experimentally [1] $\mathcal{B}(B^\pm\to\omega\pi^\pm)=\left(1.1^{+0.6}_{-0.5}\pm0.2\right)\times10^{-5}$, it is evident that $N_c^{\rm eff}=\infty$ is disfavored by the data. Measurements of interference effects in charged B decays $B^-\to\pi^-(\rho^-)\pi^0(\rho^0)$ will help determine the parameter $N_c^{\rm eff}$. For example, the ratio $R_1\equiv2\mathcal{B}(B^-\to\pi^-\rho^0)/\mathcal{B}(\bar{B}^0\to\pi^-\rho^+)$ is calculated to be 2.50 for $N_c^{\rm eff}=2$ and 0.26 for $N_c^{\rm eff}=\infty$. Hence, a measurement of R_1, which has the advantage of being independent of the Wolfenstein parameters ρ and η, will provide a decisive determination of the effective number of colors $N_c^{\rm eff}$.

4. Exclusive charmless hadronic B decays to η' and η

The CLEO collaboration has recently reported the preliminary branching ratios for the exclusive decay $B\to\eta' K$ dominated by gluonic penguin daigrams [1]:

$$\begin{aligned}\mathcal{B}(B^\pm\to\eta' K^\pm)&=\left(7.1^{+2.5}_{-2.1}\pm0.9\right)\times10^{-5},\\ \mathcal{B}(\overset{(-)}{B}{}^0\to\eta'\overset{(-)}{K}{}^0)&=\left(5.3^{+2.8}_{-2.2}\pm1.2\right)\times10^{-5}.\end{aligned}\tag{8}$$

Early theoretical estimate of the $B^\pm\to\eta' K^\pm$ branching ratio [12,13,3] lies in the range of $(1-2)\times10^{-5}$. [a] The CLEO result thus appears to be abnormally large. The question is then can the CLEO observation of $B\to\eta' K$ be accommodated in the standard model ? Do the new data imply new physics ? The theoretical interest and speculation in this subject has surged, as evidenced by the recent literature [4,14-20] that offer various interpretations on the unexpected large branching ratios.

In order to illustrate the problem clearly we choose the following parameters for calculation:

$$F_0^{BK}(0)=0.34,\quad \sqrt{3}F_0^{B\eta_0}(0)=0.254,\quad f_0=f_8=f_\pi,\quad m_s=150\,{\rm MeV},\tag{9}$$

and $N_c^{\rm eff}=2$, $\theta=-19.5°$, the $\eta-\eta'$ mixing angle. Using the renormailization scale and scheme independent effective Wilson coefficients $c_i^{\rm eff}$ discussed in

[a] The prediction $\mathcal{B}(B^\pm\to\eta' K^\pm)=3.6\times10^{-5}$ given in [12] is too large by about a factor of 2 because the normalization constant (i.e. $1/\sqrt{3}$) of the η_0 wave function was not taken into account in the form factor $F_0^{B\eta_0}$. This negligence was also made in some recent papers on $B\to\eta' K$.

Sec. II, we find

$$\mathcal{B}(B^\pm \to \eta' K^\pm) = \begin{cases} 1.4 \times 10^{-5}, & \text{for } \eta = 0.35, \ \rho = 0.08, \\ 1.6 \times 10^{-5}, & \text{for } \eta = 0.34, \ \rho = -0.12. \end{cases} \tag{10}$$

In the ensuing discussion, we will use (9) and (10) as the benchmarked values to be compared with. Since the choice of form factors and light quark masses is uncertain, one may argue that the CLEO data (8) can be fitted by choosing a small strange quark mass and/or large form factors F_0^{BK} and $F_0^{B\eta_0}$ [15]. For example, the above model estimate with $m_s = 55$ MeV or $F_0^{BK}(0) = 0.63$ will fit to the central value of $\mathcal{B}(B^\pm \to \eta' K^\pm)$. However it is dangerous to fit the parameters to a few particular decay modes. The point is that comparison between theory and experiment should be carried out using the same set of parameters for all decay channels. Indeed the measured branching ratio of $B \to \pi K$ puts a constraint on the strange quark mass and it indicates that m_s cannot be too small. In the SU(3) limit we have the relation $F_0^{B\pi^\pm} = F_0^{BK}$. Most of the existing QCD-sum-rule and quark model calculations show that $F_0^{B\pi^\pm}(0) \lesssim 0.33$ (for a review, see [20]). We shall see below that a severe constraint on $F_0^{B\pi^\pm}(0)$ can be derived from the current limit on the decay $B^+ \to \eta\pi^+$. Since SU(3) breaking is expected at most of 30% level, it is very unlikely that $F_0^{BK}(0)$ can deviate much from 0.33. Likewise, the nonet symmetry relation $\sqrt{3}F_0^{B\eta_0} = F_0^{B\pi^\pm}$ implies that $\sqrt{3}F_0^{B\eta_0}(0)$ cannot be too large than the model estimate, say 0.254 given in (9). In short, the parameters given in (9) cannot be modified dramatically without violating SU(3) symmetry relation and experimental observation of other decay modes.

Nevertheless, we can adjust the parameters in (9) slightly to improve the discrepancy between theory and experiment. The key point is that an accummulation of several small enhancement may eventually lead to a sizable enhancement. First, the current quark mass $m_s = 150$ MeV in (9) is defined at the renormalization scale $\mu = 1$ GeV. For reason of consistency, one ought to apply the small running quark mass $m_s(m_b) \simeq 105$ MeV in calculation. Second, we use $f_8/f_\pi = 1.38 \pm 0.22$ and $f_0/f_\pi = 1.06 \pm 0.03$ to take into account SU(3) breaking effects in decay constants [22]. Third, for the form factor $F_0^{B\eta_0}$ we follow [4] to use $\sqrt{3}F_0^{B\eta_0}(0) = 0.33$, which is slightly larger than the value of 0.254 obtained in [21]. Fourth, previously we employed the mixing angle $\theta = -19.5°$ so that the wave functions of the η and η' have the simple expressions: $\eta = \frac{1}{\sqrt{3}}(u\bar{u} + d\bar{d} - s\bar{s})$ and $\eta' = \frac{1}{\sqrt{6}}(u\bar{u} + d\bar{d} + 2s\bar{s})$. Here we instead use the value $\theta = -22.0° \pm 3.3°$ extracted in [22]. Applying the new value for each of the aforementioned parameters individually, we find that the branching ratio of $B^\pm \to \eta' K^\pm$ is enhanced by 62%, 37%, 19%, and 5%, respectively. Hence, the dominant enhancement comes from the running strange quark mass and SU(3) breaking in decay constants. When all new parameters are employed, we obtain (see Table I)

$$\mathcal{B}(B^\pm \to \eta' K^\pm) = \begin{cases} 3.87 \times 10^{-5}, \\ 4.04 \times 10^{-5}, \end{cases} \qquad \mathcal{B}(\overset{(-)}{B}{}^0 \to \eta' \overset{(-)}{K}{}^0) = \begin{cases} 3.70 \times 10^{-5}, \\ 3.67 \times 10^{-5}, \end{cases} \quad (11)$$

for $\eta = 0.35$, $\rho = 0.08$ (upper entry) and $\eta = 0.34$, $\rho = -0.12$ (lower entry), respectively. Note that $\mathcal{B}(B^\pm \to \eta' K^\pm)$ increases with $1/N_c^{\rm eff}$ and it becomes as large as order of 6×10^{-5} at $N_c^{\rm eff} = \infty$. However, as we have stressed before, it is very unlikely to have a large $N_c^{\rm eff}$ in the two-body charmless B decay. Therefore, we see that by adjusting the parameters in (9) within some reasonable range, the standard penguin contribution can account for the observed decay rate of $B^0 \to \eta' K^0$ but only marginally for $B^\pm \to \eta' K^\pm$. Nevertheless, the current data allow for some new contributions (but not necessarily new physics) unique to the η'. Of course, we have to await more new data to sort it out.

There are several mechanisms which are unique to the η' and may enhance the decay rate of $B \to \eta' K$. (i) The $b \to sg^*$ penguin followed by the transition $g^* \to g\eta'$ via the QCD anomaly can in principle contribute to the exclusive decay $B \to \eta' K$. This anomalous mechanism was originally advocated to explain the observed large inclusive $B \to \eta' + X$ signal[23,24]. However, since this mechanism involves a production of a gluon before hadronization, it will not play an essential role in low-multiplicity two-body exclusive decays unless the gluon is soft and absorbed in the wave function of the η'. Another possibility is that the gluon produced from the penguin diagram and the gluon emitted from the light antiquark fuse into the η' [18,19]. As the average momentum of the gluon emitted from the antiquark is in general less than 1 GeV, it is not clear if perturbative QCD is still applicable in this case. (ii) The process $b \to s + g^* g^* \to s + \eta'$ involves two gluon production in the penguin-like diagram followed by the η'-gluon anomalous interaction. The decay $b \to s + g^* g^*$ has been calculated in the literature[25]. It appears that the branching ratio arising from this mechanism[26] is less than 1×10^{-5}. (iii) A new internal W-emission contribution comes from the Cabibbo-allowed process $b \to c\bar{c}s$ followed by a conversion of the $c\bar{c}$ pair into the η' via two gluon exchanges. This new contribution is important since its mixing angle $V_{cb}V_{cs}^*$ is as large as that of the penguin amplitude and yet its Wilson coefficient $a_2^{\rm eff}$ is larger than that of penguin operators. The decay constant $f_{\eta'}^{(\bar{c}c)}$, defined by $\langle 0|\bar{c}\gamma_\mu\gamma_5 c|\eta'\rangle = i f_{\eta'}^{(\bar{c}c)} q_\mu$, has been estimated to be $|f_{\eta'}^{(\bar{c}c)}| = (50-180)$ MeV, based on the OPE, large-N_c approach and QCD low energy theorems[14]. It was claimed in [14,27]

Table I. Averaged branching ratios for charmless B decays to η' and η, where "Tree" refers to branching ratios from tree diagrams only, "Tree+QCD" from tree and QCD penguin diagrams, and "Full" denotes full contributions from tree, QCD and electroweak (EW) penguin diagrams in conjunction with contributions from the process $c\bar{c} \to \eta_0$. Predictions are for $k^2 = m_b^2/2$, $N_c^{\rm eff} = 2$, $f_{\eta'}^{(\bar{c}c)} = -15$ MeV, $\eta = 0.35$, $\rho = 0.08$ (the first number in parentheses) and $\eta = 0.34$, $\rho = -0.12$ (the second number in parentheses).

Decay	Tree	Tree+QCD	Tree+QCD+EW	Full	Exp. [1]
$B^\pm \to \eta' K^\pm$	1.41×10^{-7}	$(4.00,\ 4.18)\ 10^{-5}$	$(3.87,\ 4.04)\ 10^{-5}$	$(5.48,\ 5.69)\ 10^{-5}$	$(7.1^{+2.5}_{-2.1} \pm 0.9)\ 10^{-5}$
$B^\pm \to \eta K^\pm$	3.56×10^{-7}	$(5.48,\ 3.37)\ 10^{-7}$	$(3.40,\ 3.80)\ 10^{-7}$	$(5.05,\ 8.68)\ 10^{-7}$	$< 0.8 \times 10^{-5}$
$B^\pm \to \eta' K^{*\pm}$	2.11×10^{-7}	$(2.69,\ 1.90)\ 10^{-6}$	$(2.94,\ 2.11)\ 10^{-6}$	$(5.90,\ 3.24)\ 10^{-7}$	$< 29 \times 10^{-5}$
$B^\pm \to \eta K^{*\pm}$	5.25×10^{-7}	$(0.92,\ 1.53)\ 10^{-6}$	$(1.52,\ 2.42)\ 10^{-6}$	$(2.49,\ 3.70)\ 10^{-6}$	$< 24 \times 10^{-5}$
$B^\pm \to \eta' \pi^\pm$	1.94×10^{-6}	$(1.13,\ 1.06)\ 10^{-5}$	$(1.12,\ 1.05)\ 10^{-5}$	$(1.29,\ 1.21)\ 10^{-5}$	$< 4.5 \times 10^{-5}$
$B^\pm \to \eta \pi^\pm$	4.93×10^{-6}	$(9.57,\ 6.02)\ 10^{-6}$	$(9.82,\ 6.24)\ 10^{-6}$	$(1.04,\ 0.67)\ 10^{-5}$	$< 0.8 \times 10^{-5}$
$B^\pm \to \eta' \rho^\pm$	3.95×10^{-6}	$(1.08,\ 1.84)\ 10^{-5}$	$(1.08,\ 1.84)\ 10^{-5}$	$(1.01,\ 1.71)\ 10^{-5}$	
$B^\pm \to \eta \rho^\pm$	9.72×10^{-6}	$(1.21,\ 1.71)\ 10^{-5}$	$(1.19,\ 1.66)\ 10^{-5}$	$(1.19,\ 1.63)\ 10^{-5}$	
$B_d \to \eta' K^0$	5.09×10^{-9}	$(3.82,\ 3.80)\ 10^{-5}$	$(3.70,\ 3.67)\ 10^{-5}$	$(5.22,\ 5.19)\ 10^{-5}$	$(5.3^{+2.8}_{-2.2} \pm 1.2)\ 10^{-5}$
$B_d \to \eta K^0$	1.93×10^{-8}	$(1.23,\ 0.77)\ 10^{-7}$	$(1.97,\ 3.11)\ 10^{-8}$	$(2.49,\ 3.28)\ 10^{-7}$	
$B_d \to \eta' K^{*0}$	3.88×10^{-9}	$(2.23,\ 2.08)\ 10^{-6}$	$(2.47,\ 2.32)\ 10^{-6}$	$(2.44,\ 2.96)\ 10^{-7}$	$< 4.2 \times 10^{-5}$
$B_d \to \eta K^{*0}$	1.62×10^{-8}	$(5.71,\ 6.64)\ 10^{-7}$	$(1.23,\ 1.36)\ 10^{-6}$	$(2.26,\ 2.44)\ 10^{-6}$	$< 3.3 \times 10^{-5}$
$B_d \to \eta' \pi^0$	3.31×10^{-11}	$(4.83,\ 6.80)\ 10^{-6}$	$(4.63,\ 6.52)\ 10^{-6}$	$(5.32,\ 7.35)\ 10^{-6}$	$< 2.2 \times 10^{-5}$
$B_d \to \eta \pi^0$	6.70×10^{-9}	$(2.71,\ 3.67)\ 10^{-6}$	$(2.66,\ 3.60)\ 10^{-6}$	$(2.86,\ 3.83)\ 10^{-6}$	
$B_d \to \eta' \rho^0$	1.21×10^{-7}	$(2.07,\ 3.33)\ 10^{-6}$	$(2.01,\ 3.24)\ 10^{-6}$	$(1.79,\ 2.93)\ 10^{-6}$	
$B_d \to \eta \rho^0$	3.46×10^{-7}	$(1.28,\ 2.20)\ 10^{-6}$	$(1.16,\ 1.98)\ 10^{-6}$	$(1.11,\ 1.89)\ 10^{-6}$	$< 8.4 \times 10^{-5}$

that $|f_{\eta'}^{(\bar{c}c)}| \sim 140$ MeV is needed in order to exhaust the CLEO observation of $B^\pm \to \eta' K^\pm$ and $B \to \eta' + X$ by the mechanism $b \to c\bar{c} + s \to \eta' + s$ via gluon exchanges. However, a large value of $f_{\eta'}^{(\bar{c}c)}$ seems to be ruled out for three reasons. First, the decay constant $f_{\eta'}^{(\bar{u}u)}$ is only of order 50 MeV. Second, suppose the pseudoscalar content of $c\bar{c}$ is dominated by the η_c. Then from the data of $J/\psi \to \eta_c \gamma$ and $J/\psi \to \eta' \gamma$, one can show that $|f_{\eta'}^{(\bar{c}c)}| \geq 6$ MeV, where the lower bound corresponds to the nonrelativitsic quark estimate. (When the relativistic effect of the η' in $J/\psi \to \eta' \gamma$ is taken into account, $|f_{\eta'}^{(\bar{c}c)}|$ is larger than 6 MeV.) Even contributions from e.g., η_c', η_c'', $\cdots$ to the $c\bar{c}$ are included, it is argued in [28] that $|f_{\eta'}^{(\bar{c}c)}| \leq 40$ MeV. Third, based on the $\eta\gamma$ and $\eta'\gamma$ transition form factor data, the range of allowed $f_{\eta'}^{(\bar{c}c)}$ is estimated [29] to be $-65\,\text{MeV} \leq f_{\eta'}^{(\bar{c}c)} \leq 15$ MeV.

From Table I it is clear that for $f_{\eta'}^{(\bar{c}c)} = -15\,\text{MeV}$, which is consistent with above-mentioned constraints, the agreement between theory and experiment for $B \to \eta' K$ is substantially improved in the presence of large charm content in the η'. We conclude that no new physics is needed to account for the CLEO data of $B \to \eta' K$. We have also calculated the branching ratios of other exclusive charmless B decays involving η' and η (see Table I), where use of $f_\eta^{(c\bar{c})} = -\tan\theta f_{\eta'}^{(\bar{c}c)}$ has been made. [b] Three comments are in order. (i) The effect of $c\bar{c}$ conversion into the η' contributes destructively to $B \to \eta' K^*$. Consequently, the branching ratio of $B \to \eta' K^*$ is suppressed [c] and $\mathcal{B}(B \to \eta' K^*)/\mathcal{B}(B \to \eta K^*) \sim \mathcal{O}(10^{-1})$. If $B \to \eta' K$ is assumed to be entirely accommodated by any of aforementioned new mechamisms, the decay rate of $B \to \eta' K^*$ will be predicted to be the same order of magnitude as $B \to \eta' K$ [27,18]. (ii) Contrary to $B \to \eta(\eta')K^*$ decays, we see from Table I that $\mathcal{B}(B \to \eta' K)/\mathcal{B}(B \to \eta K) \sim \mathcal{O}(10^2)$ due to the destructive interference in the penguin diagrams of $B \to \eta K$. (iii) The electroweak penguin effects are in general very small, but they become important for $B \to \eta K$ and $B \to \eta K^*$ decays due to a large cancellation of QCD penguin contributions in these decay modes.

For $B \to \eta'(\eta)\pi(\rho)$ decays, the mechanism of $c\bar{c} \to \eta_0$ is much less dramatic since it does not gain mixing-angle enhancement as in the case of $B \to \eta'(\eta)K(K^*)$. Their branching ratios are sensitive to the light quark masses m_u, m_d and form factors such as $F_0^{B\pi}$ [5]. The current experimental limit on the decay $B^\pm \to \eta\pi^\pm$ puts useful constraints on m_q and $F_0^{B\pi}$. The predicted values presented in Table I are for $m_u(m_b) = 5$ MeV, $m_d(m_b) = 10$ MeV and $F_0^{B\pi^\pm}(0) = 0.30$. We find that even a slight increase of $F_0^{B\pi}(0)$ or decrease of m_q, say $m_u(m_b) \approx m_d(m_b)/2 \sim 3$ MeV, will make the decay rate of $B^\pm \to \eta\pi^\pm$ exceeding the present upper bound significantly. We also see that a negative ρ, which in turn implies a unitarity triangle γ in the range $90° < \gamma < 180°$, is preferred [11]. By contrast, the present experimental value of the ratio $R_2 \equiv \Gamma(B^0 \to \pi^\mp K^\pm)/\Gamma(B^\pm \to \pi^\pm K^0)$ favors a positive ρ [4]. Note that a positive ρ is also preferred by the limit on the ratio $\Delta M_s/\Delta M_d$ [30]. Clearly more data of $B^\pm \to \eta\pi^\pm$ and R_2 are needed to pin down the sign of ρ.

Acknowledgments

I would like to thank the organizers for this well run, fruitful and stimulating workshop. This work was supported in part by the National Science Council of ROC.

1. CLEO Collaboration, R. Godang *et al.*, CLNS 97-1522 [hep-ex/9711010]; S. Anderson *et al.*, CLEO CONF 97-22a (1997); D.M. Asner *et al.*, CLEO CONF 97-13; B. Behrens, talk presented at *The Second International Conference on*

[b] In the two mixing angle parametrization scheme given in [29], the decay constant $f_\eta^{(c\bar{c})}$ is much smaller: $f_\eta^{(c\bar{c})} = -\tan\theta_1 f_{\eta'}^{(\bar{c}c)}$ with $\theta_1 = -(6° \sim 9°)$.

[c] In [5] we have employed $f_{\eta'}^{(\bar{c}c)} \sim -50$ MeV. In that case, $B \to \eta' K^*$ is dominated by the process $c\bar{c} \to \eta_0$ and its branching ratio is of order 10^{-5}.

B Physics and CP Violation, March 24-27, 1997, Honolulu, Hawaii; J. Smith, talked presented at the 1997 Aspen Winter Physics Conference on Particle Physics, January, 1997.

2. A.J. Buras *et al., Nucl. Phys.* **B370**, 69 (1992); *Nucl. Phys.* **B408**, 209 (1993); R. Fleischer, *Z. Phys.* **C58**, 483 (1993); M. Ciuchini *et al., Nucl. Phys.* **B415**, 403 (1994).
3. G. Kramer, W.F. Palmer, and H. Simma, *Z. Phys.* **C66**, 429 (1995); *Nucl. Phys.* **B428**, 77 (1994).
4. A. Ali and C. Greub, DESY 97-126 [hep-ph/9707251].
5. H.Y. Cheng and B. Tseng, IP-ASTP-03-97 [hep-ph/9707316], to appear in Phys. Lett.
6. H.Y. Cheng, *Phys. Lett.* **B335**, 428 (1994).
7. A.N. Kamal and A.B. Santra, *Z. Phys.* **C72**, 91 (1996); A.N. Kamal, A.B. Santra, and R.C. Verma, *Phys. Rev.* **D53**, 2506 (1996).
8. H.Y. Cheng and B. Tseng, *Phys. Rev.* **D51**, 6295 (1995).
9. M. Neubert and B. Stech, CERN-TH/97-99 [hep-ph/9705292].
10. A. Deandrea, N. Di Bartolomeo, R. Gatto, F. Feruglio, and G. Nardulli, *Phys. Lett.* **B320**, 170 (1994).
11. H.Y. Cheng and B. Tseng, IP-ASTP-04-97 [hep-ph/9708211].
12. L.L. Chau, H.Y. Cheng, W.K. Sze, H. Yao, and B. Tseng, *Phys. Rev.* **D43**, 2176 (1991).
13. D.S. Du and L. Guo, *Z. Phys.* **C75**, 9 (1997); D.S. Du, M.Z. Yang, and D.Z. Zhang, *Phys. Rev.* **D53**, 249 (1996).
14. I. Halperin and A. Zhitnitsky, *Phys. Rev.* **D56**, 7247 (1997).
15. A. Datta, X.G. He, and S. Pakvasa, UH-511-864-97 [hep-ph/9707259]; A.L. Kagan and A.A. Petrov, UCHEP-27 UMHEP-443 [hep-ph/9707354].
16. A.S. Dighe, M. Gronau, and J.L. Rosner, *Phys. Rev. Lett.* **79**, 4333 (1997); H. Fritzsch, LMU 08/97 [hep-ph/9708348].
17. N.G. Deshpande, B. Dutta, and S. Oh, OITS-641 [hep-ph/9710354].
18. M.R. Ahmady, E. Kou, and A. Sugamoto, RIKEN-AF-NP-274 [hep-ph/9710509].
19. D.S. Du, C.S. Kim, and Y.D. Yang, BIHEP-Th/97-15 [hep-ph/9711428].
20. R. Casalbuoni *et al., Phys. Rep.* **281**, 145 (1997).
21. M. Wirbel, B. Stech, and M. Bauer, *Z. Phys.* **C29**, 637 (1985).
22. E.P. Venugopal and B.R. Holstein, hep-ph/9710382.
23. D. Atwood and A. Soni, *Phys. Lett.* **B405**, 150 (1997).
24. W.S. Hou and B. Tseng, hep-ph/9705304.
25. J. Liu and Y.P. Yao, *Phys. Rev.* **D41**, 2147 (1990); H. Simma and D. Wyler, *Nucl. Phys.* **B344**, 283 (1990).
26. G.L. Lin, private communication.
27. I. Halperin and A. Zhitnitsky, hep-ph/9705251
28. F. Yuan and K.T. Chao, *Phys. Rev.* **D56**, 2495 (1997).
29. T. Feldmann and P. Kroll, WUB 97-28 [hep-ph/9711231].
30. P. Paganini, F. Parodi, P. Roudeau, and A. Stocchi, hep-ph/9711261.

STANDARD MODEL PARAMETERS FROM LATTICE QCD

T. ONOGI
Department of Physics, Hiroshima University, 1-3-1 Kagamiyama, Higashi-Hiroshima, Hiroshima 739, JAPAN

I review progress in the determination of the strong coupling constant and the B meson decay constant in recent years.

1 Introduction

High precision calculation of matrix elements in lattice QCD is important for testing the standard model. Since a typical lattice spacing a^{-1} is around 2 - 3 GeV, discretization errors are of 20%. In order to reduce the discretization error, one usually extrapolates the result for several a's to the continuum limit assuming a linear dependence in a, but in recent years, one alternatively exploits Symanzik improvement, where the lattice discretization errors of $O(a)$ are explicitly removed by adding "irrelevant operators" to the lattice action and operators. The coefficients are determined by improved perturbation theory proposed by Lepage and Mackenzie[1]. I review the application of the Symanzik improvement to heavy quark physics such as quarkonium and heavy–light systems.

2 Heavy Quarkonium System

Quarkonium systems are good places to test the Symanzik improvement program because the effective Hamiltonian can be expanded systematically in powers of the velocity. This system was first studied by the Fermilab group [3] from which they determined the strong coupling constant. This was followed by a series of studies by the NRQCD collaboration. The most recent results can be found in [2].

The quark action can be expanded as follows

$$\begin{aligned} S &= Q^\dagger[iD_0 + \frac{\vec{D}^2}{2m_Q}]Q \\ &= Q^\dagger[-\frac{g}{8m_Q^2}(\vec{D}\cdot\vec{E} - \vec{E}\cdot\vec{D}) - \frac{\vec{D}^2}{8m_Q^3} - \frac{ig}{2m_Q}\vec{\sigma}\cdot\vec{B} \\ &\quad -\frac{ig}{8m_Q^2}\vec{\sigma}\cdot(\vec{D}\times\vec{E} - \vec{E}\times\vec{D}) + \cdots]Q. \end{aligned}$$

Table 1: $\alpha_{\overline{MS}}(M_Z)$ by various groups

Group	$\alpha^{(5)}_{\overline{MS}}(M_Z)$	a^{-1} from
NRQCD group[2]	0.1174(15)(19)	$b\bar{b}$ 1P-1S
	0.1173(21)(18)	$b\bar{b}$ 1S-2S
	0.1136(52)(16)	$c\bar{c}$ 1P-1S
FNAL[3]	0.108(4)	$c\bar{c}$ 1P-1S
KEK[4]	0.110(7)	$c\bar{c}$ 1P-1S

Here, the first and second terms are of $O(mv^2)$ and $O(mv^4)$, respectively. The expected systematic errors are 10% for $c\bar{c}$ and 4% for $b\bar{b}$. Figure 1 shows the

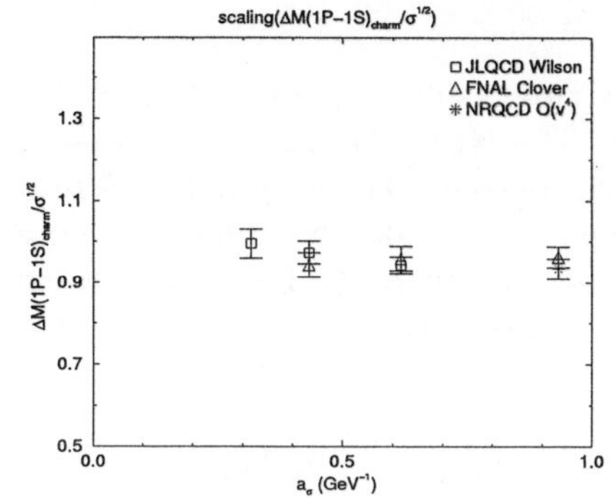

Figure 1: Scaling of 1P–1S level splitting in charmonium.

charmonium 1P–1S splitting in units of the string tension as a function of the lattice spacing. We find that the ratio has a good scaling behavior.

Using the experimental value of 1P–1S splitting or 1S–2S splitting as input, one can determine a. From improved lattice perturbation theory, the lattice coupling is then converted to $\alpha_{overline MS}(\mu)$ where μ is an arbitrary scale near $1/a$, which is finally scaled up to M_Z using the RG equation. Table 1 gives $\alpha_{\overline{MS}}(M_Z)$ from various groups. It seems that the results from bottomonium give larger values than those from charmonium. The charmonium system suffers from larger systematic errors from $O(v^4)$ corrections. The bottomonium results have much smaller statistical and perturbative errors as well as smaller $O(v^6)$ corrections, but one needs some care for the extrapolation in n_f. Figure 2 shows α_V at $\mu = 8.2$ GeV for $n_f = 0, 2$ determined from bottomonium 1P–1S and 1S–2S splitting by the NRQCD group. Here α_V is the coupling determined from the static quark potential. A linear extrapolation in

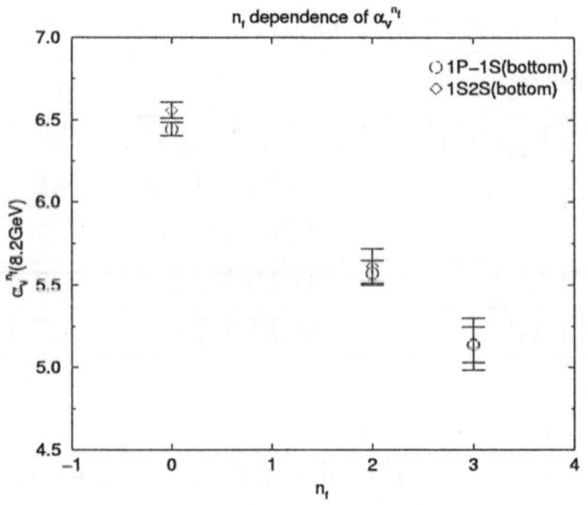

Figure 2: n_f dependence of α_V by the NRQCD group. The strong coupling constant determined from 1P–1S and 1S–2S agree when extrapolated to $n_f = 3$.

n_f is also shown. The two results from 1P–1S and 1S–2S almost agree with each other for $n_f = 3$. The NRQCD group estimates the extrapolation error by taking the differences between 1P–1S and 1S–2S results. However, a more conservative error estimate is to take the differences between the extrapolated result $\alpha^{(3)}$ and the $n_f = 2$ result $\alpha^{(2)}$. This gives an extrapolation error of 5 %. Therefore the most conservative estimate of the strong coupling constant is

$$\alpha_{\overline{MS}} = 0.117(2)(5), \tag{1}$$

where the first error is the error from statistics, perturbation, and discretization while the second error comes from the n_f extrapolation.

3 Heavy–light systems

In heavy–light systems, the heavy quark effective action can be approximated not by a velocity expansion but by a momentum expansion. The heavy quark and light quark actions are

$$\begin{aligned} S_Q &= Q^\dagger[\quad iD_0 + \frac{\vec{D}^2}{2m_Q} - \frac{ig}{2m_Q}\vec{\sigma}\cdot\vec{B} \\ &\qquad -\frac{g}{8m_Q^2}(\vec{D}\cdot\vec{E} - \vec{E}\cdot\vec{D}) - \frac{ig}{8m_Q^2}\vec{\sigma}\cdot(\vec{D}\times\vec{E} - \vec{E}\times\vec{D}) + \cdots]Q. \\ S_q &= \overline{\psi}[\quad iD_\mu\gamma^\mu - m_q + \frac{1}{2}aD^2 + \frac{1}{2}a\sigma^{\mu\nu}F_{\mu\nu} + \cdots]\psi. \end{aligned} \tag{2}$$

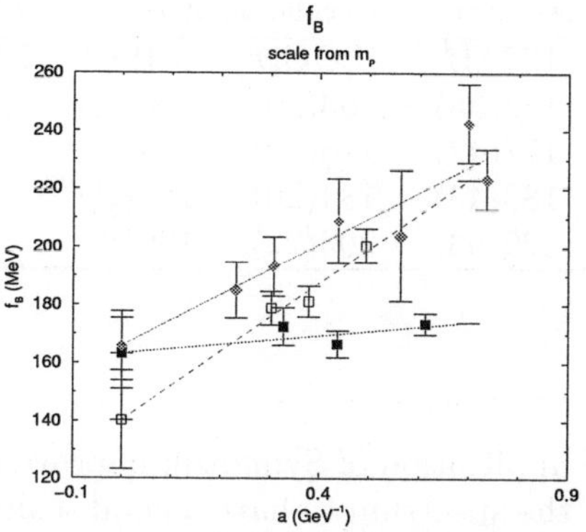

Figure 3: Continuum extrapolation of f_B from MILC Wilson (diamond), JLQCD Wilson (open square) and JLQCD ($O(a)$-improved action).

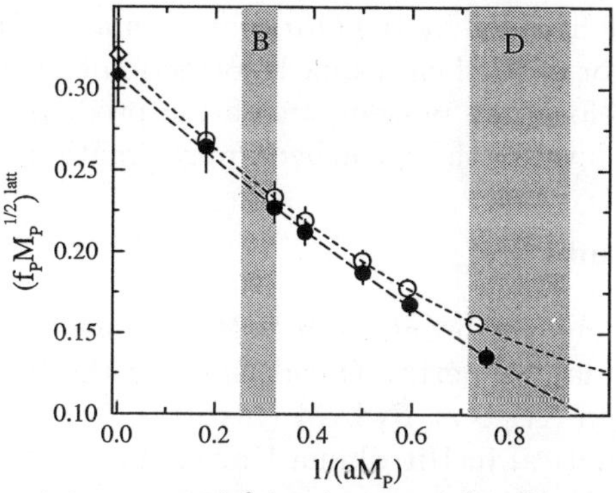

Figure 4: Comparison of f_B using $O(p^2/m_Q)$ and $O(p^3/m_Q^2)$ heavy quark actions by the Hiroshima group.

A naive estimate suggests that the use of the Wilson action contains $O(a)$ errors, which amount to 20 - 30% for $a^{-1} = 2 - 3$ GeV. The truncation errors of the heavy quark can give few % errors using the $O(p^2/m_Q)$ action. In Figure 3 we compare results for f_B using the Wilson action from MILC[5] and JLQCD[6] collaborations, and those using an $O(a)$ improved action from JLQCD. The graph shows that the naive expectation is indeed true and the $O(a)$ error can be removed with Symanzik improvement. Figure 4 compares f_B from $O(p^2/m_Q)$ and $O(p^3/m_Q^2)$ heavy quark actions by the Hiroshima group[9]. We find that the truncation error of higher order relativistic correction is less than 6% . Finally we list the most advanced calculation of the heavy–light decay constants reported this year in table 2.

Table 2: Most advanced calculation of f_B by various groups.

(MeV)	FNAL[8]	JLQCD[6]	MILC[5]	NRQCD[7]
f_B	156(35)	163(20)	$153(^{40}_{16})$	$149(^{25}_{15})$
f_{B_s}	177(42)	175(20)	$164(^{50}_{16})$	
f_D	183(43)	184(20)	$186(^{31}_{20})$	
f_{D_s}	229(53)	203(22)	$199(^{44}_{13})$	

4 Summary

We have seen that the application of Symanzik improvement to heavy quarkonium systems enabled the spectrum to have a good scaling behavior for $a^{-1} = 2-3GeV$. Using experimental values for splittings in the charmonium and bottomonium spectra, the strong coupling constant can be determined to the few percent level. In application to the heavy–light system, we saw that 1) a large lattice spacing can be handled by the $O(a)$ improvement of the light quark and 2) the truncation error of the heavy quark correction terms of $O(p^3/m_Q^2)$ is at most a few %. Therefore, the present simulation predicts the B meson decay constant at the 10% level in the quenched approximation.

5 Acknowledgments

I would like to thank my collaborators Dr.S. Hashimoto, K-I. Ishikawa, H. Matsufuru, and N. Yamada. Numerical calculations for the Hiroshima group have been done on Paragon XP/S at INSAM (Institute for Numerical Simulations and Applied Mathematics) in Hiroshima University.

References

1. G. P. Mackenzie and P. B. Mackenzie, *Phys. Rev.* D48(1993)2250.
2. C. T. H. Davies *Phys. Rev.* D56(1997)2755.
3. A. X. El-Khadra et.al, *Phys. Rev. Lett.69(1992)729.*
4. *S. Aoki et.al,* Phys. Rev. Lett.74(1995)22.
5. C. Bernard et.al, hep-lat/9709142
6. S. Aoki et al. KEK-CP-71, hep-lat/9711041 .
7. A.Ali Khan et.al, *Phys. Rev.* D56(1997)7012
8. A.X.El-Khadra et.al, FNAL Pub-97/322-T, hep-ph/9711426.
9. K-I. Ishikawa et.al, *Phys. Rev.* D56(1997)7028.

USING B_d^0, B_s^0 DECAYS AT HADRONIC B-FACTORIES TO DETERMINE CP ANGLES α AND β

C.S. KIM

Department of Physics, Yonsei University, Seoul 120-749, Korea

At the first PPPP workshop, I gave a review talk on physics of B and CP-phases. In my talk, I explained the DGRG method and our extension to extract CP angles α and γ from measurements of the decay rates of $B_{d,s}^0 \to \pi\pi$, πK, KK. There are several advantages to this extension: discrete ambiguities are removed, fewer assumptions are necessary, and the method works even if all strong phases vanish. I also talked on several other topics.

1 Introduction

Here I review an extension of the DGR method [1] which avoids most of its problems. In addition to the B_d^0 and B^+ decays used by DGR, it requires the measurement of their $SU(3)$-counterpart B_s^0 decays: $B_s^0 \to \pi^+ K^-$, $B_s^0(t) \to K^+K^-$, and $B_s^0 \to K^0 \bar{K}^0$. This overconstrains the system, which eliminates most discrete ambiguities in the extraction of α and γ.

Although challenging experimentally, this method can probably be carried out at hadron collider experiments such as BTeV and LHC-B [2]. All branching ratios are $\mathcal{O}(10^{-5})$, and all final states involve only charged pions and K's – there are no π^0's. It will be necessary to resolve B_s^0-$\overline{B_s^0}$ oscillations, but this should not be be a problem at hadron colliders: for example, BTeV expects to be able to measure values of the mixing parameter $x_s \sim 50$. The principal obstacle to obtaining precise values of α and γ is the fact that 12 measurements must be combined to extract these CP angles, as well as other quantities. If the experimental errors are large, then all sensitivity to the CP angles may be washed out. However, preliminary studies indicate that the errors will in fact be relatively small. A BTeV simulation found that the CP asymmetry of the decay $B_d^0(t) \to \pi^+\pi^-$ can be measured with an error of ± 0.05 [2]. This error includes the background, which consists of the decays $B_d^0 \to \pi K$, $B_s^0 \to \pi K$ and $B_s^0 \to KK$. Since these are precisely the decays involved in our extension of the DGR method, this suggests that these decays can be experimentally separated from one another, and their branching ratios measured with reasonable precision, say $\sim 5\%$ or less. If this is so, then we can expect that α and γ can also be extracted fairly precisely.

2 Using B_s^0 decays to determine the CP angles α and β

The DGR method involves the decays $B_d^0 \to \pi^+\pi^-$, $B_d^0 \to \pi^- K^+$, and $B^+ \to \pi^+ K^0$. The amplitudes for these decays can be written

$$\begin{aligned} A_{\pi\pi} &\equiv A\left(B^0 \to \pi^+\pi^-\right) = -\left(T + P\right), \\ A_{\pi K} &\equiv A\left(B^0 \to \pi^- K^+\right) = -\left(T' + P'\right), \\ A_{\pi K}^+ &\equiv A\left(B^+ \to \pi^+ K^0\right) = P' . \end{aligned} \tag{1}$$

Note that, since only tree and penguin terms are involved, EWP contributions are negligible. The weak phase of T is $\mathrm{Arg}(V_{ud}V_{ub}^*) = \gamma$, and similarly for T': $\mathrm{Arg}(V_{us}V_{ub}^*) = \gamma$. The $b \to s$ penguin P' is dominated by the internal t-quark, so its weak phase is $\mathrm{Arg}(V_{ts}V_{tb}^*) = \pi$. As for the $b \to d$ penguin P, if it also is dominated by the t-quark, its weak phase is $\mathrm{Arg}(V_{td}V_{tb}^*) = -\beta$. This is the assumption made by DGR, but we can relax it.

If $SU(3)$ were unbroken, the amplitudes T and T' would be related simply by the ratio of their CKM matrix elements: $|T'/T| = |V_{us}/V_{ud}|$. However, if one includes first-order $SU(3)$ breaking, there is an additional factor involving the ratio of K and π decay constants if factorization is assumed: $|T'/T| = |V_{us}|f_K/|V_{ud}|f_\pi \equiv r_u$. On the other hand, since factorization is unlikely to hold for penguin amplitudes, P and P' are not related in a simple way. However, DGR do assume that the strong phase of the penguin diagram, δ_P, is unaffected by $SU(3)$ breaking. With these assumptions, the amplitudes in Eqs. (1) can be written

$$\begin{aligned} A_{\pi\pi} &= \mathcal{T} e^{i\delta_T} e^{i\gamma} + \mathcal{P} e^{i\delta_P} e^{-i\beta} , \\ A_{\pi K} &= r_u \mathcal{T} e^{i\delta_T} e^{i\gamma} - \mathcal{P}' e^{i\delta_P} , \\ A_{\pi K}^+ &= \mathcal{P}' e^{i\delta_P} , \end{aligned} \tag{2}$$

where $\mathcal{T} \equiv |T|$, $\mathcal{P} \equiv |P|$, and $\mathcal{P}' \equiv |P'|$.

There are thus six unknown quantities in the above 3 amplitudes: $\alpha \equiv \pi - \beta - \gamma$, γ, $\mathcal{T}$, $\mathcal{P}$, $\mathcal{P}'$, and $\delta \equiv \delta_T - \delta_P$. The time-dependent, tagged B_d^0 and $\overline{B_d^0}$ decay rates to $\pi^+\pi^-$ are given by

$$\begin{aligned} \Gamma\left(B_d^0(t) \to \pi^+\pi^-\right) &= e^{-\Gamma t}\Big[\left|A_{\pi\pi}\right|^2 \cos^2\left(\frac{\Delta m}{2}t\right) + \left|\bar{A}_{\pi\pi}\right|^2 \sin^2\left(\frac{\Delta m}{2}t\right) \\ &\qquad + \mathrm{Im}\left(e^{2i\beta} A_{\pi\pi}\bar{A}_{\pi\pi}^*\right)\sin(\Delta m t)\Big] , \\ \Gamma\left(\overline{B_d^0}(t) \to \pi^+\pi^-\right) &= e^{-\Gamma t}\Big[\left|A_{\pi\pi}\right|^2 \sin^2\left(\frac{\Delta m}{2}t\right) + \left|\bar{A}_{\pi\pi}\right|^2 \cos^2\left(\frac{\Delta m}{2}t\right) \\ &\qquad - \mathrm{Im}\left(e^{2i\beta} A_{\pi\pi}\bar{A}_{\pi\pi}^*\right)\sin(\Delta m t)\Big] . \end{aligned} \tag{3}$$

From these measurements one can determine $|A_{\pi\pi}|^2$, $|\bar{A}_{\pi\pi}|^2$, and $\mathrm{Im}\left(e^{2i\beta} A_{\pi\pi} \bar{A}^*_{\pi\pi}\right)$. The rates for the self-tagging decays $B^0_d \to \pi^- K^+$ and $\overline{B^0_d} \to \pi^+ K^-$ are

$$\begin{aligned} |A_{\pi K}|^2 &= r_u^2 \mathcal{T}^2 + \mathcal{P}'^2 - 2 r_u \mathcal{T}\mathcal{P}' \cos(\delta + \gamma), \\ |\bar{A}_{\pi K}|^2 &= r_u^2 \mathcal{T}^2 + \mathcal{P}'^2 - 2 r_u \mathcal{T}\mathcal{P}' \cos(\delta - \gamma). \end{aligned} \tag{4}$$

Finally, the rates for $B^+ \to \pi^+ K^0$ and its CP-conjugate decay give

$$|A^+_{\pi K}|^2 = |\bar{A}^-_{\pi K}|^2 = \mathcal{P}'^2 \ . \tag{5}$$

This method thus allows to obtain the following six quantities:

$$A \equiv \frac{1}{2}\left(|A_{\pi\pi}|^2 + |\bar{A}_{\pi\pi}|^2\right) = \mathcal{T}^2 + \mathcal{P}^2 - 2\mathcal{T}\mathcal{P}\cos\delta\cos\alpha, \tag{6}$$

$$B \equiv \frac{1}{2}\left(|A_{\pi\pi}|^2 - |\bar{A}_{\pi\pi}|^2\right) = -2\mathcal{T}\mathcal{P}\sin\delta\sin\alpha, \tag{7}$$

$$C \equiv \mathrm{Im}\left(e^{2i\beta} A_{\pi\pi}\bar{A}^*_{\pi\pi}\right) = -\mathcal{T}^2 \sin 2\alpha + 2\mathcal{T}\mathcal{P}\cos\delta\sin\alpha, \tag{8}$$

$$D \equiv \frac{1}{2}\left(|A_{\pi K}|^2 + |\bar{A}_{\pi K}|^2\right) = r_u^2\mathcal{T}^2 + \mathcal{P}'^2 - 2r_u\mathcal{T}\mathcal{P}'\cos\delta\cos\gamma, \tag{9}$$

$$E \equiv \frac{1}{2}\left(|A_{\pi K}|^2 - |\bar{A}_{\pi K}|^2\right) = 2r_u\mathcal{T}\mathcal{P}'\sin\delta\sin\gamma, \tag{10}$$

$$F \equiv |A^+_{\pi K}|^2 = \mathcal{P}'^2 \ . \tag{11}$$

These give 6 equations in 6 unknowns, so that one can solve for α, γ, $\mathcal{T}$, $\mathcal{P}$, $\mathcal{P}'$, and δ. However, because the equations are nonlinear, there is an up to 8–fold discrete ambiguity in extracting these quantities.

If $\delta = 0$, the quantities B and E vanish, so that one is left with 4 equations in 5 unknowns. In this case one must use additional assumptions to extract information about the CP phases. Furthermore, even if $\delta \neq 0$, if one relaxes any of the assumptions described above, the method breaks down. For example, if one allows the strong phase of the P' diagram to be different from that of the P diagram, as might be the case in the presence of $SU(3)$ breaking, then one has 6 equations in 7 unknowns. And if one relaxes the assumption that the $b \to d$ penguin is dominated by the t-quark, then once again additional parameters are introduced, and the method breaks down.

All of these potential problems can be dealt with by considering additional B^0_s decays. The problems with the DGR method can be resolved by adding amplitudes which depend on the same 6 quantities, thus overconstraining the system. In this case, if one adds a parameter or two, perhaps by relaxing certain assumptions, the method will be less likely to break down.

Within $SU(3)$ symmetry, the obvious decays to consider are the $SU(3)$ counterparts to the DGR decays, namely $B_s^0 \to \pi^+ K^-$, $B_s^0(t) \to K^+K^-$, and $B_s^0 \to K^0 \bar{K}^0$. The amplitudes for these decays are completely analogous to those in Eqs. (1):

$$\begin{aligned} B_{\pi K} &\equiv A\left(B_s^0 \to \pi^+ K^-\right) = -\left(\tilde{T} + \tilde{P}\right), \\ B_{KK} &\equiv A\left(B_s^0 \to K^+ K^-\right) = -\left(\tilde{T}' + \tilde{P}'\right), \\ B_{KK}^s &\equiv A\left(B_s^0 \to K^0 \overline{K^0}\right) = \tilde{P}' . \end{aligned} \tag{12}$$

Here we have denoted the tree and penguin diagrams involving a spectator s quark by $\tilde{T}$ and $\tilde{P}$, respectively. As before, the unprimed and primed quantities denote $\Delta S = 0$ and $\Delta S = 1$ processes, respectively.

The weak phase of $\tilde{T}$ and $\tilde{T}'$ is γ, and that of $\tilde{P}'$ is π. As for $\tilde{P}$, as a first step we make the same assumptions as DGR, namely that it is dominated by the t-quark, so that its weak phase is $-\beta$. Turning to $SU(3)$ breaking, we assume factorization for the tree amplitudes, so that $|\tilde{T}'/\tilde{T}| = r_u$. The magnitudes of the $\tilde{P}$ and $\tilde{P}'$ amplitudes are unrelated to one another. However, again as a first step, like DGR we assume that they have the same strong phase, $\delta_{\tilde{P}}$. The one new assumption we make is that the relative strong phase between the tree and penguin amplitudes is independent of the flavor of the spectator quark, i.e. $\delta_s = \delta$ where $\delta_s \equiv \delta_{\tilde{T}} - \delta_{\tilde{P}}$ and $\delta \equiv \delta_T - \delta_P$. (The most likely way for this to occur is if $\delta_T = \delta_{\tilde{T}}$ and $\delta_P = \delta_{\tilde{P}}$.) This assumption is motivated by the spectator model. Under these assumptions, the amplitudes in Eqs. (12) can be written

$$\begin{aligned} B_{\pi K} &= \tilde{\mathcal{T}} e^{i\delta_T} e^{i\gamma} + \tilde{\mathcal{P}} e^{i\delta_P} e^{-i\beta} , \\ B_{KK} &= r_u \tilde{\mathcal{T}} e^{i\delta_T} e^{i\gamma} - \tilde{\mathcal{P}}' e^{i\delta_P} , \\ B_{KK}^s &= \tilde{\mathcal{P}}' e^{i\delta_P} , \end{aligned} \tag{13}$$

where $\tilde{\mathcal{T}} \equiv \left|\tilde{T}\right|$, $\tilde{\mathcal{P}} \equiv \left|\tilde{P}\right|$, and $\tilde{\mathcal{P}}' \equiv \left|\tilde{P}'\right|$.

The important point here is that three new parameters have been introduced in the above amplitudes: $\tilde{\mathcal{T}}$, $\tilde{\mathcal{P}}$, and $\tilde{\mathcal{P}}'$. However, as in the DGR method, 6 quantities can be extracted from measurements of the rates for these decays. Here, the self-tagging decays are $B_s^0 \to \pi^+ K^-$ and $\overline{B_s^0} \to \pi^- K^+$, whose rates are

$$\begin{aligned} |B_{\pi K}|^2 &= \tilde{\mathcal{T}}^2 + \tilde{\mathcal{P}}^2 - 2\tilde{\mathcal{T}}\tilde{\mathcal{P}} \cos(\delta - \alpha), \\ |\bar{B}_{\pi K}|^2 &= \tilde{\mathcal{T}}^2 + \tilde{\mathcal{P}}^2 - 2\tilde{\mathcal{T}}\tilde{\mathcal{P}} \cos(\delta + \alpha). \end{aligned} \tag{14}$$

The time-dependent, tagged B_s^0 and $\overline{B_s^0}$ decay rates to K^+K^- are given by

$$\begin{aligned}
\Gamma\left[B_s^0(t) \to K^+K^-\right] &= e^{-\Gamma t}\left[|B_{KK}|^2\cos^2\left(\frac{\Delta m_s}{2}t\right) + \left|\bar{B}_{KK}\right|^2\sin^2\left(\frac{\Delta m_s}{2}t\right)\right. \\
&\quad \left. + \,\mathrm{Im}\left(B_{KK}\bar{B}_{KK}^*\right)\sin(\Delta m_s t)\right], \\
\Gamma\left[\overline{B_s^0}(t) \to K^+K^-\right] &= e^{-\Gamma t}\left[|B_{KK}|^2\sin^2\left(\frac{\Delta m_s}{2}t\right) + \left|\bar{B}_{KK}\right|^2\cos^2\left(\frac{\Delta m_s}{2}t\right)\right. \\
&\quad \left. - \,\mathrm{Im}\left(B_{KK}\bar{B}_{KK}^*\right)\sin(\Delta m_s t)\right], \qquad (15)
\end{aligned}$$

from which the quantities $|B_{KK}|$, $\left|\bar{B}_{KK}\right|$, and $\mathrm{Im}\left(B_{KK}\bar{B}_{KK}^*\right)$ can be extracted. Finally, we turn to $B_s^0(t) \to K^0\overline{K^0}$. In principle there can be indirect CP violation in these decays. However, within the SM, this CP violation is zero to a good approximation, since both B_s^0-$\overline{B_s^0}$ mixing and the $b \to s$ penguin diagram, which dominates this decay, are real. Thus, measurements of the rates for these decays yield

$$|B_{KK}^s|^2 = |\bar{B}_{KK}^s|^2 = \tilde{\mathcal{P}}'^2 \,. \qquad (16)$$

Obviously, any violation of this equality will be clear evidence for new physics.

Therefore the above measurements yield 6 new quantities:

$$\begin{aligned}
\tilde{A} &\equiv \frac{1}{2}\left(|B_{\pi K}|^2 + \left|\bar{B}_{\pi K}\right|^2\right) = \tilde{\mathcal{T}}^2 + \tilde{\mathcal{P}}^2 - 2\tilde{\mathcal{T}}\tilde{\mathcal{P}}\cos\delta\cos\alpha, && (17)\\
\tilde{B} &\equiv \frac{1}{2}\left(|B_{\pi K}|^2 - \left|\bar{B}_{\pi K}\right|^2\right) = -2\tilde{\mathcal{T}}\tilde{\mathcal{P}}\sin\delta\sin\alpha, && (18)\\
\tilde{C} &\equiv \mathrm{Im}\left(B_{KK}\bar{B}_{KK}^*\right) = r_u^2\tilde{\mathcal{T}}^2\sin 2\gamma - 2r_u\tilde{\mathcal{T}}\tilde{\mathcal{P}}'\cos\delta\sin\gamma, && (19)\\
\tilde{D} &\equiv \frac{1}{2}\left(|B_{KK}|^2 + \left|\bar{B}_{KK}\right|^2\right) = r_u^2\tilde{\mathcal{T}}^2 + \tilde{\mathcal{P}}'^2 - 2r_u\tilde{\mathcal{T}}\tilde{\mathcal{P}}'\cos\delta\cos\gamma, && (20)\\
\tilde{E} &\equiv \frac{1}{2}\left(|B_{KK}|^2 - \left|\bar{B}_{KK}\right|^2\right) = 2r_u\tilde{\mathcal{T}}\tilde{\mathcal{P}}'\sin\delta\sin\gamma, && (21)\\
\tilde{F} &\equiv |B_{KK}^s|^2 = \tilde{\mathcal{P}}'^2 && (22)
\end{aligned}$$

Combined with the 6 quantities in Eqs. (6-11), we have 12 equations in 9 unknowns. As shown below, this allows us to solve for the CP angles, as in the DGR method, but greatly reduces the discrete ambiguities.

The CP angles can be obtained as follows. First, one finds the ratios $\tilde{\mathcal{T}}/\mathcal{T}$, $\tilde{\mathcal{P}}/\mathcal{P}$, and $\tilde{\mathcal{P}}'/\mathcal{P}'$:

$$a \equiv \frac{\tilde{\mathcal{T}}}{\mathcal{T}} = \frac{\tilde{E}}{E}\sqrt{\frac{F}{\tilde{F}}}\,, \qquad b \equiv \frac{\tilde{\mathcal{P}}}{\mathcal{P}} = \frac{\tilde{B}E}{B\tilde{E}}\sqrt{\frac{\tilde{F}}{F}}\,, \qquad c \equiv \frac{\tilde{\mathcal{P}}'}{\mathcal{P}'} = \sqrt{\frac{\tilde{F}}{F}}\,. \qquad (23)$$

Using these, we can find the values of all the magnitudes of the amplitudes. The amplitudes $\mathcal{T}$ and $\mathcal{P}$ are obtained from

$$\mathcal{T}^2 = \frac{(acD - \tilde{D}) - c(a-c)F}{a(c-a)r_u^2} \, , \quad \mathcal{P}^2 = \frac{abA - \tilde{A}}{b(a-b)} + \frac{a}{b}\frac{(acD - \tilde{D}) - c(c-a)F}{a(c-a)r_u^2} \, , \tag{24}$$

and the remaining amplitudes can be found using Eq. (23). Note that all magnitudes are positive, by definition.

We now turn to the angles. Using our knowledge of the magnitudes of the amplitudes, we have

$$\begin{aligned} \cos(\delta - \alpha) &= \frac{\mathcal{T}^2 + \mathcal{P}^2 - A - B}{2\mathcal{T}\mathcal{P}} \, , \\ \cos(\delta + \alpha) &= \frac{\mathcal{T}^2 + \mathcal{P}^2 - A + B}{2\mathcal{T}\mathcal{P}} \, , \\ \cos(\delta - \gamma) &= \frac{r_u^2\mathcal{T}^2 + F - D + E}{2r_u\mathcal{T}\sqrt{F}} \, , \\ \cos(\delta + \gamma) &= \frac{r_u^2\mathcal{T}^2 + F - D - E}{2r_u\mathcal{T}\sqrt{F}} \, . \end{aligned} \tag{25}$$

These equations can be solved to give the phases α, γ and δ up to a four-fold ambiguity. That is, if α_0, γ_0 and δ_0 are the true values of these phases, then the following four sets of phases solve the above equations: $\{\alpha_0, \gamma_0, \delta_0\}$, $\{-\alpha_0, -\gamma_0, -\delta_0\}$, $\{\alpha_0 - \pi, \gamma_0 - \pi, \delta_0 - \pi\}$, and $\{\pi - \alpha_0, \pi - \gamma_0, \pi - \delta_0\}$. Note, however, that we still haven't used the C and $\tilde{C}$ measurements. Their knowledge eliminates two of the four sets, leaving

$$\begin{aligned} &\{\alpha_0, \gamma_0, \delta_0\} \, , \\ &\{\alpha_0 - \pi, \gamma_0 - \pi, \delta_0 - \pi\} \, . \end{aligned} \tag{26}$$

These two solutions correspond to two different orientations of the unitarity triangle, one pointing up, the other down. This final ambiguity cannot be resolved by this method alone. However, within the SM it can be removed by using other measurements such as ϵ_K or the third CP angle β.

3 More Comments

For more details on this part of the talk, please see ref.[3]. I also talked on several other topics: determination of $|V_{td}/V_{ts}|$ from $B \to X_{s,d} l\bar{l}$, of $(\sin\gamma/\sin\beta)$ from $B \to \rho(\pi)\nu\bar{\nu}$, and of $|V_{ub}/V_{cb}|$ from the invariant hadronic mass distribution of $B \to X l \bar{\nu}$:

1. We propose[4] a new method to extract $\frac{|V_{td}|}{|V_{ts}|}$ from the ratio of the decay distributions $B \to X_d l\bar{l}/B \to X_s l\bar{l}$. This ratio depends only on the KM ratio $\frac{|V_{td}|}{|V_{ts}|}$ within 15% theoretical uncertainty, if the dilepton invariant mass-squared is away from the peaks of the accessible resonance states, J/ψ, ψ', *etc.* We also give a detailed analytical and numerical analysis of $B \to X_q l\bar{l}$.

2. We propose[5] a new method for the precise determination of $\left|\frac{V_{td}}{V_{ub}}\right|$ from the ratios of branching ratios $\frac{\mathcal{B}(B\to\rho\nu\bar{\nu})}{\mathcal{B}(B\to\rho l\nu)}$ and $\frac{\mathcal{B}(B\to\pi\nu\bar{\nu})}{\mathcal{B}(B\to\pi l\nu)}$. As is well known, $\left|\frac{V_{td}}{V_{ub}}\right|$ equals $\left(\frac{\sin\gamma}{\sin\beta}\right)$ in the CKM version of CP-violation within the Standard Model.

3. In order to determine the ratio of CKM matrix elements $|V_{ub}/V_{cb}|$ (and $|V_{ub}|$), we propose[6] a new model-independent method based on the heavy quark effective theory. In the forthcoming asymmetric B-experiments with microvertex detectors, BABAR and BELLE, the total separation of $b \to u$ semileptonic decays from the dominant $b \to c$ semileptonic decays would be experimentally viable.

Acknowledgments

This work is supported by the Korean Research Foundation made in the program year of 1997.

References

1. A.S. Dighe, M. Gronau and J.L. Rosner, *Phys. Rev.* D **54**, 3309 (1996).
2. S. Stone, hep-ph/9709500.
3. C.S. Kim, D. London and T. Yoshikawa, hep ph/9708356.
4. C.S. Kim, T. Morozumi and A.I. Sanda, *Phys. Rev.* D **56**, 7240 (1997); hep-ph/9706380.
5. T.M. Aliev and C.S. Kim, hep-ph/9710428.
6. C.S. Kim, *Phys. Lett.* B **414**, 347 (1997); *Nucl. Phys.* **B59**(Proc.Suppl.), 114 (1997) (hep-ph/9701316).

PHYSICS AT ASYMMETRIC B FACTORIES

MASASHI HAZUMI
Graduate School of Science, Osaka University,
1-1 Machikaneyama, Toyonaka, Osaka 560-0043, Japan

Asymmetric B factories, i.e. high-luminosity e^+e^- colliders running on the $\Upsilon(4S)$, are under construction at KEK and SLAC. Physics runs will start in 1999 in both experiments. In this article, the expected physics performance of both experiments is reported.

1 Introduction

BELLE and BABAR are two collaborations for asymmetric e^+e^- colliders which are now under construction. The BELLE detector[1] is for the KEKB collider. The design luminosity of KEKB is $L = 10^{34}$ cm^{-2}s^{-1} and the expected integrated luminosity is ~100 fb^{-1}/year[a]. The BELLE detector will be operational for physics runs from January 1999. The BABAR detector[2] is for the PEPII collider at SLAC. The design luminosity of PEPII is $L = 3 \times 10^{33}$ cm^{-2}s^{-1}, corresponding to an expected integrated luminosity of $\sim 30 fb^{-1}/year$. The BABAR detector will be operational for physics runs from July 1999.

The principal physics goal of the asymmetric e^+e^- colliders running on the $\Upsilon(4S)$ is to discover and measure the time–dependent CP asymmetry in B decays[3] by measuring the distance between the decay vertices of two B mesons. This becomes possible only by boosting the produced $B^0/\bar{B}^0$ system. Thus the electron and positron beams should have different energies. The values chosen are 8 GeV e^- × 3.5 GeV e^+ at KEKB and 9 GeV e^- × 3.1 GeV e^+ at PEPII. The Lorenz factor $\beta\gamma$ for B mesons is 0.42 at KEKB and 0.56 at PEPII in the laboratory frame. From the observed CP asymmetry, angles of the 'unitarity triangle' can be obtained. Decays of B mesons into $J/\psi K_S$, $J/\psi K_L$, $\psi(2S)K_S$, $D^{(*)}D^{(*)}$, $\pi^+\pi^-$ etc. are available for this study. In conjunction with measurements of CKM elements (from $b \to ul\nu$, $cl\nu$ transitions etc.), we are able to overconstrain the triangle, which will be a strict test of the mechanism of CP violation proposed first by Kobayashi and Maskawa in 1973[4]. In addition to the time dependent CP asymmetry which requires $B^0 - \bar{B}^0$ mixing, the expected high luminosity allows us to explore direct CP violation in decays such as $B^\pm \to \pi^0 K^\pm$. A big advantage of B physics at e^+e^- colliders, compared with hadron machines, is the high signal-to-background ratio. Thus

[a] At the beginning the luminosity will be 2×10^{33} cm^{-2}s^{-1}.

Table 1: Technologies and dimensions of BELLE and BABAR detectors.

	BELLE	BABAR
Vertexing	Double-sided silicon strips 3 layers	Double-sided silicon strips 5 layers
Tracking	Drift chamber 50 (anode) layers 3 (cathode) layers	Drift chamber 40 layers
PID	Silica Aerogel Scintillator (TOF)	DIRC
Calorimetry	CsI(Tl) ($16.1X_0$) barrel ($r = 125cm$) forward ($z = +196cm$) rear ($z = -102cm$) BGO (extreme forward)	CsI(Tl) ($16 - 17.5X_0$) barrel only ($r = 90.5cm$)
Magnet	Super-conducting inn.rad.=170cm B=1.5T	Super-conducting inn.rad.=140cm B=1.5T
K_L, μ	RPC inside return yoke (14 layers)	RPC inside return yoke (18-19 layers)

decay modes including photons and neutral pions can be studied with small background fractions. An example is the $b \to s\gamma$ transition, which opens a window for physics beyond the Standard Model such as supersymmetry. Physics of τ, charm, 2 photon processes, and B_S (on $\Upsilon(5S)$) can also be beyond the present reach. An asymmetric B factory is an exciting physics challenge with $O(10^8)$ produced B mesons.

In order to achieve the physics goals mentioned above, it is indispensable to reconstruct various *exclusive* B decays into CP eigenstates with precise vertex information. Effective flavor tagging is also essential. Since decay modes into CP eigenstates have typical branching ratios of $O(10^{-4})$ or even below, the efficiencies of exclusive reconstruction should be as high as possible. The specific requirements for the detectors are:

- Wide angular coverage (acceptance of more than 90%);
- Good resolution on the distance between B decay vertices ($\sim 100\mu$m or better);
- Good momentum resolution for charged tracks (e.g. $B \to \pi^+\pi^-$ should be distinguished from $K^\pm\pi^\mp$ through the difference in invariant masses);

- Good low p_t track finding (e.g. to tag slow π from D^*, with p_t down to several tens of MeV/c);
- e, μ tagging with high efficiency and purity (e.g. for flavor tagging and $J/\psi \to l^+l^-$);
- Good K_L tagging for $B \to J/\psi K_L$;
- $K^\pm/\pi^\pm$ separation for a wide momentum range (e.g. in the low momentum region for flavor tagging, in the high momentum region for $B \to \pi^+\pi^-/K^\pm\pi^\pm$ separation);
- Good energy and position resolution for photons with a wide momentum range (from several tens of MeV to $4 \sim 5$ GeV); and
- Stable operation for $\sim$10 years.

Table1 summarizes the technologies and dimensions of the BELLE and BABAR detectors. As is seen, most detector technologies are similar at both experiments: Double–sided silicon strip detectors (DSSDs) for vertexing, a drift chamber with a minimal amount of material for tracking, a CsI calorimeter inside a superconducting magnet, and resistive plate counters for K_L and μ identification. In some points, however, the two detectors are clearly different: in the structure of the vertex detectors and devices for particle identification. Different approaches are taken also for the data acquisition and trigger systems.

In the following sections, we will take a closer look at the expected physics performance of the asymmetric B factories.

2 Measurements on time-dependent CP violation

Monte Carlo simulation programs have been developed by both experiments based on the expected detector performance. The sensitivity for CP violating parameters has been estimated with these programs. The results are compiled in Table2[2,5,6]. It should be emphasized that a lot of decay modes are useful, thanks to the high signal-to-background ratio. Although not shown in the table, studies on additional charmonium modes for ϕ_1 measurements, such as $\chi_{c1} K_S$ and $\eta_C K_S$, seem promising and studies are in progress. Combining all the available decay modes, CP–violating phases (if not tiny) will be determined precisely.

Assuming the CP reach shown in the table, a rough prediction of the first observation of a CP asymmetry in B decays is as follows: with an integrated luminosity of 50 fb^{-1}, the entire region of the $\rho - \eta$ plane allowed at the

Table 2: Errors on the measurement of the CP-violating phases $\sin 2\phi_1$, $\sin 2\phi_2$, $\sin 2\phi_3$. Integrated luminosities of 100 fb^{-1} for BELLE and 30 fb^{-1} for BABAR are assumed, both of which are expectations with one year running.

CP-violating phase	CP Mode	$\delta \sin 2\phi$	
		BELLE	BABAR
ϕ_1 (=β)	$B^0 \to J/\psi K_S$	0.080	0.098
	$B^0 \to J/\psi K_L$	0.137	0.16
	$B^0 \to J/\psi K^*$	0.191	0.19
	$B^0 \to \psi(2S) K_S$	0.13	
	$B^0 \to D^+ D^-$		0.21
	$B^0 \to D^{*+} D^{*-}$		0.15
	$B^0 \to D^{*\pm} D^{\mp}$		0.15
ϕ_2 (=α)	$B^0 \to \pi^+ \pi^-$	0.147	0.20
	$B^0 \to \rho^\pm \pi^\mp$	$\leq$0.19	0.11
	$B^0 \to a_1 \pi$		0.24
ϕ_3 (=γ)	$B^0 \to D^0 K_S$	$\delta\phi_3 \leq 25°$	
	$B^\pm \to D^0 K^\pm$	$\delta\phi_3 \leq 15°$	
	$B^0 \to D^0 K^*$	$\delta\phi_3 \leq 27°$	

90%C.L. will be covered combining all the ϕ_1 measurements. Even with only 10 fb^{-1}, most of the 1σ allowed region in the $\rho - \eta$ plane will be covered.

For the measurements of ϕ_1 and ϕ_2, first of all a neutral B meson, which decays into a CP eigenstate or a flavor non–specific state, should be fully reconstructed. All the remaining particles are regarded as the decay products of the other B meson (denoted as B_{tag}). It is essential to determine the flavor of B_{tag} (called flavor tagging hereafter). There are two major methods which have been studied in detail: primary lepton tagging and kaon charge tagging. In the primary lepton tagging method, one tries to find a lepton which presumably comes from a semi–leptonic decay of B_{tag}. Applying a cut on the lepton momentum in the rest frame of B_{tag} (or $\Upsilon(4S)$), one can reject leptons from charmed hadrons (i.e. cascade decays) rather effectively.

In kaon charge tagging, one tries to find a charged kaon which presumably comes from the cascade decay $b \to c \to s$. As described in the previous section, it is essential to have excellent $K^\pm/\pi^\pm$ separation capability for this purpose. Kaon charge tagging generally has a higher "wrong tagging" probability (p_w) compared with that of primary lepton tagging. Thanks to the higher tagging efficiency (p_t), however, the effective tagging efficiency defined as $p_t(1-2p_w)^2$ is generally 3-4 times as large as that of the primary lepton tagging.

Table 3: The expected sensitivities for triangle-side measurements with integrated luminosities of 10 and 100 fb^{-1}. The expected statistical errors on $|V_{cb}|, |V_{ub}/V_{cb}|$ and $|V_{td}/V_{ts}|$ are compared with the presently achieved statistical and systematic errors. We assume $Br(B \to \pi l \nu) = 1.0 \times 10^{-4}$ and $Br(B \to \rho\gamma) = Br(B \to \omega\gamma) = 5 \times 10^{-6} \sim 1 \times 10^{-6}$.

side	decay mode		statistical error			syst.
	BELLE	now	BELLE		now	now
			100 fb^{-1}	10 fb^{-1}		
$\lvert V_{cb}\rvert$	$B \to D^* \ell \nu$	$B \to D^* \ell \nu$	0.3%	0.8%	5 %	4 %
$\lvert V_{ub}/V_{cb}\rvert$	$B \to \pi \ell \nu$ incl. $\ell^{\pm}$	$B \to \pi \ell \nu$ incl. $\ell^{\pm}$	5%	16%	- 15 %	- 14 %
$\lvert V_{td}/V_{ts}\rvert$	$B \to \rho(\omega)\gamma /$ $B \to K^*\gamma$	-	$10 \sim 34\%$	$31 \sim 107\%$	-	-

3 Other physics

Among many other physics topics, the sizes of the CKM matrix elements V_{cb}, V_{ub}, and V_{td} are very important. Note that these are the lengths of the sides of the unitarity triangle. Table3 shows the expected precision for the BELLE experiment.

Searches for direct CP violation in either neutral or charged B meson decays are also of crucial importance. In most cases, however, there is no way to predict the non–trivial phases from final state interactions. Therefore it is generally difficult to predict the expected physics performance precisely. From the experimental point of view, what is important will thus be to reconstruct as many decay modes as possible with high efficiency and purity. Such a B reconstruction "engineering" is not a trivial task but will be a key to finding new (even unexpected) phenomena in B decays. This is also important to test new theoretical ideas[7] within a short time.

"Traditionally" speaking, one good way to test physics beyond the Standard Model (SM) is to measure the branching fractions of penguin decays such as $b \to s\gamma$. In addition to more precise measurements of such kinds at asymmetric B factories, one is able to search for anomalous time dependent CP asymmetries which cannot be explained within the SM. An example is the ϕ_1 measurement in hadronic $b \to s$ penguin decays, which can be affected significantly by new physics[8].

4 Summary

BELLE at KEK and BABAR at SLAC plan to start physics runs in 1999. At the beginning of the next century, after a few years of running, it is anticipated that exciting results on CP violation and flavor physics will appear: time–dependent CP violation, direct CP violation, and measurements of $|V_{ub}|$, $|V_{cb}|$ and $|V_{td}|$. Some results on physics beyond the Standard Model will be also obtained.

While the choices of detector technologies are similar in the two collaborations, there also exist some differences especially in the particle identification and the structure of the vertex detectors. Each detector is expected to provide the best performance for each collider. Thus efficient and precise reconstruction of exclusive B decays will be realized.

Acknowledgments

The author expresses a special gratitude to the Organizing Committee for the invitation to this conference.

References

1. BELLE Technical Design Report, KEK Report 95-1, April 1995.
2. BABAR Technical Design Report, SLAC-R-95-457, March, 1995.
3. A. Carter and A.I. Sanda, *Phys. Rev. Lett.* **45**, (1980) 952.; *Phys. Rev.* **D23**, (1981) 1567; I.I. Bigi and A.I. Sanda, *Nucl. Phys.* **B193**, (1981) 851.
4. M. Kobayashi and T. Maskawa, Prog. Theo. Phys., 49 (1973) 652.
5. BELLE Progress Report, 1996.
6. BELLE Progress Report, 1997.
7. A good example of such new theoretical ideas is found in R. Fleischer and T. Mannel, hep-ph/9704423.
8. D. London and A. Soni, *Phys. Lett.* **B407** (1997) 61.

PQCD ANALYSIS OF INCLUSIVE HEAVY HADRON DECAYS

HOI-LAI YU
Institute of Physics, Academia Sinica, Taipei, Taiwan 115, Republic of China

We develop the perturbative QCD formalism for inclusive heavy hadron decays. Transverse degrees of freedom of partons are introduced to facilitate the factorization description of the heavy hadron decays.

In this talk, we shall derive the PQCD factorization formula for the semi-leptonic decay $B \to X_u \ell \nu$ in a rigorous way, to demonstrate how techniques of PQCD can be applied to heavy hadron decays. The factorization procedures demand the inclusion of the transverse degrees of freedom of partons at the end point of the charged lepton spectrum, where an outgoing jet is present. Hence, we have to perform the resummation of large perturbative corrections in the transverse configuration space using the technique developed in [1].

We work in the rest frame of the B meson, and choose the following light-cone components for relevant momenta,

$$P_B = (P_B^+, P_B^-, \mathbf{0}_\perp) \,, \quad p_\ell = (p_\ell^+, 0, \mathbf{0}_\perp) \,, \quad p_\nu = (p_\nu^+, p_\nu^-, \mathbf{p}_{\nu\perp}) \,, \tag{1}$$

with $P_B^+ = P_B^- = M_B/\sqrt{2}$ and $p_\nu^2 = 0$. The independent variables are identified as p_ℓ^+, p_ν^- and p_ν^+, and their relations to E_ℓ, q^2 and q_0 are $E_\ell = p_\ell^+/\sqrt{2}$, $q^2 = 2p_\ell^+ p_\nu^-$ and $q_0 = (p_\ell^+ + p_\nu^+ + p_\nu^-)/\sqrt{2}$, respectively. We define $P_b = P_B - p$ as the b quark momentum, which satisfies $P_b^2 \approx M_b^2$, M_b being the b quark mass. p is the kick from the light components inside the B meson, which has a large plus component p^+ and small transverse components $\mathbf{p}_\perp$. The b quark decays into a u quark with momentum $P_u = P_B - p - q$. We have distinguished the B meson momentum P_B from the b quark momentum P_b here.

It is more convenient to employ the scaling variables

$$x = \frac{2E_\ell}{M_B} \,, \quad y = \frac{q^2}{M_B^2} \,, \quad y_0 = \frac{2q_0}{M_B} \,, \tag{2}$$

instead of the dimensionful ones E_ℓ, q^2 and q_0. Note that the scaling variables are defined in terms of the B meson mass M_B, since we formulate the factorization according to the B meson kinematics. Using the momentum configurations defined in eq. (1), the phase space is

$$0 \le x \le 1, \quad 0 \le y \le x, \quad \frac{y}{x} + x \le y_0 \le y + 1 \,. \tag{3}$$

In the end point region with $x \to 1$ ($p_\ell^+ \to M_B/\sqrt{2}$) and $y \to 0$ ($p_\nu^- \to 0$), we have $y_0 \to 1$ ($p_\nu^+ \to 0$) and $p \to 0$. The u quark momentum then has a large minus component $P_u^- = (1 - y/x)M_B/\sqrt{2}$ but a very small plus component $P_u^+ = (1 - y_0 - y/x)M_B/\sqrt{2}$, and thus a very small invariant $P_u^2 = M_B^2(1 - y_0 + y)$, which forms an on-shell jet subprocess. The u quark travels a long distance of $\mathcal{O}(1/\Lambda_{\rm QCD})$ before it hadronizes. Besides, the B meson is dominated by soft dynamics, which is the origin of a soft function. Therefore, the important contributions are factorized into the soft (S), jet (J) and hard (H) subprocesses.

The factorization formula for the inclusive semi-leptonic decay $B \to X_u \ell \nu$ is written as

$$\frac{1}{\Gamma_\ell^{(0)}} \frac{d^3\Gamma}{dx dy dy_0} = M_B^2 \int_{z_{\rm min}}^{z_{\rm max}} dz \int d^2\mathbf{p}_\perp \times S(z, \mathbf{p}_\perp, \mu) J(z, P_u^-, \mathbf{p}_\perp, \mu) H(z, P_u^-, \mathbf{p}_\perp, \mu) , \quad (4)$$

with the momentum fraction z defined by $z = P_b^+/P_B^+ = 1 - p^+/P_B^+$ and $\Gamma_\ell^{(0)} = \frac{G_F^2}{16\pi^3}|V_{ub}|^2 M_B^5$. μ in eq. (4) is the renormalization and factorization scale. The triple differential decay rate is, of course, μ independent. Note that in the region $y \to x \sim 1$ the outgoing u quark becomes soft and eq. (4) fails. We shall show that contributions from this dangerous region are suppressed by phase space. The upper limit of z takes the value $z_{\rm max} = 1$ in our analysis. When performing the factorization according to the b quark kinematics, one must assume $z_{\rm max} = M_B/M_b$, which is greater than 1, in order to fill up the kinematic window. It has been explained[2] that $z_{\rm max} > 1$ is not allowed in perturbation theory, and is thus of nonperturbative origin. From the kinematic constraints in eq. (3) and the on-shell condition of the u-quark jet, the lower limit of z should be $z_{\rm min} = x$, instead of $z_{\rm min} = 0$.

At the tree level the convolution factors J and H are given by

$$\begin{aligned} J^{(0)} &= \delta\left(M_B^2\left[1 - y_0 + y - (1-z)(1 - \frac{y}{x}) - \frac{2\mathbf{p}_\perp \cdot \mathbf{p}_{\nu\perp}}{M_B^2} - \frac{\mathbf{p}_\perp{}^2}{M_B^2}\right]\right) , \\ H^{(0)} &\propto (P_b \cdot p_\nu)(p_\ell \cdot P_u) = ((P_B - p)\cdot p_\nu)(p_\ell \cdot P_u) \\ &\propto (x - y)\left(y_0 - x - (1-z)\frac{y}{x} + \frac{2\mathbf{p}_\perp \cdot \mathbf{p}_{\nu\perp}}{M_B^2}\right) , \end{aligned} \quad (5)$$

where the identity $J^{(0)} = \delta(P_u^2)$ has been used. Equation (4) can be regarded as an expression at the intermediate stage in the derivation of conventional factorization theorems. If the $\mathbf{p}_\perp$ dependence in J and H is negligible, the variable $\mathbf{p}_\perp$ in S can be integrated over, and eq. (4) reduces to the conventional factorization formula.

Suppose we consider higher-order corrections to eq. (4) from a gluon crossing the final state cut, and route the loop momentum ℓ through, say, the jet subprocess. Without losing generality, we approximate $J(p^+ + \ell^+, P_u^- + \ell^-, \mathbf{p}_\perp + \ell_\perp) \approx J(p^+, P_u^-, \mathbf{p}_\perp + \ell_\perp)$ according to the kinematic relations $\ell^+ < p^+$, $\ell^- < P_u^-$ and $\ell_\perp \approx \mathbf{p}_\perp$. Hence, the loop integral cannot be performed unless the dependence of J on the transverse momentum is known. This difficulty can be removed by a Fourier transformation,

$$J(p^+, P_u^-, \mathbf{p}_\perp + \ell_\perp) = \int \frac{d^2\mathbf{b}}{(2\pi)^2} \tilde{J}(p^+, P_u^-, \mathbf{b}) e^{i(\mathbf{p}_\perp + \ell_\perp)\cdot\mathbf{b}} , \tag{6}$$

where the impact parameter b (Fourier conjugate variable of $p_\perp$) measures the transverse distance travelled by the jet. Using eq. (6), the $\ell_\perp$ dependence is decoupled from the jet function, and the factor $e^{i\ell_\perp\cdot\mathbf{b}}$ is absorbed into the loop integral, which can then be performed. Therefore, an extra factor $e^{i\ell_\perp\cdot\mathbf{b}}$ is associated with each gluon crossing the final state cut in our formalism.

To further simplify the factorization formula, we neglect those terms involving $\mathbf{p}_{\nu\perp}$ in J and H. This is a good approximation for $x \to 0$ and $x \to 1$, since for $x \to 0$ contributions from transverse momenta are not important, and for $x \to 1$ the magnitude $p_{\nu\perp} = M_B\sqrt{(y_0 - x - y/x)y/x}$ vanishes. Equation (4) then becomes

$$\begin{aligned}\frac{1}{\Gamma_\ell^{(0)}}\frac{d^3\Gamma}{dx dy dy_0} &= M_B^2 \int_x^1 dz \int \frac{d^2\mathbf{b}}{(2\pi)^2} \\ &\quad \times \tilde{S}(z, \mathbf{b}, \mu)\tilde{J}(z, P_u^-, \mathbf{b}, \mu) H(z, P_u^-, \mu) .\end{aligned} \tag{7}$$

Large radiative corrections spoil the perturbation theory and must be organized. One has to resum these large corrections up to next-to-leading logarithms. The first step in resummation is to map out the leading regions of radiative corrections. In the collinear region with the loop momentum ℓ parallel to P_u and in the soft region with $\ell \to 0$ we can eikonalize the heavy b-quark line. With the eikonal approximation, the b-quark propagator is expressed as $1/v \cdot \ell$ to the order $1/M_B$ with $v = (1, 1, \mathbf{0}_\perp)$. Hence, the factorization in eq. (7) is in fact valid to $\mathcal{O}(1/M_B)$. The physics involved in this approximation is that a soft gluon or a gluon moving parallel to P_u can not explore the details of the b quark, and its dynamics can be factorized. This is consistent with HQEFT, where the b quark is treated as a classical relativistic particle carrying color. Since the large mass M_B does not appear in the eikonal propagator, the only large scale in J is P_u^-. After eikonalization procedures, one obtains the Sudakov exponent, which resums large double logarithms arising from both collinear and soft regions,

$$S = 2\int_{1/b}^{m_B} \frac{dp}{p} \int_{1/b}^{p} \frac{d\mu}{\mu} A(\alpha_s(\mu)) , \tag{8}$$

$$A = \mathcal{C}_F \frac{\alpha_s}{\pi} + \left[\frac{67}{9} - \frac{\pi^2}{3} - \frac{10}{27} n_f\right] \left(\frac{\alpha_s}{\pi}\right)^2 , \tag{9}$$

with $\mathcal{C}_F = 4/3$ the color factor.

Having summed up the double logarithms, we concentrate on the single logarithms in $\tilde{S}$, H and the initial condition $\tilde{J}(b, \mu)$. Since both the differential decay rate and the Sudakov exponent $s(P_u^-, b)$ are RG invariant, we have the following RG equations:

$$\begin{aligned} \mathcal{D}\tilde{J}(b,\mu) &= -2\gamma_q \tilde{J}(b,\mu) , \\ \mathcal{D}\tilde{S}(b,\mu) &= -\gamma_S \tilde{S}(b,\mu) , \\ \mathcal{D}H(P_u^-,\mu) &= (2\gamma_q + \gamma_S) H(P_u^-,\mu) , \end{aligned} \tag{10}$$

with

$$\mathcal{D} = \mu \frac{\partial}{\partial \mu} + \beta(g) \frac{\partial}{\partial g} . \tag{11}$$

$\gamma_q = -\alpha_s/\pi$ is the quark anomalous dimension in axial gauge, and $\gamma_S = -(\alpha_s/\pi)C_F$ is the anomalous dimension of $\tilde{S}$. Integrating eq. (10), we obtain the evolution of all the convolution factors,

$$\begin{aligned} \tilde{J}(z,P_u^-,b,\mu) &= \exp\left[-2s(P_u^-,b) - 2\int_{1/b}^{\mu} \frac{d\bar{\mu}}{\bar{\mu}} \gamma_q(\alpha_s(\bar{\mu}))\right] \tilde{J}(z,b,1/b) , \\ \tilde{S}(z,b,\mu) &= \exp\left[-\int_{1/b}^{\mu} \frac{d\bar{\mu}}{\bar{\mu}} \gamma_S(\alpha_s(\bar{\mu}))\right] f(z,b,1/b) , \\ H(z,P_u^-,\mu) &= \exp\left[-\int_{\mu}^{P_u^-} \frac{d\bar{\mu}}{\bar{\mu}} [2\gamma_q(\alpha_s(\bar{\mu})) + \gamma_S(\alpha_s(\bar{\mu}))]\right] H(z,P_u^-,P_u^-) . \end{aligned} \tag{12}$$

We shall neglect the intrinsic b dependence of the distribution function f below. If Sudakov suppression in the large-b region is strong, we may drop the evolution of f and $\tilde{J}$ in b, which is proportional to $\alpha_s(1/b)$. Hence, we assume $f(z,b,1/b) = f(z)$, $\tilde{J}(z,b,1/b) = \tilde{J}^{(0)}(z,b)$, the Fourier transform of the tree level expression in eq. (5), and $H(z,P_u^-,P_u^-) = H^{(0)}(z,P_u^-)$.

Substituting eq. (13) into (7), we derive the factorization formula for the inclusive semi-leptonic B meson decay,

$$\frac{1}{\Gamma_\ell^{(0)}} \frac{d^3\Gamma}{dx dy dy_0} = M_B^2 \int_x^1 dz \int_0^\infty \frac{bdb}{2\pi} f(z) \tilde{J}^{(0)}(z,b) H^{(0)}(z,P_u^-)$$

$$\times \exp[-S(P_u^-, b)] . \tag{13}$$

The complete Sudakov exponent S is given by

$$S(P_u^-, b) = 2s(P_u^-, b) - \frac{5}{3\beta_1} \ln \frac{\hat{P}_u^-}{\hat{b}} , \tag{14}$$

with $\hat{P}_u^- = \ln(P_u^-/\Lambda)$, which combines all the exponents in eq. (13) and includes both leading and next-to-leading logarithms. It is straightforward to observe from eq. (14) that the Sudakov form factor e^{-S} falls off quickly at large $b \sim 1/\Lambda$, where $\alpha_s(1/b) > 1$ and perturbation theory fails. Hence, the Sudakov form factor guarantees that main contributions to the factorization formula come from the small b, or short-distance, region, and the perturbative treatment is indeed self-consistent.

Since all the double logarithmic corrections have been absorbed into the jet subprocess, the soft function $\tilde{S}$ contains only soft single logarithms. These single logarithms can be summed by solving the RG equation $\mathcal{D}\tilde{S} = -\gamma_S \tilde{S}$ as shown in eq. (10). The solution has been given in eq. (13),

$$\tilde{S}(z, b, \mu) = \exp\left[-\int_{1/b}^{\mu} \frac{d\bar{\mu}}{\bar{\mu}} \gamma_S(\alpha_s(\bar{\mu})) \right] f(z, b) , \tag{15}$$

where the initial condition $f(z)$ for the evolution of $\tilde{S}$ must be determined phenomenologically. From the definition of $\tilde{S}$, it is obvious that f depends only on the properties of the bound state $|B\rangle$, but not on the particular short distance subprocess. Therefore, f is a process-independent universal function describing the distribution of the b quark inside a B meson. Since the intrinsic b dependence (the perturbative b dependence has been collected into the exponent S) is not known yet, we take the ansatz $f(z, b) = f(z) \exp(-\Sigma(b))$, which leads to $\Sigma \to 0$ as $b \to 0$ according to the definition $f(z, b = 0) \equiv f(z)$. It is also natural to assume $\Sigma > 0$ for all b from the viewpoint that the b quark is bound inside the B meson. Hence, the intrinsic b dependence provides further suppression. The nonperturbative function $f(z)$, identified as the B-meson distribution function, can be expressed as the matrix element of the b quark fields, whose first three moments are[3,4]

$$\begin{aligned}
&\int_0^1 f(z) dz = 1 , \\
&\int_0^1 f(z)(1-z) dz = \frac{\bar{\Lambda}}{m_B} + O(\Lambda_{\rm QCD}^2/m_B^2) , \\
&\int_0^1 f(z)(1-z)^2 dz = \frac{1}{m_B^2}\left(\bar{\Lambda}^2 - \frac{\lambda_1}{3}\right) + O(\Lambda_{\rm QCD}^3/m_B^3) .
\end{aligned} \tag{16}$$

Though the exponent Σ is unknown, we can nevertheless extract its leading behavior by means of an IR renormalon analysis. Note that the perturbative Sudakov factor e^{-S} in Eq. (15) becomes unreliable as the transverse distance b approaches $1/\Lambda_{\rm QCD}$. Near this end point, $\alpha_s(1/b)$ diverges, and IR renormalon contributions are significant. We reexpress the RG result of the evolution of the distribution function, which is contained in the second term of S, as

$$W = \exp\left[4\pi C_F \int \frac{d^4 l}{(2\pi)^4}\frac{v_\mu v_\nu}{(v\cdot l)^2} 2\pi\delta(l^2)\alpha_s(l_T^2)e^{il_T\cdot\mathbf{b}}N^{\mu\nu}\right] . \tag{17}$$

The tensor $N^{\mu\nu} = g^{\mu\nu} - (n^\mu l^\nu + l^\mu n^\nu)/(n\cdot l) + n^2 l^\mu l^\nu/(n\cdot l)^2$ comes from the gluon propagator in axial gauge $n \cdot A = 0$.

Substituting the identity $\alpha_s(l_T^2) = \pi \int_0^\infty d\sigma \exp[-2\sigma\beta_1 \ln(l_T/\Lambda_{\rm QCD})]$, $\beta_1 = (33-2n_f)/12$, into Eq. (17), and performing the loop integral for $n \propto (-1,1,\mathbf{0})$, we obtain

$$W = \exp\left[C_F \int_0^\infty d\sigma \left(\frac{b\Lambda_{\rm QCD}}{2}\right)^{2\sigma\beta_1} \frac{\Gamma(-\sigma\beta_1)}{\Gamma(1+\sigma\beta_1)}\right] . \tag{18}$$

It is easy to observe that the pole of $\Gamma(-\sigma\beta_1)$ at $\sigma \to 0$ gives the perturbative anomalous dimension of the distribution function appearing in Eq. (8). The extra poles at $\sigma \to 1/\beta_1, 2/\beta_1, ...$, then correspond to the IR renormalons, giving corrections of powers b^2, b^4,..., respectively. These renormalons generate singularities, which must be compensated by nonperturbative power correctons in order to have a well-defined perturbative expansion. Using the "minimal" ansatz [2], *i.e.* picking up only the leading (left-most) renormalon contribution, we parametrize the exponent $\Sigma(b)$ by $\Sigma(b) = c' m_B^2 b^2$, corresponding to the fact that the power corrections start at $O(b^2)$. Certainly, other parametrizations consistent with this fact are equally good. We postulate the following two-parameter distribution function for the B meson,

$$f_B(z) = N\frac{z(1-z)^2}{[(z-a)^2+\epsilon z]^2}\theta(1-z) , \tag{19}$$

The charged lepton spectrum for the decay $B \to X_u \ell\nu$ from the naive quark model is obtained by simply choosing $f(z) = \delta(1-z)$ and ignoring the transverse momentum dependence in $J^{(0)}$ and $H^{(0)}$. A simple calculation leads to

$$Li\frac{1}{\Gamma_\ell^{(0)}}\frac{d\Gamma}{dx} = \frac{x^2}{6}(3-2x) . \tag{20}$$

This spectrum does not fall off at the end point $x \to 1$, contradicting the observed behavior of the inclusive semi–leptonic decays of B mesons. The discrepancy implies that the tree level analysis is not appropriate, especially in the end-point region where PQCD corrections are important.

The spectrum from the parton model without Sudakov suppression is obtained by adopting $H^{(0)} = (x-y)[y_0 - x - (1-z)y/x]$ and $P_u^2 = M_B^2[1-y_0+y-(1-z)(1-y/x)]$. With integration over y_0, we derive

$$\frac{1}{\Gamma_\ell{}^{(0)}}\frac{d\Gamma}{dx} = \int_0^x dy \int_x^1 dz f(z)(x-y)(y+z-x) \, . \tag{21}$$

Finally, including Sudakov suppression and all other single logarithms, we arrive at the charged lepton spectrum of the $B \to X_u \ell \nu$ decay that takes into account both large perturbative and nonperturbative corrections,

$$\begin{aligned}\frac{1}{\Gamma_\ell{}^{(0)}}\frac{d\Gamma}{dx} &= M_B \int_0^x dy \int_0^{1/\Lambda} db \int_x^1 dz f(z)(x-y)\xi \Big[(z+y-x)J_1(\xi M_B b) \\ &\quad - \frac{2}{M_B b}\xi J_2(\xi M_B b) + \xi^2 J_3(\xi M_B b)\Big] e^{-S(P_u^-,b)} \, ,\end{aligned} \tag{22}$$

with $\xi = \sqrt{(x-y)(z/x-1)}$ and $J_1,\ J_2,\ J_3$ are the Bessel functions of order 1, 2 and 3, respectively.

We stress that our results do not violate the conclusion from HQEFT and also respect global quark–hadron duality, if they are interpreted properly. To confirm this, we identify $P_b = (M_b^2/2P_B^-, P_B^-, \mathbf{0})$ as the momentum carried by a free b quark in factorization theorems, where the minus component P_b^- has been set to P_B^-. That is, the free b quark is not at rest inside the B meson. Reexpress eq. (21) into a form similar to that in [3]:

$$\frac{1}{\Gamma_\ell{}^{(0)}}\frac{d\Gamma}{dx} = F(x)\theta\left(\frac{M_b^2}{M_B^2} - x\right) + F\left(\frac{M_b^2}{M_B^2}\right) M(x) \, , \tag{23}$$

with $F(x) = x^2(3-2x)/6$ being the quark-model prediction derived from the conventional approaches and

$$F\left(\frac{M_b^2}{M_B^2}\right) M(x) = \int_0^x dy \int_x^1 dz f(z)(x-y)(y+z-x) - F(x)\theta\left(\frac{M_b^2}{M_B^2} - x\right) \, . \tag{24}$$

The step function in eq. (23) specifies the maximal E_ℓ in the decay of a free b quark with the above momentum P_b. The function $M(x)$, representing nonperturbative corrections to the b quark decay, coincides with the shape function $S(x)$ defined in [3].

We shall show that the contribution from $M(x)$ to the total decay rate is indeed of $\mathcal{O}(1/M_B^2)$. Integrating eq. (24) over x, we obtain

$$F\left(\frac{M_b^2}{M_B^2}\right)\int_0^1 M(x)dx = \frac{1}{12}\int_0^1 dzz^4 f(z) - \int_0^{M_b^2/M_B^2} F(x)dx\ . \tag{25}$$

An arbitrary structure function f, which possesses the same moment as in eq. (16), can be expanded in terms of δ-functions:

$$f(z) = \delta(1-z) - \frac{\bar{\Lambda}}{M_B}\delta'(1-z) + \mathcal{O}(\bar{\Lambda}^2/M_B^2)\ . \tag{26}$$

Inserting eq. (26) into (25), we justify straightforwardly that the nonperturbative correction

$$\begin{aligned} F\left(\frac{M_b}{M_B}\right)\int_0^1 M(x)dx &= \int_{M_b^2/M_B^2}^1 F(x)dx - \frac{1}{12}\frac{\bar{\Lambda}}{M_B}\int_0^1 dzz^4\delta'(1-z) + \cdots \\ &= \frac{1}{3}\frac{\bar{\Lambda}}{M_B} - \frac{1}{3}\frac{\bar{\Lambda}}{M_B} + \mathcal{O}(\bar{\Lambda}^2/M_B^2) \end{aligned} \tag{27}$$

vanishes at $\mathcal{O}(1/M_B)$ as concluded in [3]. In summary, the quark-model contribution from the window between $x = M_b^2/M_B^2$ and $x = 1$, with a width of $\mathcal{O}(1/M_B)$, cancels the $\mathcal{O}(1/M_B)$ correction from the structure function, such that the nonperturbative correction is of $\mathcal{O}(1/M_B^2)$. Our result provides an explicit dynamical demonstration the validity of global quark-hadron duality.

Acknowledgments

I would like to thank the organizers for this well run, fruitful and stimulating workshop.

1. H-n. and H.L. Yu, Phys. Rev. Lett. **74**, 4388 (1995);
   H-n. Li and H.L. Yu, Phys. Lett. **B353**, 301 (1995).
2. G.P. Korchemsky and G. Sterman, Phys. Lett. **B340**, 96 (1994).
3. M. Neubert, Phys. Rev. D49, 4623 (1994).
4. H-n. Li and H.L. Yu, Phys. Rev. D53, 4970 (1996).

RECENT DEVELOPMENTS IN HEAVY QUARKONIUM PHYSICS

PYUNGWON KO
Dept. of Physics, KAIST
Taejon 305-701, Korea

Recent developments in heavy quarkonium physics are reviewed, including (i) non-relativistic QCD (NRQCD), (ii) the importance of color–octet $(Q\bar{Q})$ components in the decay and production of a physical heavy quarkonium state, (iii) limitations of the NRQCD factorization and (iv) the double humps in the $\pi\pi$ spectrum in the decay $\Upsilon(3S) \to \Upsilon(1S)\pi\pi$.

1 Introduction

Heavy quarkonium physics was one of the early applications of perturbative QCD (PQCD). In the original treatment of the decay and production of a heavy quarkonium state H, it was assumed that $(Q\bar{Q})_1$ inside H is in a definite ${}^{2S+1}L_J$ color–singlet state. Further, production and decay rates of a heavy quarkonium H are assumed to be factorized into (i) short distance (SD) parts, which are calculable in PQCD (in an expansion in $\alpha_s(M_Q)$), and (ii) long distance (LD) parts, that may be parameterized in terms of the $(Q\bar{Q})_1$ wave function and its derivatives at the origin [1].

This factorization hypothesis works well for $S-$wave and $P-$wave states to lowest order in α_s. For example, the χ_{c0} state decays into light hadrons (LH) through $\chi_{c0} \to gg$ at the parton level:

$$\Gamma(\chi_{c0} \to \text{LH}) = \frac{18\alpha_s^2(M_c)}{M_c^4} \, |R_p^{'}(0)|^2. \tag{1}$$

However, such a factorization in the CSM fails once PQCD corrections are included, which induce an infrared divergence in its decay rate [2]. This implies the failure of factorization in terms of a single wave function of $(Q\bar{Q})_1$ in H. This problem remained unsolved until 1992, when Bodwin *et al.* pointed out *the importance of the color–octet* $(c\bar{c})_8$ *component* of H (see (5)) [3].

In this decade, prompt J/ψ production at the Tevatron could be measured with the development of high resolution vertex detectors. The data was astonishing in that the CSM underestimates the data by a factor of ~ 10 for J/ψ production, even worse for $\psi^{'}$ production (by ~ 30). Possible contributions in the CSM are (i) gluon–gluon fusion ($gg \to J/\psi + g$), (ii) gluon fragmentation ($g \to J/\psi + gg$), (iii) $c-$quark fragmentation ($c \to J/\psi + c$), and (iv) cascade decays from $P-$wave charmonia ($\chi_{cJ}(1P) \to J/\psi + \gamma$), which is the most important J/ψ production mechanism. However, for the case of ψ' production at the Tevatron, one observed that $\sigma_{\rm exp} \sim 30 \times \sigma_{\rm th}$ based on CSM. Since there are no known $P-$wave states (such as $\chi_{cJ}(2P)$) that can decay into ψ', two options are avaiable: (i) a hypothetical $\chi_{cJ}(2P)$ decaying into $\psi' + \gamma$ with a suitable branching ratio [4], which can be tested in B and Υ decays [5], and (ii) color–octet gluon fragmentation: $g \to (c\bar{c})_8$ followed by $(c\bar{c})_8 \to J/\psi + {\rm soft}\ gg$ [6]. This second idea is the main theme of this talk, and the next two sections will be devoted to this issue.

Another less understood phenomenon in heavy quarkonium physics in the 1980's is the double hump in the $\pi\pi$ spectrum in $\Upsilon(3S) \to \Upsilon(1S)\pi\pi$ [a]. This issue is related with chiral dynamics of pions, rather than heavy quarkonium itself, and will be separately discussed in Sec. 4.

In this talk, we review these recent developments in brief. In Section 2, NRQCD (and the factorization formula) is introduced. In Section 3, we discuss the color–octet contribution to J/ψ production in various high energy processes. In Section 4 the double peaks in $\Upsilon(3S) \to \Upsilon(1S)\pi\pi$ are discussed, and a summary is given in Section 5.

2 NRQCD

2.1 NRQCD Lagrangian

A heavy quarkonium state H is associated with several different scales with the following hierarchical structure [7] :

$$\Lambda_{QCD} \simeq M_Q v^2 << M_Q v << M_Q. \tag{2}$$

In brief, M_Q is the heavy quark mass scale, which is the typical energy scale for decay and production of H, and PQCD in $\alpha_s(M_Q)$ becomes applicable. The typical momentum scale of Q inside H is $p \sim M_Q v \sim 1/{\rm size}$, and the typical kinetic energy scale of Q inside H is $K.E. \sim M_Q v^2 \sim$ the level splittings. Finally, the typical energy scale for nonperturbative dynamics of light quarks and gluons is characterized by $\Lambda_{\rm QCD}$. From the observed spectra of charmonium and upsilon families, one gets $v^2 \sim 0.3$ for the $\psi = (c\bar{c})$ system, and ~ 0.1 for the $\Upsilon = (b\bar{b})$ system. For sufficiently heavy Q, one has $v \sim \alpha_s(M_Q v) > \alpha_s(M_Q)$. Formally, the v^2 expansion is more important than α_s corrections [7].

In order to obtain the NRQCD Lagrangian, one integrates out the modes with momentum $> M_Q$ (no $Q/\bar{Q}$ creation/annihilation possible), and expands

[a] This puzzle has nothing to do with the limitation of the CSM

the resulting effective action in terms of the heavy quark velocity. The NRQCD Lagrangian thus posseses $SU(3)_c$ gauge symmetry, P (parity) and C (charge conjugation) symmetry, rotational symmetry, and approximate heavy quark spin symmetry. It is written in terms of heavy quark (ψ) and heavy antiquark fields (χ), and their covariant derivatives:

$$\mathcal{L}_{\rm NRQCD} = \psi^\dagger \left(iD_0 + \frac{D^2}{2M_Q} \right) \psi + \chi^\dagger \left(iD_0 - \frac{D^2}{2M_Q} \right) \chi + \mathcal{L}_{\rm light} + \delta\mathcal{L} \quad (3)$$

Gluon exchange with $k \sim p \sim M_Q v$ is included in terms of the static potential, A^0, whereas soft gluons with $k \sim M_Q v^2$ are treated in QCD multipole expansion. $\delta\mathcal{L}$ includes the correction terms of $O(M_Q v^4)$ and higher. One can determine the coefficients of the operators in the NRQCD Lagrangian by calculating the same process in QCD and NRQCD, expanding both results in powers of $k/M_Q \sim v$, and matching to the desired order in v and α_s.

In NRQCD, a physical heavy quarkonium state is represented as a superposition of various $(Q\bar{Q})_{1,8}$ and dynamical gluons: $|J/\psi\rangle = O(1)|(c\bar{c})_1[^3S_1]\rangle + O(v)|(c\bar{c})_8[^3S_1]g\rangle + O(v^2)|(c\bar{c})_8[^1S_0]\rangle + \ldots$. The relative importance of various Fock states in this equation (and of NRQCD matrix elements in the factorization formulae discussed in the next subsection) are determined by velocity scaling laws that can be derived from the field equations of motion of NRQCD [7].

2.2 NRQCD factorization

Decays of a heavy quarkonium into light hadrons are described in terms of local 4–fermion operators in the NRQCD Lagrangian, using the optical theorem:

$$\Gamma(H \to LH) = 2{\rm Im}\, \langle H|\delta\mathcal{L}_{4-{\rm fermion}}|H\rangle = \Sigma_n \frac{2{\rm Im} f_n(\Lambda)}{M_Q^{d_n-4}}\, \langle H|O_n(\Lambda)|H\rangle \quad (4)$$

This is a double expansion in v^2 (non–pert.) and $\alpha_s(M_Q^2)$ (pert.).

One can determine the NRQCD matrix elements either from experimental data [8] or from lattice QCD [9]. For χ_{c0} decay, the CSM prediction of eq.(1) is modified to

$$\Gamma(\chi_{c0} \to {\rm LH}) = \pi\alpha_s^2(M_c)\; M_c^5\; \langle O_1(^3P_0)\rangle + \frac{N_f \pi \alpha_s^2(M_c)}{12M_c^3}\; \langle O_8(^3S_1)\rangle \quad (5)$$

where

$$\langle O_1\rangle = |\langle 0|\chi^\dagger(-\frac{i}{2}D\cdot\sigma)\psi\chi_{c0}\rangle|^2, \quad (6)$$

$$\langle O_8\rangle = \langle \chi_{c0}|\psi^\dagger \sigma T^a \chi \cdot \chi^\dagger \sigma T^a \psi|\chi_{c0}\rangle. \quad (7)$$

The second term comes from the annihilation of the color–octet $(c\bar{c})_8$ ($S-$wave) in the physical χ_{c0} state. The IR divergence in the first term at $O(\alpha_s^3)$ is cancelled by another divergence in the color–octet matrix element in the second term. Similar cancellations were shown to occur also in $P-$wave charmonium production [3].

The inclusive cross section for heavy quarkonium (H) production is:

$$\Sigma_X d\sigma(1+2 \to H(P)+X) = \frac{1}{4E_1E_2v_{12}} \frac{d^3P}{(2\pi)^3 2E_P} \Sigma_{mn} C_{mn} \langle 0|O_{mn}^H|0\rangle, \quad (8)$$

or,

$$d\sigma = \Sigma_n d\hat{\sigma}(Q\bar{Q}[n]+X)\langle 0|O^H(n)|0\rangle. \quad (9)$$

The C_{mn} take into account the short distance parts, which are calculable using PQCD in $\alpha_s(M_Q)$. The matrix element $\langle 0|O_{mn}^H|0\rangle$ is the VEV of a four–fermion operator of NRQCD, and is proportional to the inclusive transition probability of the perturbative $Q\bar{Q}[n]$ state into the physical heavy quarkonium state H. In terms of NRQCD operators,

$$O_{mn}^H = \chi^\dagger \mathcal{K}_m^{'\dagger} \psi \; \mathcal{P}_H \; \psi^\dagger \mathcal{K}_n \chi, \quad (10)$$

where the $\mathcal{K}$'s are products of a spin matrix, a color matrix and a polynomial in the covariant derivative, D. The projection operator $\mathcal{P}$ is defined as

$$\mathcal{P}_H \equiv \Sigma_S |H(P=0,S)\rangle\langle H(P=0),S|, \quad (11)$$

where the sum is over soft hadron states S with $E < \Lambda$.

3 Color–octet mechanism in J/ψ production

3.1 ψ' anomaly at the Tevatron

As discussed in the introduction, the color–octet gluon fragmentation $g \to (c\bar{c})_8[^3S_1]$ followed by $(c\bar{c})_8[^3S_1] \to J/\psi$ + (soft gluons) may explain the large excess of ψ's at the Tevatron [6]. Qualitatively, one has

$$\begin{aligned} &\text{CSM} \quad \sim \alpha_s^3 v^3 \quad \text{(SD suppressed, LD enhanced)} && (12)\\ &\text{COM} \quad \sim \alpha_s v^7 \quad \text{(SD enhanced, LD suppressed)} && (13) \end{aligned}$$

Cho and Leibovich included the color–octet $^1S_0$ and ^3P_J contributions as well as the color–octet $^3S_1$, and obtained the following constraints on the corresponding three NP parameters from the Tevatron data [10]:

$$\langle 0|O_8^{J/\psi}(^1S_3)|0\rangle = (6.6 \pm 2.1) \times 10^{-3}\ \text{GeV}^3$$

$$\frac{\langle 0|O_8^{J/\psi}(^3P_J)|0\rangle}{M_c^2} + \frac{\langle 0|O_8^{J/\psi}(^1S_0)|0\rangle}{3} = (2.2 \pm 0.5) \times 10^{-2}\ \text{GeV}^3$$

It is very important to check the idea of the color–octet mechanism and NRQCD factorization in other processes. In view of this, it is crucial to observe that the color–octet matrix elements $\langle 0|O_8^H(^{2S+1}L_J)|0\rangle$ are universal, i.e. process–independent. Therefore, one can determine these matrix elements from a (set of) process(es), and then apply them to other processes, and test NRQCD factorization. A lot of work has been done along this line. To name only a few that were discussed at this workshop, hadroproduction of $h_c(^1P_1)$ [11], $B \to J/\psi + X$, $Z^0 \to J/\psi + X$ [12], $e^+e^- \to J/\psi + X$ at CLEO [13].

3.2 $B \to J/\psi + X$

The relevant effective Hamiltonian for this decay is

$$H_{eff} = \frac{G_E}{\sqrt{2}}\, V_{cb}V_{cq}^* \, [C_1(\bar{c}c)_{V-A}(\bar{q}b)_{V-A} + C_2(\bar{c}t^a c)_{V-A}(\bar{q}t^a b)_{V-A}]\,, \qquad (14)$$

with $C_1(m_b) \approx 0.13$ and $C_2(m_b) \approx 2.21$ in the leading logarithmic approximation. The CSM prediction to the lowest order in α_s is

$$\Gamma(B \to J/\psi + X)_{\rm csm} = \frac{3\langle 0|O_1^{J/\psi}(^3S_1)|0\rangle}{3M_c^2}\, C_1^2 \left(1 + \frac{8M_c^2}{M_b^2}\right) \hat{\Gamma}_0, \qquad (15)$$

with $\hat{\Gamma}_0 \equiv |V_{cb}|^2 \frac{G_F^2}{144\pi}\, M_b^3 M_c \, \left(1 - \frac{4M_c^2}{M_b^2}\right)^2$. Using $\langle 0|O_1^{J/\psi}(^3S_1)|0\rangle$ determined from $J/\psi \to l^+l^-$, we get $B(B \to J/\psi + X)_{\rm csm} = 0.23\%$ compared with the most recent CLEO data $B(B \to J/\psi + X) = (0.80 \pm 0.08)\%$. Higher order corrections in α_s and relativistic corrections are not that important [8]. The color–octet $^3S_1$ contribution

$$\Gamma(b \to (\bar{c}c)_8[^3S_1]+s \to J/\psi+X) = \frac{\langle 0|O_8^{J/\psi}(^3S_1)|0\rangle}{2M_c^2}\, C_2^2 \left(1 + \frac{8M_c^2}{M_b^2}\right) \hat{\Gamma}_0 \qquad (16)$$

is enhanced because of large Wilson coefficients [8], $C_2 \approx 17C_1$. Thus, we get a moderate increase in the branching ratio to 0.58%. Color–octet $^1S_0$ and ^3P_J states also contribute at the same order of v^2 expansion, dependent on $\langle 0|O_8^{J/\psi}(^1S_0)|0\rangle$ and $\langle 0|O_8^{J/\psi}(^3P_J)|0\rangle$. J/ψ polarization in $B \to J/\psi + X$ [14] may provide another useful constraint on color–octet matrix elements. Similar analyses have been done for $B \to \chi_c + X$, and B decays into $D-$wave charmonium states [15]. Here the available phase space is rather small so that the parton model description may be a poor approximation.

3.3 J/ψ Photoproduction

Inelastic J/ψ photoproduction ($\gamma + p \to J/\psi + X$) is described as parton level subprocess $\gamma + g \to J/\psi + g$ in PQCD and the CSM [16]. This process is advocated as a nice probe of the gluon distribution function inside the proton.

PQCD corrections (in the CSM) to the leading order results have been computed by M. Krämer [17]. The impact of these radiative corrections is that the scale dependence in α_s and the structure functions is reduced. However, PQCD is out of control for $z > 0.8$ and $P_T^2 < 1$ GeV^2 at HERA energy ($\sqrt{s_{\gamma p}} = 100$ GeV. Therefore one needs cuts, $z < 0.8$ and $P_T^2 > 1$ GeV^2, when one employs the PQCD correction to J/ψ photoproduction.

Color–octet contributions were considered by various groups [18 19]. It was shown that the spectrum $d\Gamma/dz$ in the high $z > 0.9$ region blows up, where z is defined as $E_{J/\psi}/E_\gamma$ in the proton rest frame. This phenomenon was originally taken to be a signal that the determination of color–octet matrix elements from the Tevatron data on J/ψ production are inconsistent with J/ψ photoproduction data.

However, this may not be the case, because NRQCD may break down near $z = 1$, as discussed by Beneke et al. [20]. In the lowest order in v^2, one ignores $\Delta M \equiv M_H - 2M_Q$, *i.e.* one sets $M_{J/\psi} \approx 2M_c$. This is not valid near the phase space boundary, since one probes the dynamics in detail: $\Delta E < \Delta M$, especially, when the matrix element has mainly a support near the phase space boundary. This is the case for J/ψ photoproduction at large $z(> 0.8 \sim 0.9)$. In such cases, predictions become sensitive to ΔM, or to momentum carried away by light hadrons during hadronization. One can summarize this effect in terms of a universal shape function, as in $b \to s\gamma$ in HQET. This shape function shifts the unphysical partonic boundary of phase space to the hadronic one that is physically more sensible. Therefore, the problem with J/ψ photoproduction at $z > 0.9$ is probably less serious.

4 $\pi\pi$ spectrum in $\Upsilon'' \to \Upsilon\pi\pi$

The $M_{\pi\pi}$ spectra in $\psi' \to J/\psi\pi\pi$ and $\Upsilon' \to \Upsilon\pi\pi$ can be understood in terms of the QCD multipole expansion and the low energy theorem for pions, which dictates the following amplitude:

$$\mathcal{M} = A \, \epsilon \cdot \epsilon' \, \left[q^2 + C M_\pi^2 \right], \tag{17}$$

with $q^2 \equiv (p_1 + p_2)^2 = M_{\pi\pi}^2$. This amplitude predicts a peak in the high $M_{\pi\pi}$ region in agreement with data from ARGUS and CLEO. However, the double hump in $\Upsilon(3S)(k, \epsilon) \to \Upsilon(1S)(k', \epsilon')\pi(p_1)\pi(p_2)$ could not be understood with the above amplitude. There are several proposals to describe this phenomenon, but none of them were successful. In Refs.[21], the above amplitude was modified as

$$\mathcal{M} = A\,\epsilon\cdot\epsilon^{'}\ \left[q^2 + BE_1E_2 + CM_\pi^2\right], \tag{18}$$

and the phase shift information for $I = 0$, S and D waves was added. The double humps could then be fitted, and various angular distributions were predicted [21]. The predictions agree with the newest CLEO data except the $\cos\theta_\pi^*$ distribution. It turns out that inclusion of terms higher in pion momenta, such as $(E_1 + E_2)q^2$ and so on, improves the agreement with the data. A more systematic study in ChPT is in progress [22].

5 Conclusion

NRQCD provides a theoretical framework in which one can study perturbative (in $\alpha_s(M_Q)$) and nonperturbative (in v^2) aspects of heavy quarkonium physics. In particular, the role of the $|(Q\bar{Q})_8 g\rangle$ Fock state in heavy quarkonium production and decay can be rigorously formulated in NRQCD. One can fit the J/ψ and $\psi^{'}$ production rates at the Tevatron via the color–octet mechanism. In order to check this idea at other processes, PQCD corrections are essential. This part has been calculated only recently by Petrelli *et al.* [23]. A complete phenomenological analysis including this new result has not yet been performed, however. Finally, the double hump in the $M_{\pi\pi}$ spectrum can be explained in terms of an amplitude that satisfies the low energy theorem for pions, once the $D-$wave dipion amplitude is properly included.

Acknowledgements

I thank the organizing committee of this workshop, especially Manuel Drees and Kaoru Hagiwara, for their efforts to make this workshop successful and exciting. I am also grateful to Prof. H.S. Song, Jungil Lee, Seung Won Baek, Sun Myong Kim and S. Chakravarty for enjoyable collaborations on some material presented in this talk. This work was supported in part by KOSEF through CTP at Seoul National University, and by the Ministry of Education through the Basic Science Research Institute, Contract No. BSRI-97-2418.

References

1. G. Gubernia, J.H. Kühn, R.D. Peccei and R. Rückl, Nucl. Phys. **B174**, 317 (1980).

2. R. Barbieri, M. Caffo, R. Gatto and E. Remiddi, Phys. Lett. **95 B**, 93 (1980); Nucl. Phys. **B 192**, 61 (1981).
3. G.T. Bodwin, E. Braaten and G.P. Lepage, Phys. Rev. **D46**, 1914 (1992); Phys. Rev. **D 46**, 3703 (1992).
4. D.P. Roy and K. Sridhar, Phys. Lett. **339**, 141 (1994).
5. P. Ko, Phys. Rev. **D 52**, 3108 (1995).
6. E. Braaten and S. Fleming, Phys. Rev. Lett. **74**, 3327 (1995).
7. G.T. Bodwin, E. Braaten and G.P. Lepage, Phys. Rev.**D 51**, 1125 (1995); Erratum-ibid. **D 55**, 5853 (1997).
8. P. Ko, J. Lee and H.S. Song, Phys. Rev. **D 53**, 1409 (1996).
9. G.T. Bodwin, D.K. Sinclair and S. Kim, Phys. Rev. Lett. **77**, 2376 (1996).
10. P. Cho and A.K. Leibovich, Phys. Rev. **D 53** (1996) 150; P. Cho and A.K. Leibovich, Phys. Rev. **D 53** (1996) 6203.
11. K. Sridhar, in these Proceedings.
12. Jungil Lee, in these Proceedings.
13. S.W. Baek, in these Proceedings.
14. S. Fleming, O.F. Hernandez, I. Maksymik and H. Nadeau, Phys. Rev. **D 55**, 4098 (1997).
15. P. Ko, J. Lee and H.S. Song. Phys. Lett. **B 395**, 107 (1997).
16. E. Berger and D. Jones, Phys. Rev. **D** , (1981).
17. M. Krämer, Nucl. Phys. **B 459**, 3 (1996).
18. M. Cacciari and M. Krämer, Phys. Rev. Lett. **76** , 4128 (1996).
19. P. Ko, J. Lee and H.S. Song, Phys. Rev. **D 54**, 4312 (1996).
20. M. Beneke, I.Z. Rothstein and M.B. Wise, **B 408**, 373 (1997).
21. S. Chakravarty and P. Ko, Phys. Rev. **D 48**, 1205 (1993); S. Chakravarty, S.M. Kim and P. Ko, Phys. Rev. **D 48**, 1212 (1993); S. Chakravarty, S.M. Kim and P. Ko, Phys. Rev. **D 50**, 389 (1994).
22. Saebyuk Bai and P. Ko, work in progress.
23. A. Petrelli, M. Cacciari, M. Greco, F. Maltoni and M.L. Mangano, hep-ph/9707223; F. Maltoni, M.L. Mangano and A. Petrelli, hep-ph/9708349.

PHYSICS AT THE ENERGY FRONTIER

KEN-ICHI HIKASA
Department of Physics, Tohoku University,
Aoba-ku, Sendai 980-8578, Japan

After a brief overview of our present position in understanding the fundamental interactions, theoretical and experimental constraints on the Higgs boson and its future perspective are discussed. The case without Higgs boson is also treated.

1 Where Do We Stand?

As there are four P's in the title of this workshop, there exist four fundamental forces in Nature: gravity, strong, electromagnetic, and weak interactions. My talk will be concerned with the last three of them.

1.1 Three Forces

A remarkable achievement in (particle) physics in the second half of this century is the unified understanding of these three forces. The strong, electromagnetic, and weak interactions can be understood as gauge interactions. Gauge interactions are characterized by their universality: the form and strength of the interactions only depends on the 'charges' of the particle. For instance, all quarks and leptons couple to the W boson with equal strength. It is often said that universality is the consequence of gauge symmetry. In fact, the symmetry alone is not sufficient to guarantee universality. Renormalizability of the theory is also essential.

Universality derives from the fact that the coupling of a particle with a gauge field arises only from the covariant derivative $D_\mu = \partial_\mu + igA_\mu$ in the kinetic term. Renormalizability forbids non-minimal couplings with the field strength $F_{\mu\nu}$. Renormalizability also ensures that the kinetic term is universal. These constraints ensures universality of the gauge forces.

The overall strength of the interaction is fixed by the gauge coupling g, the only parameter characterizing the interaction. There is an independent gauge coupling per a simple gauge group. As the Standard Model describing the three interactions has gauge group $SU(3) \times SU(2) \times U(1)$, three couplings determine the strength of the three interactions. These couplings are not independent in grand unified theories in which the gauge group is a large simple group containing $SU(3) \times SU(2) \times U(1)$.

1.2 Three Phases

Although the gauge theory description of the three forces is now universally accepted, it took a long time before this understanding was reached. In fact, non–Abelian gauge theories were proposed [1] by the president of APCTP in 1954, but only in the late 1970's the Standard Model was established as the theory of the fundamental interactions. This is essentially because the three forces look very different. In fact, it can be said that all these three forces are in a different 'phase' of the theory.

The electromagnetic interaction is in the Coulomb phase. The gauge particle is the massless photon, and the electromagnetic force mediated by the exchange of the photon is a long range one. Because of the weak coupling $\alpha \simeq 1/137$, perturbation expansion works very well.

The strong interaction is in the confinement phase. All observable states (hadrons) are singlets under the color gauge transformation and thus composite states. Elementary quarks and gluons are confined and cannot be observed singly. Because the gluons are confined, there is no observable long range force and the range of the interaction is determined by the confinement scale (more precisely the mass of the lightest hadron). The effective strong gauge coupling becomes smaller at shorter distances (asymptotic freedom), contrary to the electromagnetic coupling. This property, which is the result of the gauge dynamics itself, is related to confinement. There is no need of extra interactions to confine quarks and gluons. *The confining force is the strong gauge force itself.* The strong gauge coupling is reasonably small at high energies, which makes perturbative calculations possible. Perturbation theory ceases to work at low energies, which is the reason why we could not see the simplicity of the strong interactions for quite a while.

The weak interaction is in the Higgs phase. The interaction is mediated by the massive gauge bosons $W^{\pm}$ and Z^0, which acquire masses from the spontaneous breakdown of the gauge symmetry. The couplings have similar sizes as the electromagnetic coupling, and perturbation expansion again gives a good approximation.

1.3 Three Forces are Not Enough

Since the $SU(2) \times U(1)$ gauge couplings can be considered to be in the perturbative regime, the gauge interactions themselves cannot break the gauge symmetry. This is in contrast with the strong interactions, where the gauge force is responsible for confinement. If the electroweak gauge symmetry were not broken, the weak gauge bosons as well as the quarks and leptons would remain massless. Therefore, even if we ignore gravity, the three forces cannot be a consistent whole. *New extra forces are needed to explain our world.* Something which breaks the electroweak gauge symmetry (Higgs sector) and something which generates the fermion masses (Yukawa sector) are missing from our experiments.

1.4 The Standard Model or ...

In the Standard Model, introduction of a single Higgs field φ (an $SU(2)$ doublet) with all possible renormalizable interactions does everything needed. These interactions include the Higgs self interaction $\lambda(\varphi^\dagger\varphi)^2$ and the Yukawa interaction $f\varphi^\dagger\bar{\psi}_R\psi_L + \text{h.c.}$: a very economical model. Unlike the gauge interactions, we have no guiding principle to constrain the Yukawa couplings. Each quark or lepton thus has an independent coupling to the Higgs field. This leads to the many parameters in the Standard Model.

It should be stressed that we have no experimental evidence for these interactions whatsoever. In this sense, *only one third of the Standard Model has been verified.* We hope that we may get some clues on these new forces at next generation energy–frontier colliders.

There is a different approach to the electroweak symmetry breaking. If a new strong gauge interaction exists, it may dynamically break the symmetry and the introduction of a Higgs field may be circumvented. This is the underlying idea for the technicolor scenario for SU(2) × U(1) symmetry breaking. Unfortunately, we know of no realistic model of this type to incorporate the observed pattern of quark and lepton masses, in spite of many ideas and attempts for almost two decades. This disappointing fact may be saying that the scenario does not work, but it can as well be that we are just not clever enough.

1.5 Standard Model Parameters

Although there are quite a few parameters (around 18) in the Standard Model as listed in Tables 1 and 2, only one of them is dimensionful (at classical level). It is the Higgs mass term $\mu^2\varphi^\dagger\varphi$. This implies that all masses in the theory are proportional to this parameter, or equivalently the Higgs vacuum expectation value (or whatever you choose). Since the particle mass is given by the product of the vev and the coupling, the coupling of any elementary SM particle to the Higgs boson is proportional to the mass of the particle. If observed, this would be termed as a second kind of universality, the equalness of the mass–coupling ratio.

If one proceeds to the quantum level, however, other dimensionful couplings arise by dimensional transmutation: renormalization introduces a new

Table 1: Standard model parameters at the fundamental and phenomenological levels. We have no precise knowledge of the underlined parameters.

	Lagrangian level	Observational level
Gauge sector	g_s	$\Lambda_{\rm QCD}$ or $\alpha_s(m_Z)$
	$\left.\begin{matrix} g \\ g' \end{matrix}\right\}$	$\left\{\begin{matrix} \alpha \\ \sin^2\theta_W \text{ or } m_Z \end{matrix}\right.$
Higgs sector	$\left.\begin{matrix} \mu^2 \\ \lambda_H \end{matrix}\right\}$	$\left\{\begin{matrix} G_F \\ \underline{m_H} \end{matrix}\right.$
Yukawa sector	f_ℓ (3 physical parameters)	$m_e,\ m_\mu,\ m_\tau$
	$\left.\begin{matrix} f_u \\ f_d \end{matrix}\right\}$ (10 physical parameters)	$\left\{\begin{matrix} m_u,\ m_c,\ m_t \\ m_d,\ m_s,\ m_b \\ (\lambda,\ A,\ \underline{\rho},\ \underline{\eta}) \text{ or } \underline{V_{ij}} \end{matrix}\right.$

mass scale. Perhaps the only phenomenologically relevant such mass parameter is the QCD scale. At present we do not have any explanation for its ratio to the weak scale. Renormalization also introduces a theoretical problem known as the naturalness problem. Solutions to this problem generally imply new physics beyond the standard model *at TeV scale*, which is the target of the next generation colliders.

2 The Higgs Boson

2.1 Present Experimental Constraints

When the LEP experiments started in 1989, the experimental constraint on the Higgs boson was almost nonexistent except for nuclear transition limits, and any mass larger than ≈10 MeV was allowed. The situation changed drastically in 1989–90 to increase the limit by more than three orders of magnitude. The most recent limit comes from the search for $e^+e^- \to ZH^0$ at LEP II, giving $m_H \gtrsim 77\,\text{GeV}$ if the four experiments are combined. [8]

In the framework of the minimal Standard Model, it is possible to obtain information on the Higgs mass from electroweak precision measurements, as-

Table 2: Current values of the standard model parameters.

parameter	value	accuracy	source
$\alpha_s(m_Z^2)$	0.119 ± 0.002	0.02	RPP97 [2]
α	$1/137.0359895(61)$	4.5×10^{-8}	RPP96 [3]
G_F	$11.6639(2)\,\mathrm{TeV}^{-2}$	2×10^{-5}	RPP96 [3]
m_Z	$91.1867(20)\,\mathrm{GeV}$	2.5×10^{-5}	LEPEWG [4]
m_e	$510.99906(15)\,\mathrm{keV}$	3×10^{-7}	RPP96 [3]
m_μ	$105.658389(34)\,\mathrm{keV}$	3×10^{-7}	RPP96 [3]
m_τ	$1777.00(28)\,\mathrm{GeV}$	2×10^{-4}	RPP96 [3]
m_u	(2–8) MeV		RPP96 [3]
m_d	(5–15) MeV		RPP96 [3]
m_s	(100–300) MeV		RPP96 [3]
m_c	(1.0–1.6) GeV		RPP96 [3]
m_b	(4.1–4.5) GeV		RPP96 [3]
m_t	$174.1 \pm 5.4\,\mathrm{GeV}$	0.03	RPP97 [2]
λ	0.2196(23)	8×10^{-3}	RPP97 [2]
A	0.82 ± 0.04	0.05	RPP97 [2]
$\sqrt{\rho^2 + \eta^2}$	0.36 ± 0.09	0.25	RPP97 [2]
η	≈ 0.3		RPP97 [2]
m_H	$> 70.7\,\mathrm{GeV}$	(95%CL)	ALEPH [5]
m_{ν_e}	$< 4.35\,\mathrm{eV}$	(95%CL)	Belesev [6]
m_{ν_μ}	$< 0.17\,\mathrm{MeV}$	(90%CL)	RPP96 [3]
m_{ν_τ}	$< 24\,\mathrm{MeV}$	(95%CL)	RPP96 [3]

Light quark masses are current masses, charm and bottom masses are $\overline{\mathrm{MS}}$ masses, and m_t is from direct measurements. RPP96 [3] gives a conservative ν_e mass limit of 15 eV.

suming that no new particles or interactions are involved. At the tree level, the electroweak interactions describing *e.g.* Z production and decay are determined by the three parameters α, G_F and m_Z. All of them are known with an accuracy of 10^{-5} or better. At the one–loop level, however, other parameters affect the process. Of these remaining parameters, the lepton masses are precisely measured. The quark masses (and mixings) enter through hadronic vacuum polarization, which can be calculated with a reasonable accuracy from low energy e^+e^- data. The strong gauge coupling is also known with a good accuracy. Now the top quark mass is directly measured to a few % also. So,

m_H is the only uncontrolled parameter and a precision measurement of one observable (other than those described above) can yield information on m_H.

The mass of the W boson is one example. We now have measurements of m_W from both $p\bar{p}$ colliders and LEP II. The world average in summer 1997 is[7] $m_W = 80.40 \pm 0.08\,\text{GeV}$. This central value corresponds to a Higgs mass of around 100 GeV, whereas $m_H \sim 1\,\text{TeV}$ is about two standard deviations away.

There are various other precision measurements at the Z resonance including the Z width, forward–backward and left–right asymmetries, τ polarization and so on. Combining all these data, the LEP Electroweak Working Group gives[4] $m_H = 115^{+116}_{-66}$ GeV with the one–sided 95%CL upper limit of 420 GeV.

The preference for small m_H has been noted since a couple of years ago. Actually, the claim was not very reliable since it resulted mainly from the large value of R_b which itself was inconsistent with the Standard Model. This year, the low value seems genuine, although it is too early to exclude a heavy Higgs of $\sim 1\,\text{TeV}$ with confidence.

2.2 Theoretical Prejudices or Limits on the Higgs Mass

The mass of the Higgs boson in the Standard Model is proportional to the vev times the Higgs self coupling λ (its square root, to be exact). The Higgs mass is thus an independent parameter of the theory in principle; in this sense there is no theoretical limit on m_H. However, if m_H is larger than $\approx 1\,\text{TeV}$, perturbation theory breaks down and the Higgs sector becomes strongly interacting (this will be discussed further in the last section).

Renormalization group arguments can constrain the Higgs mass more tightly. The running of the effective Higgs self coupling λ is mainly controlled by λ itself and the top Yukawa coupling. The Yukawa coupling tends to reduce λ at higher energies and can eventually drive it to a negative value. On the other hand, λ pulls itself to the other direction. For the top mass around 175 GeV, there is a delicate balance between the two effects. For example, if one assumes that the minimal Standard Model is valid up to 10^{19} GeV, m_H has to be between 160 and 180 GeV, below which the vacuum becomes unstable, and above which the self coupling diverges. This bound, however, should not be taken too seriously, since it is rather difficult to imagine that no new physics appears until such a distant scale.

2.3 The Case of Minimal Supersymmetry

In the Minimal Supersymmetric Standard Model (MSSM), the Higgs self coupling is not an independent parameter but is related to the electroweak gauge couplings. This leads to a strong constraint on the Higgs mass, although the

Table 3: Lightest Higgs mass (in GeV) in the MSSM when the other Higgses are heavy.

$\tan\beta$	1.1	4	50
$m(\tilde{t}) = 250\,\text{GeV}$	42	90	100
$m(\tilde{t}) = 1\,\text{GeV}$	91	121	128

existence of two Higgs doublets in the theory means that a useful limit can be given only for the lightest neutral Higgs boson h^0. After including an important one-loop effect by top–stop loops, m_h has to be smaller than 120–130 GeV. An approximate expression for m_h in the case where the other Higgs bosons are much heavier is

$$m_H \simeq m_Z^2 \cos^2 2\beta + \frac{3G_F m_t^4}{\sqrt{2}\pi^2} \log \frac{m_{\tilde{t}}^2}{m_t^2} \quad (m_A \gg m_Z),$$

where $\tan\beta$ is the ratio of the two Higgs vevs. Some illustrative values are listed in Table 3.

A linear collider with $E_{\rm cm} = 300$–$500\,\text{GeV}$ can thus kill the MSSM if no Higgs is found in its reach.

2.4 Detection of the Standard Model Higgs

At electron–positron colliders, it is a rather easy task to discover a Higgs boson if the mass is within reach. The production mechanisms are $e^+e^- \to ZH$ via $s-$channel Z exchange (important at lower energies) and $e^+e^- \to \nu\bar{\nu}H$ via WW fusion (important at higher energies). The present limit of $\sim$80 GeV at LEP II may be improved to 350 GeV at $E_{\rm cm} = 500\,\text{GeV}$ and 700 GeV at $E_{\rm cm} = 1\,\text{TeV}$. It is also important to note that essentially all decay final states of the Higgs can be studied at e^+e^- machines.

At the LHC the production is thorough gluon–gluon fusion via a top quark loop. The knowledge of m_t reduces the uncertainty in the cross section calculation. Although the cross section is large, the detection is not entirely trivial. For $m_H \gtrsim 130\,\text{GeV}$, $H \to ZZ^{(*)} \to \ell\bar{\ell}\ell\bar{\ell}$ gives a clear signal. For $m_H \lesssim 130\,\text{GeV}$, it is claimed that the $\gamma\gamma$ decay mode should be detectable with a good detector.

3 Higgs Physics

Discovery of the Higgs boson is only the starting point for a wealth of physics. With the Higgs boson, we can study the heretofore unknown Yukawa and Higgs sectors. Linear colliders are particularly suited for this purpose. For example, branching ratio measurements for various Higgs decay modes can be done quite accurately [9,10]. Here let us summarize what kind of physics is important to study.

1. *Quantum numbers:* The Higgs boson is the 'radial vibration' of the vacuum. It thus has the same quantum numbers as the vacuum $J^{PC} = 0^{++}$. This can be examined using production and decay angular distributions.

2. *Saturation of vev:* The Higgs coupling to a gauge boson pair is fixed by the gauge coupling and the gauge boson mass. If there are more than one Higgs fields with vev, the coupling is smaller than expected. This may be studied using the cross sections $\sigma(ZH)$, $\sigma(\nu\bar{\nu}H)$, $\sigma(e^+e^-H)$, or $\Gamma(H \to WW)$.

3. *Doublet structure:* The doublet nature of the SM Higgs boson may be checked by comparing $B(WW)$ with $B(ZZ)$, or $\sigma(\nu\bar{\nu}H)$ with $\sigma(e^+e^-H)$.

4. *Yukawa couplings:* The top Yukawa coupling may be measured from $\sigma(t\bar{t}H)$ (for a rather light Higgs) or $B(H \to t\bar{t})$ (for a heavy Higgs), the bottom Yukawa from $B(bb)/B(WW)$. If there are two doublets generating the quark masses, the Higgs branching ratio to up–type and down–type quarks differs from the minimal case. For instance, in the MSSM 10% deviation from the canonical value is expected [11] even if the pseudoscalar Higgs A^0 is as heavy as 400 GeV.

5. *Higgs self coupling:* This is the ultimate test of the Higgs sector. It may be possible using $\sigma(ZHH)$ or $\sigma(\nu\bar{\nu}HH)$.

4 Strong Higgs Sector

4.1 What Goes Wrong If There Is No Higgs

As can be seen from the fact that we have no accurate knowledge of the Higgs boson mass, observables in our hands are quite insensitive to the presence of a Higgs boson. The Glashow model is almost as good as the Weinberg-Salam model! Absence of the Higgs boson reveals itself only at high energies with heavy particles. Let us consider the scattering of two W bosons, for example the process $W^+W^- \to ZZ$. It proceeds via $t-$ and $u-$channel W exchange and via the four–point contact interaction. The form of the vertices are prescribed by gauge symmetry and their strengths are determined by the electroweak gauge couplings.

If one calculates these Feynman graphs, one finds that the amplitude is proportional to $E^2_{\rm cm}$. If one looks more closely, this occurs only when the weak bosons are longitudinally polarized (helicity 0), corresponding to the Nambu–Goldstone degree of freedom. Moreover, the angular dependence of this E^2 piece is flat, implying that only the $J = 0$ partial wave has a problem. Eventually, partial wave unitarity is violated at $\approx 1\,$TeV, which in fact means the *appearance of a strong interaction at TeV scale.* In the Standard Model, the problematic piece is cancelled by another Feynman diagram, $s-$channel Higgs exchange.

4.2 Higgsless Standard Model

If we take the Higgs boson out of the Standard Model, it becomes a non-renormalizable theory. However, it is possible to construct a no–Higgs version of the Standard Model in which the SU(2) × U(1) symmetry is nonlinearly realized. It allows for a systematic 'low-energy' expansion and is called the electroweak chiral Lagrangian. As long as we are interested in processes below the TeV scale, this theory is as good as the Standard Model.

This theory can be regarded as an effective field theory written in terms of only the light physical degrees of freedom and contains the most general interaction terms allowed by the gauge symmetry. This framework breaks down when a new particle threshold is crossed, at which the introduction of these new degrees of freedom would be required.

The *main character* of the electroweak chiral Lagrangian is the Nambu-Goldstone fields w^a ($a = 1,\, 2,\, 3$) representing the longitudinal component of the weak bosons. The Lagrangian is written in terms of

$$U = \exp\bigl(i\tau^a w^a/v\bigr) \;\in\; \mathrm{SU}(2)$$

on which the gauge symmetry is nonlinearly realized

$$U \to V_L U V_Y^\dagger$$

where $V_L \in SU(2)$, $V_Y = \exp(i\tau^3\theta_Y) \in U(1)_Y$. The transformation matrix V_Y may be promoted to a global $\mathrm{SU}(2)_R$, which ensures $\rho = m_W^2/m_Z^2 \cos^2\theta_W = 1$ (custodial symmetry).

A systematic expansion is possible in terms of the number of derivatives. The custodially symmetric terms up to dimension-four are

$$\begin{aligned}\mathcal{L} &= \frac{1}{4}\mathrm{tr}(D_\mu U^\dagger D^\mu U) + \alpha_1 g g' \,\mathrm{tr}(\mathcal{W}_{\mu\nu} U \mathcal{B}^{\mu\nu} U^\dagger) \\ &\quad -i\alpha_2 g' \,\mathrm{tr}(D_\mu U^\dagger D_\nu U \mathcal{B}^{\mu\nu}) - i\alpha_3 g \,\mathrm{tr}(D_\mu U^\dagger \mathcal{W}^{\mu\nu} D_\nu U)\end{aligned}$$

where $\mathcal{W}_{\mu\nu} = \tau^a W^a_{\mu\nu}$, $\mathcal{B}_{\mu\nu} = \tau^3 B_{\mu\nu}$, and

$$D_\mu U = \partial_\mu U + \frac{1}{2} i g \tau^a W^a_\mu U - \frac{1}{2} i g' U \tau^3 B_\mu \,.$$

The quark/lepton part of the Lagrangian is to be added to this.

The unitary gauge with no unphysical scalars corresponds to $U = 1$. The dimension–two term becomes the W and Z mass terms and the parameter v fixes the mass scale. The dimension–four terms give gauge–boson two–, three–, and four–point vertices. Only the first term contains two–point functions. The parameter α_1 is proportional to the Peskin–Takeuchi S parameter ($S = -16\pi\alpha_1$), and is tightly constrained by the LEP/SLC precision measurements ($|\alpha_1|$ should be less than 0.005–0.01). (The other two Peskin-Takeuchi parameters T, U break the custodial symmetry and can be written in a similar form in the chiral Lagrangian.)

The other two parameters α_2 and α_3 can be measured in gauge boson pair production such as $e^+e^- \to W^+W^-$. Their effect becomes more prominent at higher energies because they break the subtle gauge cancellation between the neutrino exchange and γ/Z exchange diagrams. Although LEP II is too close to the threshold (and not high luminosity) for a meaningful measurement of these parameters, a linear collider with $E_{\rm cm} = 500\,\mathrm{GeV}$ or higher can determine them precisely. With a luminosity of $50\,\mathrm{pb}^{-1}$ at $E_{\rm cm} = 500\,\mathrm{GeV}$, $\alpha_{2,3}$ can be measured with an accuracy of one percent.[12] This accuracy is good enough to exclude a 'techni-ρ' resonance up to 3 TeV or so.

5 Final Word

Our next step in extending the energy frontier is eagerly awaited.

References

1. C. N. Yang and R. L. Mills, Phys. Rev. **96**, 191 (1954)
2. Particle Data Group, 1997 update (http://pdg.lbl.gov).
3. Particle Data Group, R. M. Barnett *et al.*, Phys. Rev. D **54**, 1 (1996).

4. The LEP Collaborations, the LEP Electroweak Working Group, and the SLD Heavy Flavor Group, CERN-PPE/97-154 (1997).
5. ALEPH Collaboration, R. Barate *et al.*, Phys. Lett. B **412**, 155 (1997).
6. Belesev *et al.*, Phys. Lett. B **350**, 263 (1995).
7. See *e.g.*, L. Taylor, hep-ex/9712016.
8. T. Mori, these proceedings.
9. M. D. Hildreth, T. L. Barklow, and D. L. Burke, Phys. Rev. D **49**, 3441 (1994).
10. I. Nakamura and K. Kawagoe, Phys. Rev. D **54**, 3634 (1996).
11. J. Kamoshita, Y. Okada, and M. Tanaka, Phys. Lett. B **391**, 124 (1997); Y. Okada, KEK-TH-508 (1997).
12. M. Tanabashi, A. Miyamoto, and K. Hikasa, in *Workshop on Physics and Experiments with Linear Colliders*, Morioka-Appi, Iwate, 1995, edited by A. Miyamoto, Y. Fujii, T. Matsui, and S. Iwata (World Scientific, Singapore, 1996), Vol. 2, pp. 493–501.

SUPERSYMMETRY PHENOMENOLOGY AT COLLIDERS

XERXES TATA
Department of Physics and Astronomy, University of Hawaii,
Honolulu, HI 96822, USA

We briefly overview SUSY phenomenology at current and future colliders.

1 What do we expect to find in a SUSY World?

Supersymmetry (SUSY) is a novel symmetry that inter–relates bosons and fermions and thus provides a level of synthesis never before achieved. [1] In a SUSY World, for each multiplet of SM particles (quarks, leptons, gauge bosons and Higgs bosons), there must be a corresponding multiplet of sparticles (squarks, sleptons, gauginos and Higgsinos) with the same mass and internal quantum numbers, but with spin differing by $\frac{1}{2}$. This feature follows directly from the supersymmetry algebra and excludes SUSY as an exact symmetry of nature: our experimental colleagues assure us that a bosonic particle with the mass and charge of an electron could not have escaped detection.

Supersymmetry must, therefore, be a broken symmetry. Herein lies all the problem. We are quite ignorant of the mechanism that underlies SUSY breaking. We expect that some dynamics leads to spontaneous SUSY breakdown — probably in a sector of the theory we do not know about, so that this breaking has also to be communicated to SM particles and their spartners — and that this breaking manifests itself in the form of *soft supersymmetry breaking* masses and couplings of the weak scale effective theory.

Within the simplest SUSY extension of the SM, dubbed the MSSM, [a] these soft terms consist [b] of a mass for each $SU(3) \times SU(2) \times U(1)$ multiplet of squarks, sleptons and the two Higgs sector weak doublets (9+6+2 masses), a mass for the gauginos of each factor of the gauge group (3 masses) together with nine trilinear (dubbed A-terms) and one bilinear scalar coupling (usually denoted by B). A single supersymmetry–conserving Higgsino mass parameter μ together with the SM and soft parameters completely specifies the model. It

[a] It is possible to introduce a new multiplicative quantum number $R-$parity which is defined to be $+1$ (-1) for SM (SUSY) particles and only allow interactions which conserve R. While it is not compulsory [2] to do so on phenomenological grounds, it is customary to include $R-$parity conservation as part of the definition of the MSSM. With this assumption the lightest SUSY particle (LSP) is stable.

[b] Allowing inter–generation mixing between squarks or $R-$parity violation greatly increases the number of parameters.

is important to note that while the introduction of soft SUSY breaking terms indeed destroys the unwanted equality of particle and sparticle masses, *it does not alter the dimension four (gauge and Yukawa) interactions of sparticles* which, together with sparticle masses, determine many broad features of SUSY phenomenology. Thus, despite our ignorance of how SUSY is broken, the picture has some predictive power.

The size of the parameter space makes the sort of phenomenological analyses that are essential for the planning of future experimental facilities very difficult. It is, therefore, customary to make further assumptions (usually about symmetries of high energy dynamics) that relate the various soft parameters. But even in a SUSY GUT, where there is just one gaugino mass and at most two scalar masses per matter generation, the number of free parameters is uncomfortably large. An economic framework that has proven very useful for phenomenological analyses is the so–called minimal supergravity (mSUGRA) GUT model where gravity communicates SUSY breaking to the superpartners of SM particles. Its characteristic feature is that just as gaugino masses unify to a common mass $m_{1/2}$, all scalar masses and scalar trilinear couplings also take on universal values, m_0 and A_0, at an ultra–high scale $M_X \sim M_{GUT} - M_{Planck}$. The diverse masses and couplings that enter the effective Lagrangian used for phenomenology at the weak scale are determined from the GUT scale parameter set, $(m_0, m_{1/2}, \mu_0, B_0, A_0)$ plus the SM parameters, by renormalization group evolution. Remarkably, this also drives the squared mass of one of the Higgs fields negative over a substantial range of parameters, leading to the observed breakdown of electroweak symmetry. We can thus eliminate (the weak scale parameter) μ^2 in favour of M_Z^2 from the parameter list above. It is also customary to eliminate B in favour of $\tan\beta = \frac{v_u}{v_d}$, the ratio of the two Higgs field VEVs. While it is very interesting that the correct pattern of electroweak breaking arises rather naturally in this framework, this should not be regarded as an *explanation* of it, since there are parameter regions where colour is also broken. The mSUGRA model is consistent with all phenomenological constraints from colliders, FCNC searches, proton decay, *etc.* and, if we assume $R-$parity is conserved, the LSP is a viable candidate for cosmological dark matter.

We caution the reader that the mSUGRA model is based on various assumptions [1] about physics at the high energy scale. These assumptions may well prove to be incorrect. It turns out that these assumptions can be directly tested [3] in experiments at future colliders. We should also mention that there are other models which make *different* assumptions about physics beyond the weak scale and lead to different predictions for sparticle masses and mixing patterns. [4] Examples include models where gauge interactions, not gravity, communicate SUSY breaking, or models in which squarks and quark matrices are somehow aligned to avoid FCNC. Measurement of sparticle properties will help us decide which, if any, of the many scenarios that have been considered is realized in nature. But first, we have to find the sparticles.

2 Discovering Sparticles at Colliders

Although we will adopt the mSUGRA framework for this discussion, our general conclusions about the reach of different facilities are probably insensitive to this, as long as $R-$parity is conserved, and coloured sparticles are substantially heavier than uncoloured ones. Heavy sparticles rapidly decay into SM particles and a lighter sparticle, which again decays until this decay cascade terminates in a stable LSP, which from cosmological arguments is taken to be neutral. Since $R-$parity conservation requires that SUSY particles are produced in pairs, every SUSY event contains two LSPs which escape the experimental apparatus undetected. Thus missing energy and an imbalance of transverse momentum ($\not{E}_T$) are regarded as hallmarks of SUSY events.

The visible debris that accompanies the $\not{E}_T$ depends on what is produced. The production of heavy gluinos and squarks leads to events with hard primary jets, together with softer jets and isolated leptons from the decays of daughter charginos and neutralinos, and $\not{E}_T$. These events may also contain real Higgs or W and Z bosons. Gluino couplings are democratic in flavour so that their production also leads to a substantial rate for central $b-$jets which may be tagged using microvertex detectors. In contrast to gluino and squark production which essentially *always* results in jet activity, the production of sleptons and sneutrinos, or charginos and neutralinos, can also lead to hadronically quiet events if these sparticles decay via their leptonic modes (in much the same way that W and Z bosons can decay purely leptonically).

2.1 e^+e^- Colliders

Sparticles with non–vanishing $SU(2)\times U(1)$ quantum numbers will be produced with cross sections with $R \sim 1$ except for mass and spin factors. Current searches for sparticles at LEP focus on the lightest visibly decaying sparticle which, of course, will directly decay to the LSP. Selectron (and smuon or stau) pair production will thus be signalled by the observation of an acollinear e^+e^- (muon or tau) pair together with $\not{E}_T$. Chargino production will be signalled by the production of $4j + \not{E}_T$ or $2j + \ell + \not{E}_T$ events. These event topologies are very characteristic, and should be readily distinguishable from WW (and other SM) background. A non-observation of such events above background allows experimentalists at LEP to infer bounds[5] that are near to the kinematic limit for charginos[6] ($m_{\widetilde{W}} > 85$ GeV if $m_{\tilde{\nu}} > 200$ GeV), and just a bit below for selectrons ($m_{\tilde{e}} > 70$ GeV). The clean environment at electron positron colliders makes this search quite straightforward, so that a Linear Collider (LC) should be able to probe[3,7] charginos and sleptons as heavy as about 90% of the beam energy.

2.2 Hadron Colliders

Tevatron: At hadron colliders gluinos and squarks will be the most copiously produced sparticles if they are kinematically accessible. A non–observation of multijet plus $\not{E}_T$ events above SM expectation has enabled[5] the CDF and D0 experiments to infer bounds $m_{\tilde{g}} \geq 160$ GeV (260 GeV, if $m_{\tilde{q}} = m_{\tilde{g}}$) on their masses. The reach in this channel is background limited. Their production and cascade decay can also lead to multilepton event topologies (particularly like sign dilepton and trilepton channels) for which the background is considerably smaller. Despite their small cross sections, the D0 and CDF experiments are already sensitive to these striking topologies, and have announced[8] gluino and squark mass limits that are not very far below those obtained via the $\not{E}_T$ search. Multilepton signals for SUSY will likely play an important role in the Main Injector (MI) era.

In models (with universal scalar and gaugino masses) where coloured sparticles are significantly heavier than colourless ones, it is possible that gluinos and squarks may become kinematically inaccessible. In this case, electroweak production of charginos and neutralinos, followed by their subsequent leptonic decays, $\widetilde{W}_1 \to \ell\nu\widetilde{Z}_1$, $\widetilde{Z}_2 \to \ell\bar{\ell}\widetilde{Z}_1$, which results in extremely striking isolated trilepton events free from jet activity, could provide the maximum reach for SUSY. Tevatron experiments have already attained[5,8] sensitivity to this channel which will be important at the MI, and certainly at the proposed TeV33 upgrade of the Tevatron. We warn the reader that the leptonic branching fraction of $\widetilde{Z}_2$ becomes very small for some range of parameters, so that there may be no observable signal even if chargino masses are close to their current bound.[9]

Another complication[10] occurs if $\tan\beta$ happens to be very large. In this case, the tau and bottom Yukawa coupling effects become important. The decays of charginos and neutralinos into τ leptons begin to dominate the branching fraction, reducing the cross section for the various multilepton signals discussed above. For very large values of $\tan\beta$ it may even be that decay to real staus dominate these decays. Prospects for SUSY searches at the Tevatron in this difficult case are under investigation.

Large Hadron Collider Many studies[11,12] have shown that the LHC should be able to search for gluinos and squarks with masses as large as 2 TeV via the $\not{E}_T$ and single lepton channels provided that $R-$parity is conserved. This provides a reasonable safety margin beyond the "upper limit" on sparticle masses expected from fine–tuning arguments. Direct searches for sleptons should also be possible[13] if these have masses smaller than 250-300 GeV, a range favoured by dark matter considerations. Moreover, as long as $\tan\beta$ is not very large, and gluinos and squarks are lighter than 1 TeV, there must be observable SUSY signals in several multilepton (plus multijet) plus $\not{E}_T$ channels.[12] This conclusion is probably generic to all models where $R-$parity is conserved, and where coloured sparticles are heavier than their colour neutral cousins.

If $R-$parity is not conserved, the LSP decays and we lose the SUSY source of $\not{E}_T$. Experimentally, the most difficult case is when the LSP decays hadronically: not only is the $\not{E}_T$ reduced, but the jets from LSP decay make it more difficult for leptons in SUSY events to remain isolated. Assuming LSP decay is the only significant effect of $R-$violating interactions, a recent study[14] has shown that the LHC reach for gluinos and squarks (via multilepton channels) still extends beyond 1 TeV.

3 Beyond SUSY Searches

If supersymmetry is discovered at future colliders, we would like to be able to identify it as such (as opposed to some other new physics), and then proceed to see if we can unravel details of the underlying model. A comparative study of the TeV33, LHC and LC capabilities for this was one of the tasks at the last Snowmass Workshop.[15]

The JLC Collaboration[3] pioneered a program about five years ago and showed that with the availability of polarized electron beam, it should be possible to measure chargino and slepton masses to within 1-2% if these sparticles directly decay to the LSP. Subsequently, it was shown[7] that a similar precision is possible even if sparticles cascade decay to the LSP. These techniques were used by the LC Group at Snowmass[15] to show that a systematic program of measurements that will allow us to proceed from the experimental data toward the underlying theory appears possible.

The optimal procedure for SUSY exploration at the LHC[15,16] as well as the precision of the measurements depends on the underlying parameters. For example, for the low mass "comparison point" studied by all Snowmass Groups, the LHC Group relied on the enormous gluino production cross section to make extremely stringent cuts that not only eliminate the SM background, but also serve to isolate the SUSY decay chain $\tilde{g} \to \bar{b}\tilde{b}_1 \to \bar{b}b\widetilde{Z}_2 \to \bar{b}b\ell\bar{\ell}\widetilde{Z}_1$. From the end point of the $m_{\ell^+\ell^-}$ distribution, they argued that $m_{\widetilde{Z}_2} - m_{\widetilde{Z}_1}$ can be determined to within 50 MeV, limited just by the systematic error! They then used the partial reconstruction technique to reconstruct this decay sequence and argued that for this case study, $m_{\tilde{g}} - m_{\tilde{b}_1}$ could be determined to within 10%. Measurements at the LHC generally yield mass differences from which masses and underlying parameters can be obtained within a particular framework. We see from Table 1 that at least in some cases, mSUGRA parameters can be extracted with comparable precision at both the LHC and the LC.

Table 1: Quoted precision with which mSUGRA parameters were extracted in the Snowmass Study. The LHC group extracted $\tan\beta$ from the LEP measurement of the Higgs mass, so that this value may be regarded as a bit of a cheat.

	LHC	NLC
δm_0	$^{+13}_{-8}$ GeV	2.7 GeV
$\delta m_{1/2}$	0.7 GeV	$^{+2.5}_{-1.0}$ GeV
$\delta \tan\beta$	0.05	$^{+.17}_{-.31}$
signμ	yes	yes

We conclude with the following remarks from this study.

- The LHC reach extends over all the "interesting parameter space". The reach of LCs is limited by the beam energy. Within the mSUGRA framework, a LC with $\sqrt{s} \sim 1.2$ TeV has a similar SUSY reach as the LHC. For SUSY purposes, a LC should be designed to allow an upgrade of energy to at least 1, and preferably, 1.5 TeV. A LC is also an excellent facility for a detailed study of the Higgs sector.

- It appears possible to have a systematic program of how to proceed if SUSY is discovered at a LC. It is hard to envision how to have such a program in place for the LHC, since how to proceed appears to depend on what is found.

- It should be possible to exclude/confirm specific models with a small number of parameters at both facilities.

- If theoretical data turn out to be incompatible with all simple models, the theoretical direction in which to proceed may be more apparent from the direct measurements that should be available at LCs

- Precision measurements of couplings that are possible at LCs may be able to probe SUSY radiative corrections and so provide information about sparticles beyond the kinematic reach.[17]

- It may be useful to archive LHC data for subsequent reanalysis if $T_{LHC} < T_{LC}$.

We should keep in mind that all our speculations of what lies beyond the SM may prove incorrect. But if the machines and experiments perform as advertised, the important thing is that we will find something. As long as this turns out to be the case, life should be very interesting regardless of what we find.

Acknowledgments

This research was supported in part by the US Department of Energy Grant No. DE-FG-03-94ER40833.

References

1. For pedagogical details and a comprehensive set of references, see X. Tata, *Lectures presented at IX J. A. Swieca Summer School*, hep-ph/9706307 (1997).
2. P. Roy, these Proceedings.
3. T. Tsukamoto *et al.* *Phys. Rev.* D**51**, 3153 (1995).
4. M. Peskin in *Workshop on Physics and Experiments with Linear Colliders*, A. Miyamoto *et al.* Editors, (World Scientific,1996).
5. C. Dionisi, presented at *XIII Int. Symp. on Lepton and Hadron Interactions, Hamburg, 1997.*
6. ALEPH Collaboration, R. Barate *et al.* hep-ex/9710012 (1997).
7. H. Baer, R. Munroe and X. Tata, *Phys. Rev.* D**54**, 6735 (1996).
8. K. Maeshima in *Proc. 28 Int. Conf. on High Energy Physics*, A. Wroblcski and Z. Adjuk, Editors (World Scientific, 1997); N. K. Mondal, *ibid.*
9. H. Baer *et al.* *Phys. Rev.* D**52**, 1565 (1995).
10. H. Baer *et al.* *Phys. Rev. Lett.* **79**, 986 (1997).
11. ATLAS and CMS Technical Proposals.
12. H. Baer *et al.* *Phys. Rev.* D**52**, 2746 (1995) and *Phys. Rev.* D**53**, 6241 (1996).
13. H. Baer *et al.* *Phys. Rev.* D**49**, 3283 (1994).
14. H. Baer *et al.* *Phys. Rev.* D**55**, 1466 (1997).
15. SUSY Working Group and LHC and NLC Subgroup Reports in *Proc. 1996 Snowmass Summer Study*, D. Cassel *et al.* Editors.
16. I. Hinchliffe *et al.* *Phys. Rev.* D**55**, 5520 (1996).
17. M. Nojiri, these Proceedings.

THE STARTING UP OF A NEUTRINO PROJECT IN TAIWAN

C.Y. CHANG [a], S.C. LEE, and H.T. WONG [b]
Institute of Physics, Academia Sinica, Taiwan R.O.C.

There may be a unique opportunity to perform a long baseline reactor neutrino oscillation experiment in Taiwan, profiting from power reactors which provide an abundant supply of neutrinos and the local mountainous geography where an underground laboratory can be built. A Taiwan-China Collaboration has been formed to intensely pursue a detector R&D program and a pilot experiment at the power reactors, the first step towards this goal. Our current efforts and future plans are presented.

1 INTRODUCTION

Phenomenal growth in basic and applied research has been taking place in the Asia Pacific region in the past decade [1]. As the economy of the region further strengthens, many projects in different fields have been initiated, and resources and positions were made available. Research directions and traditions are now being explored and defined. A wide spectrum of future particle physics projects in the Far East has been planned or discussed [2].

Activities in experimental particle physics started in Taiwan in the early 90's. By now, there are more than 10 faculty-level physicists participating in a number of international experiments (L3, CDF, Fermilab E871, Belle, AMS, RHIC-PHOBOS), working on various hardware, online software and data analysis projects. A neutrino group was started in 1996, looking into the possibilities and opportunities in reactor neutrino experiments to be performed in Taiwan.

2 TAIWAN NEUTRINO GROUP

The principal objectives of the neutrino group are to initiate a program in experimental neutrino physics and astro-particle physics, as well as to build up a qualified experimental team for future projects.

The eventual goal is to conduct a world-standard particle physics experiment in Taiwan. A good local research program is complementary to the participation in international projects, and is essential to the solid foundation of international collaborative basic research in the country. As the field grows

[a] Also at University of Maryland, U.S.A.
[b] Contact Person : htwong@phys.sinica.edu.tw

Table 1: Power Reactor Plants in Taiwan and their Parameters

Plant	Type	No. Cores	Power /core GW	Location	Status
I	Boiling Water	2	1.78	North Shore	Operational
II	Boiling Water	2	2.90	North Shore	Operational
III	Pressurized Water	2	2.78	South Shore	Operational
IV	Advanced Boiling Water	2	4.1	North Shore	Comp. 2004

and matures, the community has to acquire the experience of conceiving, formulating, fighting for, and then implementing and coordinating a big project. A local program also serves as a base for training young students and scientists, as well as promoting the subject to the general public.

Given the above implications and considerations, the field of choice is reactor neutrino. There are operational power reactors in Taiwan with parameters shown in Table 1. The mountainous landscape dotted with mines and tunnels makes the construction of an underground laboratory conceivable. The proximity of the reactor locations and the possible underground sites (Nuclear Power Plants I, II and IV are all about 20-30 km from Taipei city) to the city infrastructures and comfort provides an additional advantage. Since the project is started afresh, there is more flexibility to optimize various parameters like the reactor-detector distance and the detector target mass.

Starting 1996, a collaboration has been built up to investigate the feasibility of performing a long baseline reactor neutrino oscillation experiment in Taiwan. At present, the Collaboration comprises more than 40 scientists with diversified expertise from Taiwan (Academia Sinica, Institute of Nuclear Energy Research, National Taiwan University, National Tsing Hua University, National Central University, National Chiayi Teachers' College and Nuclear Power Plant II), China (Institute of High Energy Physics, China Institute of Atomic Energy, Nanjing University, Shandong University, University of Science and Technology at Hefei) and the United States (University of Maryland).

The Collaboration has been intensely pursuing efforts on two fronts : 1) an R&D program on liquid scintillators, and 2) the preparation of a pilot experiment using a detector with 1 ton target mass to be put inside the Nuclear Power Plant II at about 30 m from the core. These projects are funded and will serve as important learning exercise and team-building efforts for our new Collaboration. It is expected that this experience will equip us with the necessary experimental techniques in this field and will help us to build up a qualified team to make major contributions in future projects. The scientific aspects of these efforts are discussed in further details in the next section.

First investigations will soon be carried out towards the conception and formulation of the "next" project. This "next" project will have to be coupled to the world agenda and will probably be international. There are a few alternatives :

- an O(1 km) + O(10 tons) detector probing the atmospheric neutrino anomaly region (down to $\Delta m^2 \sim 10^{-3}$ eV^2 in the $\bar{\nu_e} \rightarrow \bar{\nu_x}$ channel) – whether a second generation experiment in this region is justified and of interest will depend critically on the results [3] from the Chooz [4] and Palo Verde [5] experiments.

- an O(10 km) + O(1 kilo-ton) detector probing the large mixing solution of the solar neutrino puzzle (down to $\Delta m^2 \sim 10^{-5}$ eV^2 in the $\bar{\nu_e} \rightarrow \bar{\nu_x}$ channel) - this is very ambitious and will require several years of internationally-coordinated R&D efforts to define the scientific program and the practical detector design. A proposal (Kam-LAND [2]) already exists along this line, utilizing the existing underground site of the Kamiokande detector.

- being a new group, we are also leaving our options open to other interesting avenues.

3 CURRENT ACTIVITIES

3.1 R&D Program on Liquid Scintillator

The principal objective of the current R&D program is to develop a large scale (> 1 m in the initial phase) detector system of liquid scintillator loaded with ^{10}B, and with pulse-shape discrimination (PSD) (α/γ separation) capabilities.

Electron anti-neutrinos from the reactor interact with protons from the liquid scintillator via the standard reaction :

$$\bar{\nu_e} + p \rightarrow e^+ + n \quad .$$

The emitted neutron will be thermalized and captured by ^{10}B with a typical delay of less than a few μs :

$$n + {}^{10}B \rightarrow {}^7Li^* + \alpha \quad (Q = 2.3 \text{ MeV})$$

$${}^7Li^* \rightarrow {}^7Li + \gamma \quad (E_\gamma = 480 \text{ keV}).$$

The large cross section ($\sigma \sim 4000$ barn) and high isotopic abundance (20%) makes ^{10}B an interesting isotope to provide thermal neutron capture. In particular, liquid scintillator can be produced with a concentration up to the range of 1% ^{10}B by weight (or up to 5% ^{10}B if the isotopic abundance is further enriched). Indeed, though expensive, commercial liquid scintillator with enriched-^{10}B loading[8] does exist. This leads to a much shorter capture time (a few μs) with >99% capture efficiency compared to other good isotopes adopted in previous and current reactor neutrino experiments where the capture time is of the range of 30-100 μs: ^{3}He in Goesgen (inhomogeneous detector)[6], ^{6}Li in Bugey (0.1% by weight)[7], and ^{157}Gd in Chooz[4] and Palo Verde[5] (0.1% by weight). The capture time in unloaded liquid scintillation is typically 180 μs. The capture time distribution for neutrons with an initial energy of 0.5 keV in a liquid scintillation with different amount of ^{10}B by weight, based on a full MCNP simulation[9], is depicted in Figure 1, showing a mean capture time of 2.3 μs for a ^{10}B concentration of 1%.

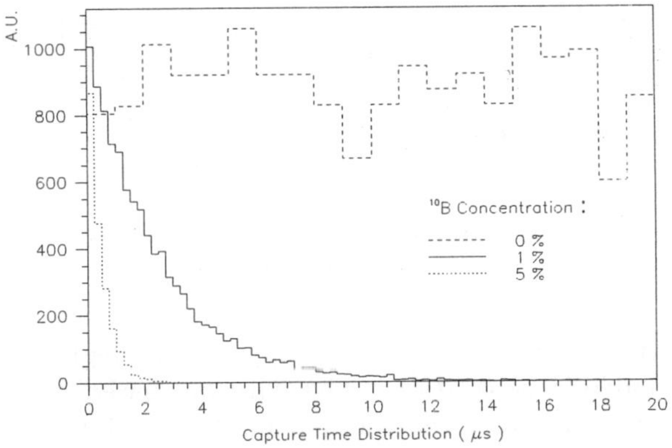

Figure 1: Full simulation of the capture time distribution of neutron with 0.5 keV initial energy in a liquid scintillator medium with 0%, 1% and 5% ^{10}B by weight. The mean capture time is 2.3 μs for a ^{10}B concentration of 1%..

Modest success has been achieved and the liquid scintillator developed (a pseudocumene + trimethyl-borate based solution) has similar performance in terms of light yield, attenuation length and PSD capabilities as the commercial product[8], whereas the cost is much reduced. They are compared in Table 2, while Figure 2 displays their measured attenuation length as a function of wavelength.

PSD capabilities are powerful means for identifying the $e^{+} - \alpha$ signatures. The measured "PSD spectrum" (partial to total charge collected) of the home-

Table 2: Comparison between the commercial and home–made liquid scintillator

	Commercial [8]	Home–made
Mean attenuation length	4 m	3.2 m
Relative light yield	1.00	1.06
PSD (relative figure of merit) ¶	0.9	1.0
% ^{10}B by weight	4.4% (90% enriched)	0.4% (natural)
Cost	USD 4600 /l	USD 60 /l

¶ On α's from ^{241}Am.

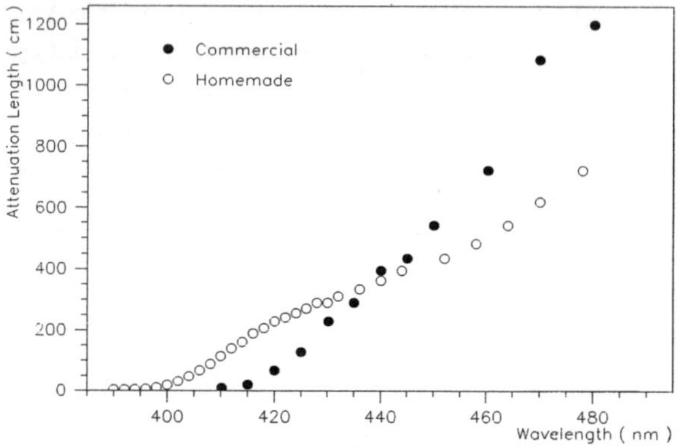

Figure 2: Measured attenuation length of the homemade liquid scintillator as compared to the commercial one.

made liquid scintillator at 1.0% ^{10}B concentration is depicted in Figure 3, demonstrating good α/γ separation. The data are based on measurements of the $\alpha + ^7$Li signatures from the thermal neutron capture of ^{10}B with an electron-equivalence light output of only 50 keV, as compared to the γ pulses from a 60 keV ^{241}Am X-ray source.

3.2 Pilot Experiment at Reactor with a CsI(Tl) Crystal Calorimeter

While the boron-loaded liquid scintillator described above offers an interesting R&D project, the physics potential of a new liquid scintillator experiment *at* a reactor remains rather limited [10]. The "pilot" experiment will be performed in Nuclear Plant II at a site about 30 m from one core and about 100 m from the other one. Apart from being a learning platform for the new Collaboration, it should preferably be able to explore new physics and to improve over previous measurements, with a technically manageable detector realizable in a relatively

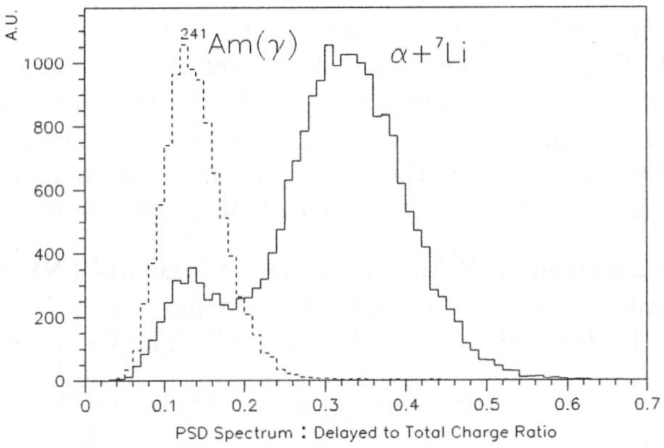

Figure 3: Measured PSD spectrum of the homemade liquid scintillator loaded with 1.0% ^{10}B by weight : γ from ^{241}Am (60 keV) and $\alpha +^{7}$ Li from the neutron capture of ^{10}B (50 keV).

short time-scale.

Almost all previous reactor neutrino experiments were based on liquid scintillator techniques to study ($\bar{\nu}_e$ p) interactions, but with different neutron capture isotopes. An experiment capable of high quality gamma detection has never been attempted. Meanwhile, in the past few years, big electromagnetic calorimeter systems (with mass up to 40 tons of crystals in the case of $B-$factory detectors) have been built for high energy physics experiments, using CsI(Tl) crystals with photo-diodes readout. The CsI-crystal production technology is well matured and the cost has been reduced enormously due to this large demand. It becomes realistic and affordable to build a CsI detector with mass of the order 1 ton for a reactor neutrino experiment.

Since a CsI detector is a new approach to neutrino experiments, it does offer a wide spectrum of physics opportunities with a rather modest investment of cost, manpower and time, well suited for a starting group. One can envisage using CsI crystals both as active target and as a gamma detector where passive target materials are inserted in between.

In the "active target" configuration, where the CsI(Tl) target mass will be about 500 kg, one can study:

1. **Neutrino-Electron Scattering** : The cross section for the process

$$\bar{\nu}_e + e^- \rightarrow \bar{\nu}_e + e^-$$

gives information on the electroweak parameters (g_V, g_A, and $\sin^2\theta_W$), and is sensitive to small neutrino magnetic moments [11]. It is the only system where the interference effects between Z and W exchanges can be studied. A 500 kg-CsI crystal calorimeter will have much more target electrons than all previous experiments [12], and therefore can potentially provide a better measurement of $\sin^2\theta_W$ in the MeV energy range, and improve the sensitivity in the search for neutrino magnetic moments. The signature will be a single hit out of the several hundred crystals.

2. **Neutrino Charged (CC) and Neutral Currents (NC) on ^{127}I and ^{133}Cs** : Neutrino interactions on nuclei have not been studied experimentally beyond a few light ones (^{1}H, ^{2}H, ^{12}C). The CC interactions

$$\bar{\nu}_e + {}^{127}\mathrm{I}\ ({}^{133}\mathrm{Cs}) \rightarrow e^+ + {}^{127}\mathrm{Te}^*\ ({}^{133}\mathrm{Xe}^*)$$

have signatures of 2 back-to-back 511 keV γs from the positron annihilation, plus the characteristic γ-lines from the daughter nuclei, while the NC interactions

$$\bar{\nu}_e + {}^{127}\mathrm{I}\ ({}^{133}\mathrm{Cs}) \rightarrow \bar{\nu}_e + {}^{127}\mathrm{I}^*\ ({}^{133}\mathrm{Cs}^*)$$

will have the characteristic γ-lines as signatures. Investigations of these interactions will provide important input to considerations of future deployment of even larger scale CsI calorimeters as neutrino detectors.

In the "passive target" scenario, virtually any target material can be inserted in between individual CsI modules. The physics menu includes:

1. **Neutrino CC and NC on Deuteron** : The interactions

$$\mathrm{CC} : \bar{\nu}_e + d \rightarrow e^+ + n + n \text{ , and}$$

$$\mathrm{NC} : \bar{\nu}_e + d \rightarrow \bar{\nu}_e + p + n$$

have been observed before [13]. Improved measurements will be of interest, especially since the NC reaction is the detection channel adopted by the forthcoming SNO experiment [14] for solar neutrino detection. A measurement of the CC/NC ratio provides a complementary method to search for neutrino oscillations. In a realistic experiment, the CsI crystals will be put in some optimized positions in a tank filled with about 500 kg of heavy water (D_2O) and good neutron capture isotopes (like gadolinium or cadmium) either as dissolved salts or as inserted sheets. The CC signatures will be rather spectacular: back-to-back 511 keV γs followed by two separate bursts of high energy γs produced by neutron capture on Gd or Cd. The NC detection will rely on a single γ-burst.

2. **Neutral Current Excitation on ^{10}B and ^{11}B** : If a compact boron-rich object like B_4C is used as passive target, characteristic γ lines (2.11, 4.45, 5.02 MeV for ^{10}B, 3.59, 5.16 MeV for ^{11}B) will be emitted by the excited daughter nuclei following NC interactions:

$$\bar{\nu}_e + {}^{10}\mathrm{B} \to \bar{\nu}_e + {}^{10}\mathrm{B}^* \text{ , and}$$

$$\bar{\nu}_e + {}^{11}\mathrm{B} \to \bar{\nu}_e + {}^{11}\mathrm{B}^* \text{ .}$$

There are theoretical works [15] suggesting that these cross sections are sensitive to the axial isoscalar component of NC interactions and the strange quark content of the nucleon.

3. **NC Excitation of Other Nuclei** [16]: If the CsI calorimeter proves itself to be optimal for studying NC excitations, where the experimental signatures are characteristic γ lines, one can insert other passive materials to measure their cross sections, and turn the experiment into a longer–term program. Theoretical activities to evaluate their cross sections may be organized in parallel to "cover the field".

4 OUTLOOK

The Asia Pacific region, with its rising economic prosperity and solid basic education, is now sufficiently mature to move into the arena of international collaborative basic research ("Big Science"). This is a new "parameter space" of scientific experience which the region will have to explore. The P^{4}97 Workshop serves as a testament to this movement.

A Taiwan and mainland China collaboration has been formed to intensely study the feasibility of performing a long baseline reactor neutrino experiment in Taiwan. A "pilot" yet independent experiment close to a reactor core using CsI(Tl) as detector is now being prepared. This is a pioneering "foundation" effort for Taiwan, and the importance of its outcome lies besides, if not beyond, neutrino physics. The Collaboration will make a rigorous attempt towards the realization of this goal.

References

1. For further details, see Nature **383**, 11 (1996).
2. "The Future of Particle Physics in the Far East", K. Nakamura, in *Procs. of XVI Int. Workshop on Weak Interactions and Neutrinos 1997*, to be published in Nucl. Phys. (Procs. Suppl.) **B** (1997).

3. M. Apollonio et al., Chooz Collaboration, submitted to Phys. Lett. **B** (1997).
4. H. de Kerret et al., Chooz Proposal (1993);
5. F. Boehm et al., Palo Verde Proposal (1993);
6. G. Zacek et al., Goesgen Collaboration, Phys. Rev. **D 34**, 2612 (1986).
7. B. Achkar et al., Bugey Collaboration, Nucl. Phys. **B 434**, 503 (1995).
8. Liquid Scintillator BC523A, Bicron, U.S.A.
9. J. F. Briesmeister ed., "MCNP - A General Monte Carlo N-particle Transport Code, Version 4A", LA-12625-M (1993).
10. For an overview of current status of reactor neutrino experiments, see, for example, "Neutrino Interactions at Nuclear Reactor", P. Vogel, in *Neutrino '94 Procs.*, Nucl. Phys. (Procs. Suppl.) **B 38**, 204 (1995).
11. B. Kayser et al., Phys. Rev. **D 20**, 87 (1997);
P.Vogel and J.Engel, Phys. Rev. **D 39**, 3378 (1989).
12. F. Reines, H.S. Gurr and H.W. Sobel, Phys. Rev. Lett. **37**, 315 (1976);
G.S. Vidyakin et al, JETP Lett. **55**, 206 (1992);
A.I. Derbin et al., JETP Lett. **57**, 769 (1993);
C. Broggin et al., Nucl. Instrum. Methods **A311**, 319 (1992).
13. E. Pasierb et al., Phys. Rev. Lett. **43**, 96 (1979).
14. G.T. Ewan et al. SNO Proposal (1987).
15. T. Suzuki, Y. Kohyama, and K. Yazaki, Phys. Lett. **B 252**, 323 (1990);
J. Bernabéu et al., Nucl. Phys. **B 378**, 131 (1992).
16. H.C. Lee, Nucl. Phys. **A 294**, 473 (1978) ;
T.W. Donnelly and R.D. Peccei, Phys. Rep. **50**, 1 (1979).

RESULTS FROM SUPER–KAMIOKANDE

M. NAKAHATA
Kamioka Observatory, ICRR, University of Tokyo

Super–Kamiokande has been taking data since April 1996. Preliminary results on atmospheric neutrinos and solar neutrinos are presented. The results re-confirmed the atmospheric neutrino anomaly and the solar neutrino problem. A possible solution of these problems could be neutrino oscillations.

1 Introduction

Super–Kamiokande was constructed to investigate neutrino oscillations through atmospheric neutrinos and solar neutrinos, and search for nucleon decays with unprecedented sensitivity.

Currently unresolved issues about neutrinos are masses and oscillations. If neutrinos have masses, it gives a strong clue for physics beyond the standard model. Furthermore, neutrinos may have played an important role in astrophysics through dark matter. Although neutrino oscillations are sensitive to only mass differences of neutrino species, much smaller mass ranges can be investigated than in direct search experiments. The probability of neutrino oscillations is given by the following formula:

$$P(\nu_\alpha \to \nu_\beta) = sin^2 2\theta sin^2(1.27\frac{L}{E}\Delta m^2),$$

where θ is the neutrino mixing angle, L is the distance from a neutrino source to a detector in units of kilometer, E is the energy of neutrinos in units of GeV, and Δm^2 is the difference of mass squares in units of eV^2. As seen in the equation, one needs longer distance and smaller neutrino energy for investigating small Δm^2 . Neutrino oscillation experiments with a baseline from the Sun to the Earth or across the Earth, namely solar neutrino and atmospheric neutrino experiments, are best suited for searching for small Δm^2 . The MSW mechanism[1] of solar neutrino oscillations has given a tool to investigate small mixing angles for the range $\Delta m^2 = 10^{-7} - 10^{-4}$ eV^2.

In this report, preliminary results from Super-Kamiokande on atmospheric neutrinos and solar neutrinos are presented.

2 Super-Kamiokande Detector

Super-Kamiokande is a water Cherenkov detector located in the Kamioka mine in Japan. The detector consists of inner and outer detectors. In the inner detector, 11,146 photomultiplier tubes(PMT), each 20 inch in diameter, are uniformly placed facing inward on a 0.707 m grid on the entire surface with dimensions 33.8 m in diameter by 36.2 m high, which contains 32,000 metric tons of water. The total photocathod surface area of all PMT's is 40 % of the surface of the inner detector. A 4π solid angle outer detector surrounds the inner-detector. The outer detector is also a water Cherenkov counter with 1,885 sets consisting of a wavelength shifter plate and an 8 inch PMT. The thickness of the outer detector is 2 meters. The outer detector is designed to reduce external gamma rays from surrounding rocks and to tag cosmic ray muons. The fiducial volume of the detector is defined to be more than 2 m from the detector wall, which amounts to 22 ktons of water. The data taking of Super-Kamiokande was started in April 1996.

The read–out electronics (ATM modules) was designed to detect low energy events without dead time. When a signal is detected in a PMT, it is held in two analog buffers at the front–end part of the ATM modules, one for pulse height information and another for timing information. The PMT signal is discriminated with a threshold of 0.2 p.e. and a logical signal of 200 nsec width is made, which is called HIT signal. The HIT signals of all PMTs are summed and a HITSUM signal is made. A global trigger of the detector is issued when the HITSUM signal is greater than 29 PMT hits, which corresponds to 5.7 MeV equivalent electron energy. The trigger signal is distributed to all PMT channels and the PMT data in the analog buffers are digitized.

3 Atmospheric Neutrinos

Atmospheric neutrinos are the decay products of hadronic showers produced by primary cosmic ray interactions in the atmosphere. In recent years, the double-ratio of neutrino events classified by lepton types, $R \equiv (\mu/e)_{DATA}/(\mu/e)_{MC}$, has been studied to approximate the atmospheric neutrino flavor ratio ν_μ/ν_e and to cancel uncertainties in the neutrino flux and cross sections. The measurements of R in underground experiments [2] are summarized in Table 1 together with preliminary results from Super-Kamiokande.

The results of NUSEX and Frejus gave R close to one. Results of Kamiokande, IMB, Soudan-II and Super-Kamiokande gave a significantly smaller R value.

Data of 20.1 kt·yr was accumulated in Super-Kamiokande between May 1996 and June 1997. The atmospheric neutrino events are classified into fully contained (FC) events and partially contained (PC) events. The FC events are selected by requiring no hits in the outer detector. The PC events are selected essentially by requiring a single outer detector hit cluster close to an exit point of the particles observed in the inner detector. The vertex positions of PC

Experiment	exposure (kt·yr)	events	R
NUSEX	0.74	50	$0.96 \,^{+0.32}_{-0.28}$
Frejus	2.0	200	$1.00 \pm 0.15 \pm 0.08$
Kamiokande sub-GeV	7.7	482	$0.60 \,^{+0.06}_{-0.05} \pm 0.05$
Kamiokande multi-GeV	6/8.2	233	$0.57 \,^{+0.08}_{-0.07} \pm 0.07$
IMB	7.7	610	$0.54 \pm 0.05 \pm 0.11$
Soudan-II	2.83	331	$0.61 \pm 0.14 \,^{+0.05}_{-0.07}$
Super-K sub-GeV	20.1	1453	$0.64 \pm 0.04 \pm 0.05$
Super-K multi-GeV	20.1	444	$0.60 \pm 0.05 \pm 0.05$

Table 1: Measurements of R in atmospheric neutrino experiments.

Ring numbers	DATA(20.1 kt·yr)	Monte Carlo
1 ring	1453	1551.6
e-like	718	593.9
μ-like	735	957.8
2 ring	436	480.5
$\geq$3 ring	97	116.1
Total	1986	2148.3

$$\frac{(\mu/e)_{DATA}}{(\mu/e)_{MC}} = 0.635^{+0.034}_{-0.033}(stat.) \pm 0.053(sys.)$$

Table 2: Summary of sub–GeV events compared with a Monte Carlo simulation.

events are required to be in the fiducial volume. In total, 2591 FC events and 156 PC events were observed in the 20.1 kt·yr data sample. The FC and PC events are sub-divided into sub–GeV and multi–GeV events. Event summaries of them are shown in Table 2 and 3.

The selection of sub–GeV events in the FC sample is that (1) the visible energy (E_{vis}) is less than 1.33 GeV, (2) the momentum is greater than 100 MeV/c for $e-$ like events and 200 MeV/c for $\mu-$like, and (3) the maximum pulse height of a single PMT is less than 200 p.e., which rejects events with a particle stopping very close to the detector wall. Multi–GeV events are required to have $E_{vis} > 1.33$ GeV. The number of sub–GeV and multi–GeV events are 1986 and 605 in the FC sample. Then, the events are analyzed

FC (20.1 kt·yr)	DATA	Monte Carlo
1 ring	288	304.8
e-like	149	120.5
μ-like	139	184.3
2 ring	207	221.0
$\geq$3 ring	110	133.0
Total	605	658.8
PC (18.1 kt·yr)	156	210.7

$$\frac{(\mu/e)_{DATA}}{(\mu/e)_{MC}} = 0.604^{+0.065}_{-0.058}(stat.) \pm 0.067(sys.)$$

Table 3: Summary of multi–GeV events compared with a Monte Carlo simulation.

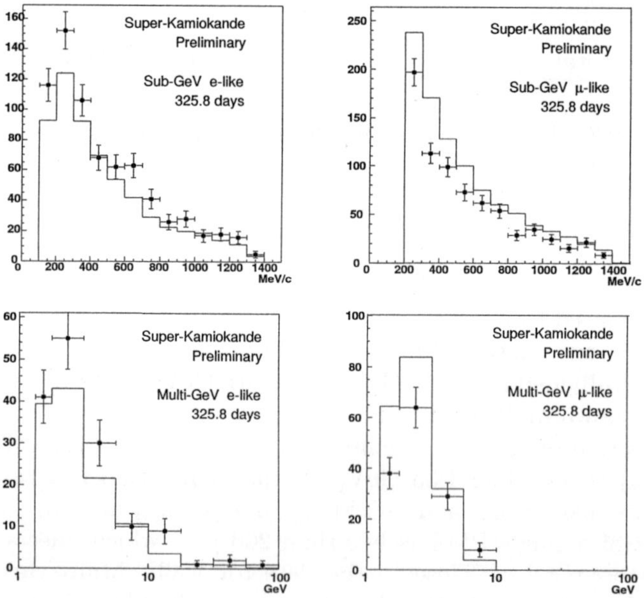

Figure 1: Momentum distribution of single ring events. The obtained data are shown in data points and the solid histograms show the Monte Carlo expectations.

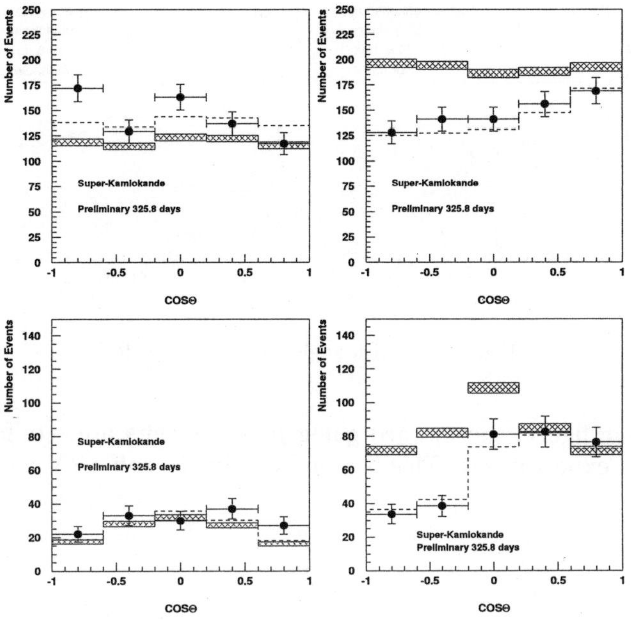

Figure 2: Zenith angle distributions of $e-$like and $\mu-$like atmospheric neutrino data in Super-Kamiokande. Upper and lower figures are for sub–GeV and multi–GeV data samples, respectively. $cos(\theta)$=1 corresponds to the downward going direction. The hatched bands show MC expectations and dotted histograms show the ones assuming $\nu_\mu \leftrightarrow \nu_\tau$ oscillations with $\Delta m^2 = 5 \times 10^{-3}$ eV^2 and $\sin^2 2\theta$ =1.0.

through a Cherenkov ring number counting program and a particle identification program. The misidentification probability of the μ/e separation is estimated to be 1% and 2% in the sub–GeV and multi–GeV samples, respectively. The observed number of single ring $e-$like and $\mu-$like events are 718 (149) and 735 (139) in the sub– (multi–)GeV data sample. All PC events are assumed to be μ-like events. The observed events are compared with a Monte Carlo simulation which uses the flux calculation by Honda et al. [3] and a neutrino interaction program based on accelerator neutrino experiments. [4] Nuclear effects, the Pauli principle and a mass difference between μ and electron are also taken into account in the Monte Carlo simulation. The results of R obtained by using single ring events are shown in Table 2 and 3. The obtained R is significantly smaller than unity and it reconfirms the results of Kamiokande and IMB. The momentum distribution of the single ring events is shown in Fig.1 together with the Monte Carlo simulation. The zenith angle distributions of observed $e-$like and $\mu-$like events are shown in Figure 2.

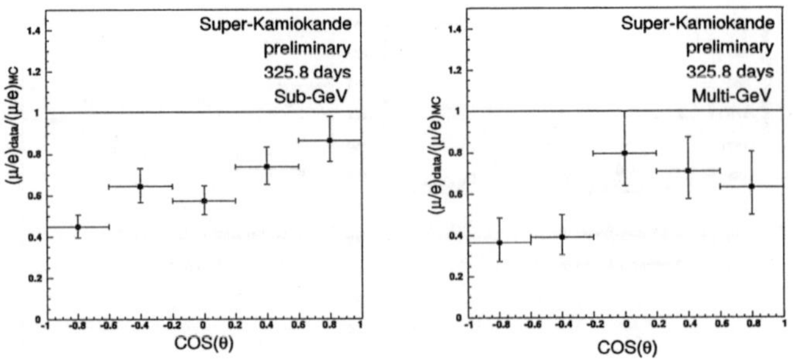

Figure 3: Zenith angle dependence of R.

As seen in the figure, the upward going μ–like events are significantly fewer than the MC expectations. This is seen both in the sub–GeV and multi–GeV samples.

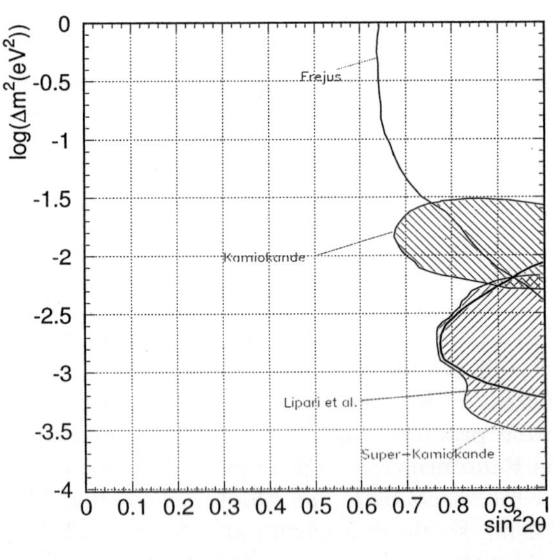

Figure 4: Allowed regions of neutrino oscillation parameters for $\nu_\mu \leftrightarrow \nu_\tau$ oscillations with 90 % C.L. obtained by atmospheric neutrino data in Super–Kamiokande and Kamiokande. The regions excluded by Frejus and IMB are also shown.

The zenith angle dependence of R is shown in Fig.3. R is not only significantly smaller than unity but also distorted with respect to the zenith angle. The distortion of the zenith angle distribution strongly indicates an anomaly in the atmospheric neutrinos, because the systematic error of up/down asymmetry of the detector is negligibly small and the shape of the angular distribution does not depend on uncertainties in neutrino cross sections, nuclear effects etc. A possible explanation of the anomaly is neutrino oscillations. Using the zenith angle information together with energy information of the events, an allowed region of neutrino oscillation parameters is obtained as Fig.4. The overlap between the allowed regions of Super–Kamiokande and Kamiokande is around $\Delta m^2 = 5\times10^{-3}$ eV^2 with large mixing angle. The dotted histograms in Fig.2 show expected zenith angle distributions assuming neutrino oscillations of $\nu_\mu \leftrightarrow \nu_\tau$ with parameters of $\Delta m^2 = 5\times10^{-3}$ eV^2 and $\sin^2 2\theta$ =1.0. The neutrino oscillation hypothesis reproduces the observed zenith angle distributions quite well.

4 Solar neutrinos

Results of current solar neutrino experiments[6] and a comparison with the standard solar model (SSM)[5] are shown in Table 4. The flux ratio, data/SSM, is

Experiment	method	flux	Data/SSM(BP95)
^{37}Cl	ν_e^{37}Cl	2.54±0.14±0.14 SNU	0.27±0.02
GALLEX	ν_e^{71}Ga	$69.7\pm6.7^{+3.9}_{-4.5}$ SNU	0.51±0.06
SAGE	ν_e^{71}Ga	73^{+10}_{-11} SNU	$0.53^{+0.07}_{-0.08}$
Kamiokande	νe scat.	$(2.80\pm0.19\pm0.33)\times10^6$ /cm^2/sec	0.42±0.06
Super-K.	νe scat.	$(2.37\ ^{+0.06}_{-0.05}\ ^{+0.09}_{-0.07})\times10^6$ /cm^2/sec	$0.358\ ^{+0.017}_{-0.013}$

Table 4: Results of running solar neutrino experiments.

27 % in the ν_e^{37}Cl experiment and almost half in the Gallium experiments. The νe scattering experiments for ^{8}B neutrinos give data/SSM =~0.4. A detailed study of the relative differences in the flux ratios indicates that astrophysical solutions have difficulties in explaining the solar neutrino problem(SNP).[7] An analysis of neutrino oscillation scenarios shows that possible solutions of SNP are a small mixing solution ($\Delta m^2 \sim 0.5\times10^{-5}$ eV^2 and $\sin^2 2\theta \sim 6\times10^{-3}$), a large mixing solution ($\Delta m^2 \sim 1\times10^{-5}$ eV^2 and $\sin^2 2\theta \sim 0.6$), and a vacuum solution ($\Delta m^2 = 10^{-11} - 10^{-10}$ eV^2 and $\sin^2 2\theta \sim 1$).[7] The high statistics data of Super–Kamiokande enable us to discuss not only the absolute flux value but

also short time variations of the flux, such as a day/night effect, and the shape of the energy spectrum. Actually, a day/night difference of the flux is expected in the large mixing solution and a distortion of the energy spectrum is expected in the small mixing and the vacuum solutions. Those checks purely depend on properties of neutrinos and are free from any ambiguities of the SSM.

Super–Kamiokande has analyzed 374 days' data taken from June 1996 to October 1997. Low energy events which have no outer detector hits and visible energy less than ~170 MeV (1000 p.e.) are selected at the first stage of data analysis. The vertex positions and directions of the selected events are reconstructed by using the charge and timing data from the hit PMT's. The events with reconstructed vertex in the fiducial volume are selected for further analysis. At this stage, the main source of background in the lower energy range is the radioactivity in the detector, such as Radon in water and gamma rays from PMT glass. On the other hand, the higher energy range is dominated by spallation background produced by cosmic ray muon interactions. To reduce the lower energy background, additional cuts are applied using the event topology and directionality with respect to the nearest wall from the vertex position. The spallation background is reduced by using timing and spatial correlation with energetic muons. The events thus obtained in the energy range from 6.5 MeV to 20 MeV are used for the solar neutrino analysis.

The Super–Kamiokande detector is calibrated by using an electron linear accelerator (LINAC). It is used to calibrate the absolute energy scale, angular resolution and vertex position resolutions. The LINAC, located ~13 m away from the Super–Kamiokande detector, injects mono–energetic electrons with a tunable energy ranging from 5 MeV to 16 MeV. The beam is transported into the detector through a 10 cm diameter beam pipe, while the beam energy is determined by a system of bending magnets and collimators. The uncertainty of the absolute energy scale is estimated to be ~ 0.45 % at 8 MeV using the LINAC calibration data.

The direction cosine of the events with respect to the direction of the sun ($\cos\theta_{sun}$) is shown in Fig.5. A clear enhancement of signals in the direction of the sun is seen in the angular distribution. The flux obtained by these signals is $(2.37^{+0.06}_{-0.05}{}^{+0.09}_{-0.07}) \times 10^6$ /cm^2/sec. Comparing the flux with the SSM by Bahcall and Pinsonneault, the flux ratio, data/SSM, is $0.358^{+0.009}_{-0.008}{}^{+0.014}_{-0.010}$.

The solar neutrino data are sub–divided into day–time and night–time data and the flux of each data set is obtained to be $(2.30 \pm 0.08\ ^{+0.09}_{-0.07}) \times 10^6$ /cm^2/sec for day-time and $(2.44 \pm 0.08\ ^{+0.10}_{-0.07}) \times 10^6$ /cm^2/sec for night-time.

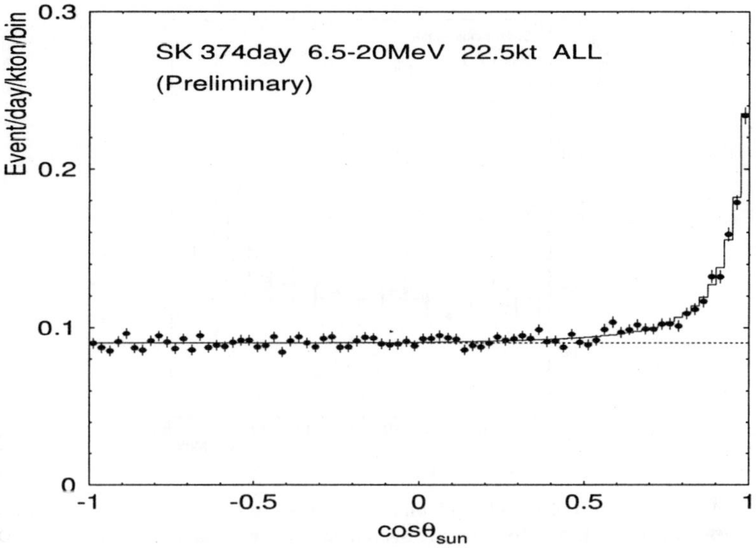

Figure 5: Angular distribution with respect to the direction of the Sun.

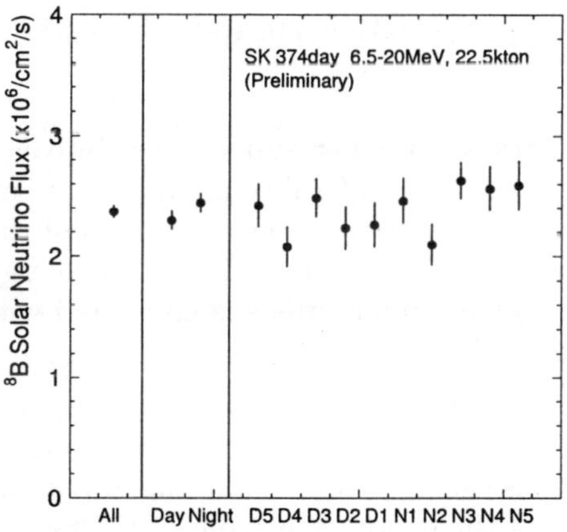

Figure 6: Solar neutrino flux of all data, day and night data, and 5 sub–divided data sets in day and night. D5 and N5 corresponds to downward and upward solar neutrinos, respectively.

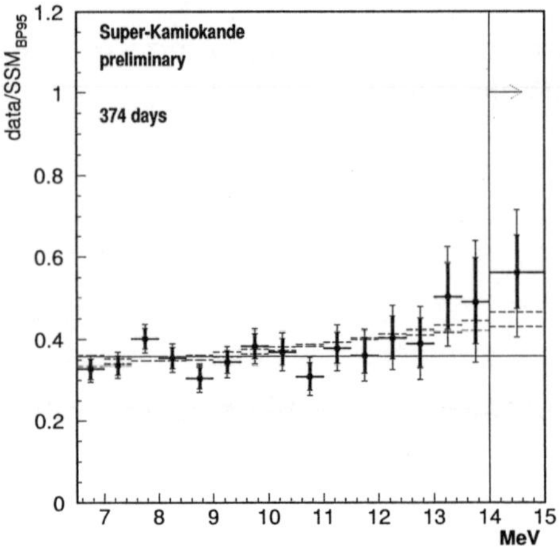

Figure 7: Energy spectrum of recoil electrons normalized to the SSM prediction. Inner error bars are statistical errors and outer bars are systematic errors. Dashed and dotted histograms are the expected spectra of typical small mixing and vacuum solutions, respectively.

The difference between day and night fluxes is

$$\frac{Day - Night}{Day + Night} = -0.031 \pm 0.024(stat.) \pm 0.014(sys.).$$

The day and night data are further sub–divided into 5 data sets according to the zenith angle of the sun. The flux of each data set is shown in Fig.6. No significant day/night difference nor any variation of the flux during night–time is seen in the current solar neutrino data. The energy spectrum of recoil electrons from solar neutrino interactions is compared with a prediction from the SSM as shown in Fig.7

Within statistical and systematic errors, the spectrum is consistent with the SSM prediction. However, by reducing these errors by taking more data and perform more precise energy calibration, the Super-Kamiokande is able to discuss the distortion of the energy spectrum in the near future.

Excluded regions of neutrino oscillation parameters are obtained by using the current analysis of day/night difference and spectral shape as shown in Fig.8. The lower half of the large mixing solution is excluded by the day/night analysis.

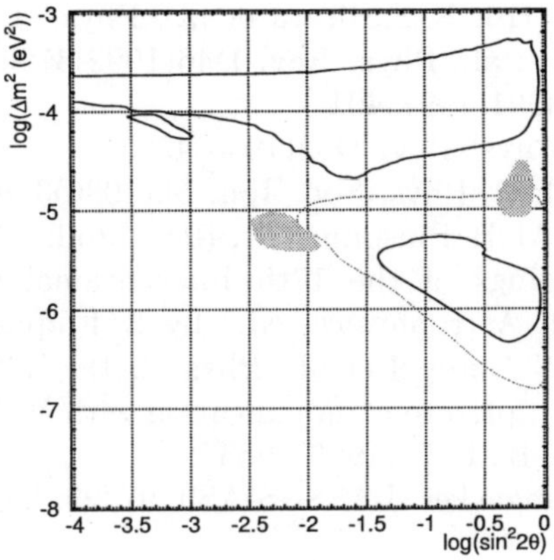

Figure 8: 95 % C.L. excluded regions of MSW solutions. The solid regions are excluded by the spectral shape analysis and the dotted region is excluded by the day/night analysis. The shaded regions are large mixing and small mixing solutions given by Hata and Langacker.

5 Conclusions

Preliminary results of 326 days' atmospheric neutrino data have been presented. The obtained double ratio, $(\mu/e)_{DATA}/(\mu/e)_{MC}$, is significantly smaller than unity. A strong distortion of the zenith angle distribution is observed in μ-like events. The atmospheric neutrino anomaly could be due to neutrino oscillations of $\nu_\mu \leftrightarrow \nu_\tau$ with $\Delta m^2 \sim 5 \times 10^{-3}$ eV^2 and $\sin^2 2\theta \sim 1.0$.

The preliminary result for the solar neutrino flux obtained from 374 days' data sample is $(2.37^{+0.06}_{-0.05}\,{}^{+0.09}_{-0.07}) \times 10^6$ /cm^2/sec, which is 36 % of the prediction from the SSM. No significant day/night difference is observed in the current data sample. A precise energy calibration by LINAC will enable us to discuss the distortion of the energy spectrum in near future.

References

1. L. Wolfenstein, Phys. Rev. D17(1978)2369; S. P. Mikheyev and A. Yu. Smirnov, Nuovo cimento 9C(1986)17; H. A. Bethe, Phys. Rev. Lett. 56(1986)1305.
2. References are in order: M. Aglietta et al., Europhys. Lett. 8(1989)611; Ch.Berger et al., Phys. Lett. B245(1990)305; K. S. Hirata et al., Phys.

Lett. B205(1988)416; K.S. Hirata et al., Phys. Lett. B280(1992)146; R. Becker-Szendy et al., Phys. Rev. D46(1992)3720; W. W. M Allison et al., Phys. Lett. B391(1997)491.

3. M. Honda et al., Phys. Rev. D52(1995)4985.
4. M. Nakahata et al., J. Phys. Soc. Jpn. 55(1986)3786.
5. J. N. Bahcall and M. H. Pinsonneault, Rev. Mod. Phys. 67(1995)781.
6. K. Lande Proceedings of the 17th International Conference on Neutrino Physics and Astrophysics, ed. by K. Enqvist et al., World Scientific(1996)25; W. Hampel et al., Phys. Lett. B357(1996)384; Results of SAGE I+II+III, private communication with V. N. Gavrin; Y. Fukuda et al., Phys. Rev. Lett. 77(1996)1683.
7. N. Hata and P. Langacker, LASSNS-AST 97/29, hep-ph/9705339.

SOME PHENOMENOLOGICAL ASPECTS OF NEUTRINO PHYSICS

SANDIP PAKVASA
Department of Physics & Astronomy, University of Hawaii
Honolulu, HI 96822 USA

I concentrate on two topics. One is techniques to distinguish amongst various oscillation scenarios from atmospheric neutrino data; and the other is the Borexino solar neutrino detector and its capabilities.

The current high level of interest in neutrino properties is well justified. Neutrino properties (such as masses, mixings, magnetic moments etc.) are of interest for a variety of reasons: (i) in their own right as fundamental parameters and (ii) as harbingers of new physics beyond the standard model (if e.g. $m_i \neq 0, \theta_i \neq 0, \mu_i \neq 0$ etc.).

I will not review here the kinematic limits on masses but concentrate on the current evidence for mixing and oscillations. First we summarize some salient features of neutrino oscillations. For two flavor mixing (say ν_e and ν_μ), the standard forms for survival probability and conversion probability are given by

$$\begin{aligned} P_{ee}(L) &= 1 - \sin^2 2\theta \sin^2\left(\frac{\delta m^2 L}{4E}\right) \\ P_{e\mu}(L) &= \sin^2 2\theta \sin^2\left(\frac{\delta m^2 L}{4E}\right) \end{aligned} \tag{1}$$

for a neutrino starting out as ν_e. Here θ is the mixing angle, $\delta m^2 = m_2^2 - m_1^2$, $L = ct$ and the ultra-relativistic limit $E_i \approx p + \frac{m_i^2}{2p}$ has been taken. Although these formulae are usually derived in plane wave approximation with $p_1 = p_2$, it has been shown that a careful wave-packet treatment yields the same formulae [1]. When the argument of the oscillating term $(\frac{\delta m^2 L}{4E})$ is too small, no oscillations can be observed. When it is much larger than one, then due to the spread of E at the source or finite energy resolution of the detector, the oscillating term effectively averages out to 1/2.

There are some obvious conditions to be met for oscillations to take place. As the beam travels, the wave packet spreads and the mass eigenstates separate. If the width Δx remains greater than the separation, then oscillations will occur; but if the separation is greater then two separate pulses of ν_1 (mass m_1) and ν_2 (mass m_2) register in the detector with intensities $\cos^2\theta$ and $\sin^2\theta$

separated by $\Delta t = (\delta m^2/2E^2)(L/c)$. In principle, the intensities as well as oscillation expressions should reflect the slightly different decay widths for different mass eigenstates but this is of no practical importance [1]. The same expressions remain valid if the mixing is with a sterile neutrino with no weak interactions. With 3 flavors mixing, the mixing matrix can have a phase (*á la* Kobayashi and Maskawa) and the oscillations have a CP non-conserving term leading to

$$P_{\alpha\beta}(L) \neq P_{\beta\alpha}(L), \quad P_{\alpha\beta}(L) \neq P_{\bar{\alpha}\bar{\beta}}(L) \tag{2}$$

etc. Some possibilities for observing CP violating effects in Long Baseline experiments were discussed here [2] by Dr. Koike and by Dr. Sato. An old observation which has become relevant recently is the following: it is possible for neutrinos to be massless but not be orthogonal [3]. For example, with three neutrino mixing we have

$$\begin{aligned} \nu_e &= U_{e1}\nu_1 + U_{e2}\nu_2 + U_{e3}\nu_3 \\ \nu_\mu &= U_{\mu 1}\nu_1 + U_{\mu 2}\nu_2 + U_{\mu 3}\nu_3 \end{aligned} \tag{3}$$

Suppose $m_1 = m_2 = 0$ but m_3 is non-zero and $m_3 > Q$ where Q is the energy released in $\beta-$decay or π-decay producing ν_e and ν_μ beams. Then ν_e and ν_μ will have zero masses but will not be orthogonal:

$$< \nu_e \mid \nu_\mu >= -U_{e3}^* U_{\mu 3} \neq 0 \tag{4}$$

(Scenarios similar to this are realized in combined fits [4] to solar and LSND neutrino anomalies). Incidentally, the "ν_e" and "ν_μ" produced in Z decay will not be massless and will be nearly orthogonal! This example illustrates the fact that neutrino flavor is not a precise concept and is process dependent.

1 Atmospheric Neutrinos

The cosmic ray primaries produce pions which on decays produce ν_μ's and ν_e's by the chain $\pi \to \mu\nu_\mu$, $\mu \to e\nu_e\nu_\mu$. Hence, one expects a ν_μ/ν_e ratio of 2:1. As energies increase the $\mu's$ do not have enough time (decay length becomes greater than 15-20 km) and the ν_μ/ν_e ratio increases. Also at low energies the ν flux is almost independent of zenith angle; at high energies due to competition between π-decay and π-interaction the famous "sec (θ)" effect takes over. Since the absolute flux predictions are beset with uncertainties of about 20%, it is better to compare predictions of the ratio (which may have only a 5% uncertainty) ν_μ/ν_e to data in the form of the famous double ratio $R = (\nu_\mu/\nu_e)_{data}/(\nu_\mu/\nu_e)_{mc}$.

For the so-called "contained" events which for Kamiokande and IMB correspond to visible energies below about 1.5 GeV, the weighted world average (before Super–Kamiokande) is $R = 0.64 \pm 0.06$ [5]. This includes all the data from IMB, Kamiokande, Frejus, Nusex and Soudan. As we heard from Dr. Nakahata, the new SuperK results are completely consistent with this [5]. It may be worthwhile to recall all the doubts and concerns which have been raised about this anomaly (i.e. deviation of R from 1) in the past and their resolution. (i) Since initially the anomaly was only seen in Water Cerenkov detectors, the question was raised whether the anomaly was specific to water Cerenkov detectors. Since then, it has been seen in a tracking detector i.e. SOUDAN II. (ii) Related to the above was the concern whether e/μ identification and separation was really as good as claimed by Kamiokande and IMB. The beam tests at KEK established that this was not a problem [6]. (iii) The ν_e and ν_μ cross–sections at low energies are not well known; however $e - \mu$ universality should hold apart from known kinematic effects. (iv) If more $\pi^{+'}$s than $\pi^{-'}$s are produced, then even though the ratio of 2/1 is preserved there is an asymmetry in $\bar{\nu}_e/\nu_e$ versus $\bar{\nu}_\mu/\nu_\mu$. Since ν cross-sections are larger than $\bar{\nu}$ cross-sections, the double ratio R would become smaller than 1 [7]. However, to explain the observed R, $\pi^{+'}$s would have to dominate over $\pi^{-'}$s by 10 to 1, which is extremely unlikely and there is no evidence for such an effect. (v) Cosmic ray muons passing through near (but outside) the detector could create neutrals (especially neutrons) which enter the tank unobserved and then create $\pi^{0'}$s faking "e" like events [8]. Again this effect reduces R. However, Kamiokande plotted their events versus distance from the wall and did not find any evidence for more "e" events near the walls [9]. (vi) Finally, the measurement of the μ flux at heights of 10–15 km to tag the parent particles as suggested by Perkins was performed by the MASS collaboration [10]. This should help decrease the uncertainty in the expected (ν_μ/ν_e) flux ratio even further. It seems that the anomaly is real and does not have any mundane explanation. The new data from SuperK that we just hear about extends the anomaly to higher energies than before and shows a clear zenith angle dependence as well. This rules out most explanations offered except for the ones based on neutrino oscillations.

If the atmospheric neutrino anomaly is indeed due to neutrino oscillations, as seems more and more likely, one would like to establish just what the nature of oscillations is. There have been several proposals recently. One is to define an up–down asymmetry for μ's as well as e's as follows:

$$A_\alpha = (N^d_\alpha - N^u_\alpha)/(N^d_\alpha + N^u_\alpha) \tag{5}$$

where $\alpha = e$ or μ, d and u stand for down-coming ($\theta_Z = 0$ to $\pi/2$) and up-coming ($\theta_Z = \pi/2$ to π) respectively. A_α is a function of E_ν. The comparison of $A_\alpha(E_\nu)$ to data can distinguish various scenarios for ν–oscillations rather easily[11]. This asymmetry has the advantage that the absolute flux cancels out and that statistics can be large. It can be calculated numerically or analytically with some simple assumptions. One can plot A_e versus A_μ for a variety of scenarios: (i) $\nu_\mu - \nu_\tau$ (or $\nu_\mu - \nu$ sterile) mixing, (ii) $\nu_\mu - \nu_e$ mixing, (iii) three neutrino mixing (iv) massless ν mixing etc. Oscillations of massless neutrinos can occur in models of flavor violating couplings to gravity and Lorentz invariance violation[12]. However, in both these cases the dependence of oscillations on the distance is very different from the conventional oscillation: $\frac{\delta m^2 L}{4E}$ is replaced by $\frac{1}{2}\delta f \phi EL$ or by $\frac{1}{2}\delta v EL$. Here $\delta f = 2\delta\gamma = 2(\gamma_2 - \gamma_1)$ is the small number parameterizing the flavor violating coupling to gravity, ϕ the gravitational potential and $\delta v = v_2 - v_1$ is the difference between the two maximum speeds of the velocity eigenstates when Lorentz invariance is violated.

The general features of the asymmetry plot are easy to understand. For the $\nu_\mu - \nu_\tau$(or $\nu_\mu - \nu_{st}$) case, A_μ increases with energy, and A_e remains 0; for $\nu_\mu - \nu_e$ mixing, A_e and A_μ have opposite signs; the three neutrino cases interpolate between the above two; for the massless case the energy dependence is opposite and the asymmetries decrease as E_ν is increased; when both ν_μ and ν_e mix with sterile ν's, both A_μ and A_e are positive etc. With enough statistics, it should be relatively straightforward to determine which is the correct one. As we heard, preliminary indications point to $\nu_\mu - \nu_\tau$ as the culprit.

There is also another suggestion which can in principle distinguish $\nu_\mu - \nu_\tau$ from $\nu_\mu - \nu_{st}$ mixing. Consider the total neutral current event rate divided by the total charged current event rate; the ratio is essentially the n.c. cross section divided by the c.c. cross section. With $\nu_\mu - \nu_{st}$ oscillations the ratio remains unchanged since ν_{st} has neither n.c. nor c.c. interactions and the numerator and denominator change equally; (the $\nu_\mu - \nu_e$ case is even simpler: nothing changes); however, in $\nu_\mu - \nu_\tau$ case the denominator decreases and the ratio is expected to increase by $\left(\frac{1+r}{P+r}\right) \approx 1.5$, (here $r = N^0_{\nu e}/N^0_{\nu\mu} \approx 1/2$ and $P = 1/2 = \nu_\mu$ survival probability). Of course, it is difficult to isolate neutral current events; but it is proposed to select $\nu N \to \nu\pi^0 N$ and $\nu N \to \ell\pi^\pm N$ events and the Kamiokande data seem to favor $\nu_\mu - \nu_\tau$ over $\nu_\mu - \nu_{st}$ or $\nu_\mu - \nu_e$ [13].

If we scale L and E each by the same amount, say ~ 100, we should again see large effects. Hence, upcoming through–going $\mu' s$ which correspond to $E \sim 100$ GeV on the average, with path lengths of $L \gtrsim 2000$ km should be depleted. There are data from Kolar Gold Fields, Baksan, Kamiokande, IMB, MACRO, SOUDAN and now SuperK. It is difficult to test the event rate for ν_μ depletion since there are no $\nu_e' s$ to take flux ratios and the absolute flux predictions have

20% uncertainties. However, there should be distortions of the zenith angle distribution and there seems to be some evidence for this [14].

2 Solar Neutrinos

The data from four solar neutrino detectors (Homestake, Kamiokande, SAGE and Gallex) have been discussed extensively [15]. The SuperK data are consistent with those from Kamiokande but increase the statistics by an order of magnitude in one year [5]. To analyze these data one makes the following assumptions: (i) the sun is powered mainly by the *pp* cycle, (ii) the sun is in a steady state, (iii) neutrino masses are zero and (iv) the $\beta-$decay spectra have the standard Fermi shapes. Then it is relatively straightforward to show using these data with the solar luminosity that the neutrinos from ^7Be are absent or at least two experiments are wrong [16]. ^7Be is necessary to produce ^8B and the decay of ^8B has been observed; and the rate for $^7Be + e^- \rightarrow \nu + Li$ is orders of magnitude greater than $^7Be + \gamma \rightarrow ^8 B + p$ and hence it is almost impossible to find a "conventional" explanation for this lack of ^7Be neutrinos. The simplest explanation is neutrino oscillations.

Assuming that neutrino oscillations are responsible for the solar neutrino anomaly, there are several distinct possibilities. There are several different regions in $\delta m^2 - sin^2 2\theta$ plane that are viable: (i) "Just-so" with $\delta m^2 \sim 10^{-10} eV^2$ and $\sin^2 2\theta \sim 1$ [16], (ii) MSW small angle with $\delta m^2 \sim 10^{-5} eV^2$ and $\sin^2 2\theta \sim 10^{-2}$ and (iii) MSW large angle with $\delta m^2 \sim 10^{-7} eV^2$ (or $\delta m^2 \sim 10^{-5} eV^2$) and $\sin^2 2\theta \sim 1$ [17]. The "just-so" is characterized by a strong distortion of the ^8B spectrum and large real–time variations of the flux, especially for the ^7Be line; MSW small angle also predicts a distortion of the ^8B spectrum and a very small $^7Be\nu$ flux, and MSW large angle predicts day–night variations. These predictions (especially spectrum distortions) will be tested in the SuperK as well as SNO detectors. In particular SNO, in addition to the spectrum, will be able to measure the NC/CC ratio thus acting as a flux monitor and reducing the dependence on solar models.

The only way to directly confirm the absence of ^7Be neutrinos is by trying to detect them with a detector with a threshold low enough in energy. One such detector under construction is Borexino, which I describe below [18].

Borexino is a liquid scintillator detector with a fiducial volume of 300 t; with energy threshold at 0.25 MeV, energy resolution of 45 keV and spatial resolution of $\sim$ 20 cm at 0.5 MeV. The PMT pulse shape can distinguish between α's and β's. Time correlation between adjacent events of up to 0.3 nsec is possible. With these features, it is possible to reduce backgrounds to a low enough level to be able to extract a signal from ^7Be $\nu_e^{\prime}s$ via $\nu - e$ scattering.

Radioactive impurities such as ^{238}U, ^{232}Th and ^{14}C have to be lower than 10^{-15}, 10^{-16} g/g and $10^{-18}(^{14}C/^{12}C)$ respectively. In the test tank CTF (Counting Test Facility) containing 6 t of LS, data were taken in 1995–96 and these reductions of background were achieved. As of last summer, funds for the construction of full Borexino have been approved in Italy (INFN), Germany (DFG) and the U.S. (NSF); and construction should begin soon. The Borexino collaboration includes institutions from Italy, Germany, Hungary, Russia and the U.S..

With a FV of 300 t, the events rate from ^7Be ν's is about 50 per day with SSM, and if ν_e's convert completely to $\nu_\alpha(\alpha = \mu/\tau)$ then the rate is reduced by a factor $\sigma_{\nu\mu e}/\sigma_{\nu ee} \sim 0.2$ to about 10 per day, which is still detectable. Since the events in a liquid scintillator have no directionality, one has to rely on the time variation due to the $1/r^2$ effect to verify the solar origin of the events. If the solution of the solar neutrinos is due to "just so" oscillations with $\delta m^2 \sim 10^{-10} eV^2$, then the event rate from ^7Be ν's shows dramatic variations with periods of months.

Borexino has excellent capability to detect low energy $\bar{\nu}_e'$s by the Reines–Cowan technique: $\bar{\nu}_e + p \to e^+ + n, n + p \to d + \gamma$ with 0.2 msec separating the e^+ and γ. This leads to possible detection of terrestial and solar $\bar{\nu}_e'$s. The terrestial $\bar{\nu}_e'$s can come from nearby reactors and from ^{238}U and ^{232}Th underground. The Geo-thermal $\bar{\nu}_e'$s have a different spectrum and are relatively easy to distinguish above reactor backgrounds. Thus one can begin to distinguish amongst various geophysical models for the U/Th distribution in the crust and mantle. Solar $\bar{\nu}_e'$s can arise via conversion of ν_e to $\bar{\nu}_\mu$ inside the sun when ν_e passes thru a magnetic field region in the sun (for a Majorana magnetic moment) and then $\bar{\nu}_\mu \to \bar{\nu}_e$ by the large mixing enroute to the earth [19].

3 Three Neutrino Mixing

In addition to the atmospheric and solar neutrino anomalies, there are also the LSND observations (as we heard from Dr. Kim) [20] which require $\nu_e \nu_\mu$ mixing with $\delta m^2 \sim 0(1)$ eV^2 and $\sin^2 2\theta \sim (0)(10^{-3})$. With the atmospheric anomaly requiring ν_μ mixing with a $\delta m^2 \sim 5.10^{-3}$ eV^2 and solar neutrinos a δm^2 in the range $10^{-5} - 10^{-7}$ eV^2(or 10^{-10} eV^2) for ν_e mixing; it is clear that one needs 4 neutrino states to mix in order to account for the three separate $\delta m^{2'} s$. There have been two proposals to account for the three effects with just three flavors. One was by Acker and Pakvasa[21], which uses the same $\delta m^2 \sim 5.10^{-3}$ with large $\nu_e - \nu_\mu$ mixing to account for both solar and atmospheric neutrinos; and a small mixing with $\nu_\tau(\delta m^2 \sim 1$ eV$^2)$ to account for the LSND. The other, by Cardall and Fuller [22] employs a δm^2 of ~ 0.3 eV^2 to account for both atmospheric and

LSND with solar neutrinos driven by either MSW ($\delta m^2 \sim 10^{-5}$ eV^2 or "just so" ($\delta m^2 \sim 10^{-10}$ eV^2). At the moment, both of these are disfavored: by the CHOOZ results [23] which saw no $\nu_e - \nu_\mu$ oscillations at a δm^2 of $5.10^{-3} eV^2$ with large mixing, and by the SuperK data which requires a δm^2 of 5.10^{-3} eV^2. It thus seems inescapable that the three anomalies together require four light neutrino states; and thus at least one sterile neutrino.

4 Conclusion

The only conclusion I can draw is that we have seen possible evidence for neutrino oscillations and within the next 3–4 years, data (from Super–Kamiokande, SNO, Borexino; the Long, Short and Intermediate Baseline Experiments, CHOOZ and Palos Verde; LSND and Karmen) will tell us more precisely the parameters of the neutrino mass matrix.

Acknowledgments

I thank Manuel Drees and Kaoru Hagiwara for the organization, the APCTP and the Seoul National University for the hospitality and J. Flanagan, J.G. Learned, and R.S. Raghavan for discussions.

This research was supported in part by the US Department of Energy Grant No. DE-FG-03-94ER40833.

References

1. C.W. Kim and A. Pevsner, *Neutrinos in Physics and Astrophysics*, Harwood, 1994 and references therein.
2. M. Koike (These Proceedings), J. Sato (These Proceedings).
3. B. W. Lee, S. Pakvasa and H. Sugawara, *Phys. Rev. Lett.* 38, 937 (1977); S. B. Treiman, F. Wilczek and A. Zee, *Phys. Rev.* D16, 152 (1977).
4. K. S. Babu, J. Pati, and F. Wilczek, *Phys. Lett.* B359, 351 (1997).
5. M. Nakahata (These Proceedings).
6. S. Kasuga et al., *Phys. Lett.* B374 , 238 (1996).
7. L. Volkova, *Phys. Lett.* B316, 178 (1993).
8. O. G. Ryazhzkaya, *JETP Lett.* 61, 237 (1995).
9. Kamiokande Collaboration, Y. Fukuda et al., *Phys. Lett.* B388, 397 (1996).
10. R. Bellotti et al,. *Phys. Rev.* D53, 35 (1996).
11. J. Flanagan, J. G. Learned and S. Pakvasa, (*Phys. Rev.* D., in press), hep-ph/9709438.

12. S.L. Glashow et al., *Phys. Rev.* D56, 2433 (1997).
13. F. Vissani and A. Smirnov, hep-ph/9710565.
14. J.G. Learned (private communication).
15. *Proceedings of Neutrino96*; June 1996, Helsinki, ed. J. Malaampi and M. Roos (World Scientific) to be published.
16. N. Hata and P. Langacker, *Phys. Rev.* D56, 6107 (1997), and references therein.
17. J. N. Bahcall, Lectures at SLAC Summer Institute on Particle Physics, Aug. 1997 (to be published); hep-ph/9711358.
18. C. Arpesella et al., *The Borexino Proposal* Vol. 1 and 2., ed. G. Bellini and R.S. Raghavan (Univ. of Milan) 1991; *Ultralow backgrounds in a large volume underground detector*; G. Alimonti et al. (*Nucl. Inst. and Methods*, in press).
19. R.S. Raghavan et al. *Phys. Rev.* D44, 3786 (1991); R.S. Raghavan et al. Bell Lab Report (to be published).
20. H.J. Kim (these proceedings).
21. A. Acker and S. Pakvasa, *Phys. Lett.* B397, 209 (1997).
22. C. Cardall and G. Fuller, *Phys. Rev.* D53, 4421 (1996).
23. The CHOOZ Collaboration, M. Apollonio et al., hep-ex/9711002.

NEUTRINO TELESCOPES AND THE HANUL PROJECT IN KOREA

YUNSIL HO

Department of Physics, Columbia University, New York, NY 10027, U.S.A.

A new neutrino telescope project, HANUL (High energy Astrophysics NeUtrino Laboratory) is to explore high energy neutrinos from various cosmic accelerators. The proposed detector is a muon magnetic spectrometer with Cherenkov detectors to measure the sign and momentum of the upward going muons generated by charged current interactions of neutrinos inside the Earth. The area covered by active detectors is about 1,000 m^2 and the detectors are to be made of many small modules. We will discuss what is unique about this new project compared with other neutrino telescope projects, along with recent developments in this field.

1 Introduction

The study of neutrinos in the energy region between a few GeV to multi-TeV is a rapidly growing field at the interface of particle physics and astrophysics. These neutrinos may be from galactic or extra-galactic sources, and also from the core of the Sun and the Earth via a SUSY mechanism. In our new project HANUL, we will detect upward going muons just like the other high energy "neutrino telescopes" (NTs) Baikal, AMANDA, and NESTOR, but we measure the muon momentum and charge via a magnetic spectrometer.

Qualitatively, any events with muon momentum greater than 1 TeV/c are potential candidates for neutrinos from extra-galactic sources. If there is an excess in the number of events in a certain muon momentum interval around nadir angle of 0 degree and there is no excess in the same momentum interval around some large angle such as 20 degree or more, then we have a possible signal for the annihilation of WIMPs (neutralinos) in the core of the Earth. The momentum distribution of these muons then tells us the mass of the neutralino. Depletion in the number of muon events in certain angular and energy intervals should be a signal for the disappearance of μ-neutrinos for specific oscillation parameters.

The HANUL detector system will be made of small modules, to make it possible to also do high energy gamma ray physics in the energy range of 1 TeV to 100 TeV.

2 HANUL as a Neutrino Telescope

The idea of building large scale neutrino telescopes is to detect high energy neutrinos from cosmic sources. The scientific goals of the NTs[1] are 1) HE neutrino astronomy, 2) study of HE cosmic ray muon properties, 3) investigation of neutrino interactions beyond the range of existing and planned accelerators, 4) neutrino oscillation, 5) search for dark matter, and 6) search for HE gamma-ray point sources.

DUMAND pioneered the development of large underwater detectors, which uses natural sea water as a medium for Cherenkov radiation produced by charged leptons from neutrino interactions. PMTs on strings are deployed in the water to detect the Cherenkov light signal. There are several neutrino telescopes in operation and more are being planned. Baikal, AMANDA, and NESTOR use or plan to use water/ice as medium for Cherenkov radiation.

Several other experiments are capable of detecting high energy neutrinos, like MACRO, Super-K, etc. MACRO, an underground detector, uses liquid scintillator and streamer tubes, and Super-K, an underground detector, uses the water Cherenkov technique. But these detectors can only measure the muon direction, once high energy muons pass through the detectors, and they only know these are above a certain energy, E_{max}. Maximum energies they measure are about 1 GeV for MACRO, and 10 GeV for Super-K. HANUL, a surface detector which uses a magnetic spectrometer, can measure muon momentum from 1 GeV/c to a few TeV/c. The higher E_{max} is, the easier it is to separate galactic and extra-galactic neutrinos from atmospheric neutrinos. For example, if E_{max} is 1 GeV, this separation is not easy, while for $E_{max} = 1$ TeV it should be much easier. HANUL, with E_{max} of about 1 TeV, is a neutrino telescope like other gigantic detectors. Rates for various sources are yet to be measured.

The status of the Baikal and AMANDA experiments is as follows. The Baikal NT is being deployed in the Siberian Lake Baikal, at a depth of 1.1 km about 3.6 km from shore. Currently NT-96, a neutrino telescope with an array of 96 optical modules, is operating. This array will be expanded to NT-200, which will consist of 192 optical modules. The optical modules are grouped in pairs along strings, with each PMT alternatively looking upward and downward. There are 7 strings symmetrically surrounding a central string. With about 200 days of running with the NT-96, they have the first two clean neutrino candidates with $E_{\mu} > 6$ GeV.

AMANDA A and AMANDA B have been deployed in the deep Antarctic ice, and are taking data. AMANDA A consists of 80 optical modules on 4 strings, positioned between 800 m to 1,000 m, where an unexpected large amount of remnant bubbles were found. These bubbles limit the scattering length of light to tens of centimeters, making it difficult to reconstruct muon tracks. Later a redesigned deeper array of about 300 optical modules, AMANDA B, was deployed in bubble-free ice between 1.5 km to 2 km in 1996/7. They also have candidates of upward going muons and data analysis is in progress.

3 HANUL Detector

Each module is made of three main components: a rectangular shaped magnet, water Cherenkov detectors, and tracking chambers (high precision drift tubes and scintillation detectors), as shown in Figure 1. The magnets are rectangular-shaped soft steel of 40 m^2 effective area and 1 m thickness. They are made of two halves, each with 50 cm thickness, with scintillation detectors placed in between. The resultant transverse momentum kick of the magnet is ~ 480 MeV/c for an average magnetic field of 1.6 Tesla. The momentum resolution due to multiple scattering is 20 %. There are five sets of tracking chambers, placed above and below each Cherenkov detector and in the gap between the magnets.

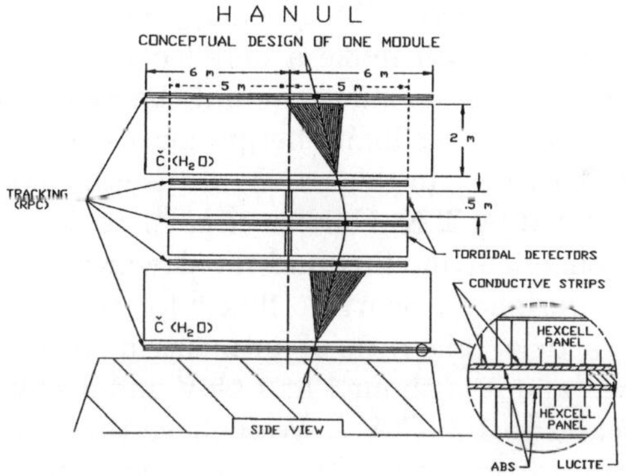

Figure 1: Conceptual design of one module of the HANUL detector.

There are two rectangular water Cherenkov detectors of 60 m^2 area and 1.5 m height placed above and below the magnets. 2 inch PMTs are to be placed in a 50 cm by 50 cm grid on both the top and bottom planes of each water Cherenkov detector. We measure the time and pulse area of all the PMT hits, and a coincidence of the two Cherenkov detectors and some of tracking chambers above and below the magnet provides the first level trigger. These two Cherenkov detectors are sandwiched between the tracking chambers and measure the direction and position of tracks. Together with the TOF (time of flight) information and position and direction measurements of tracking chambers, we should be able to determine the direction (up/down) of the muons.

An underground detector has a clear advantage over a surface detector when studying upward going muons. For example, MACRO demonstrated that they can cleanly separate upward going muons from downward going muons, when the cosmic ray flux is down by 10^6 compared to the surface flux. However, setting up an underground laboratory is costly, it is difficult to change or put more detectors later, and it eliminates the possibility of doing gamma ray physics. Therefore, we prefer to carry out this experiment on the surface. We will construct two smaller prototypes, one on the surface and the other in a mine. If the result from the prototype doesn't meet our goal, we will build an underground laboratory. Our current belief is that we should be able to carry out the experiment above ground.

4 Atmospheric Neutrinos

The only known cosmic sources of high energy neutrinos (above GeV) are atmospheric neutrinos. We will make a detailed measurement of atmospheric neutrinos as we measure the momentum and charge of muons. Only then will it become possible to discover high energy neutrinos from new sources other than atmospheric neutrinos. In this paper, we use the most recent atmospheric neutrino flux calculated by T.K. Gaisser *et al.* [3]. In our calculation we assume that the upward going neutrino flux and the downward going neutrino flux are symmetric. The atmospheric neutrino fluxes for various zenith angles can be found in the same article. The ratio of antineutrino flux to neutrino flux is also given in that paper, and it is about 1 at 1 GeV and decreases to 0.5 as the neutrino energy increases to 100 GeV [3]. We believe that the experimental ability to separate neutrinos from antineutrinos by measuring μ^-'s and μ^+'s is crucial to understanding these atmospheric neutrino fluxes. Once the atmospheric neutrino backgrounds are well understood, it should be easy to discover small important signals which may be hidden under these backgrounds. This is the unique feature of HANUL compared to other existing or planned experiments.

In order to translate our measurements into physics, we calculate the effective neutrino aperture $G(E_\nu, \theta_\nu)$ of our detector. The number of muon events $N(E_\mu, \theta_\mu)$ is the product of $G(E_\nu, \theta_\nu)$ and the (angle and energy dependent) neutrino flux.

In calculating the neutrino aperture $G(E_\nu, \theta - \nu)$, we generated charged current neutrino events of a given neutrino energy inside the earth under the detector. Then the produced muons are tracked through the earth using a GEANT based transport routine, where the earth is considered to be composed of O_2 52%, Si 32.5%, Ca 6%, Na 1.5%, Fe 2%, and Al 4%[4]. All relevant physical processes such as pair production, Bremsstrahlung, photonuclear contributions, the dE/dx energy loss, and multiple scattering of the muon toward the detector are taken into account. The muon energy at the bottom of the magnet must be greater than 1 GeV to ensure that the track goes through the magnets in a module.

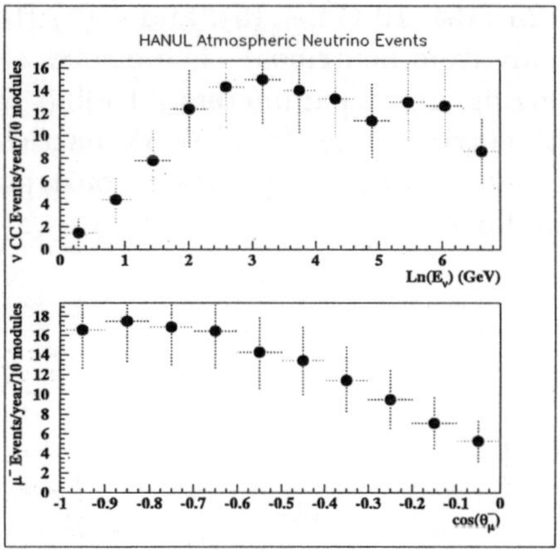

Figure 2: Upward going atmospheric neutrino and antineutrino E_ν and θ_μ distributions.

Figure 2 shows the parent neutrino energy distribution for the detected muons, and the detected muon angular distribution for neutrinos and antineutrinos, assuming a detector with fiducial area of 800 m^2. We expect to have 178 upward going muon events above 1 GeV in a year, 128 μ^- events and 50 μ^+ events respectively.

5 Cold Dark Matter Sources

If the halo of the Milky Way consists of WIMPs, then hundreds to thousands of WIMPs must pass through every square centimeter of the Earth's surface each second. If we assume a neutralino as a specific WIMP candidate, then one can calculate neutrino fluxes from WIMP annihilation. These calculations are very complex and the results are very model dependent. In this paper, we used the neutrino fluxes from the core of the Earth for several neutralino masses of 40 GeV, 60 GeV, 120 GeV, 250 GeV, 500 GeV, and 1000 GeV calculated by N. Fornengo *et al.*[5]. HANUL is an optimum experiment to search for neutralino if its mass is around a few hundred GeV. It is easier for HANUL to isolate possible signals from atmospheric neutrino backgrounds than for other neutrino telescope experiments, since HANUL can measure muon momenta below 1 TeV with good momentum resolution (about 20%) and has the ability to separate neutrinos from antineutrinos. The ratio of neutrino fluxes to antineutrino fluxes is different for signal and background.

The total event rates of neutrinos and antineutrinos we expect to observe in a 5 year run are 127 (84), 843 (556), 25 (16), 10 (7), 7 (5), and 4 (3) (the numbers in parentheses are event rates only from neutrinos). These rates can easily change by a factor of 10 by changing input parameters. Unlike the atmospheric cosmic background, the antineutrino fluxes from SUSY particles are the same as the neutrino fluxes. This fact can be an important constraint in determining if the signal is a statistical fluke.

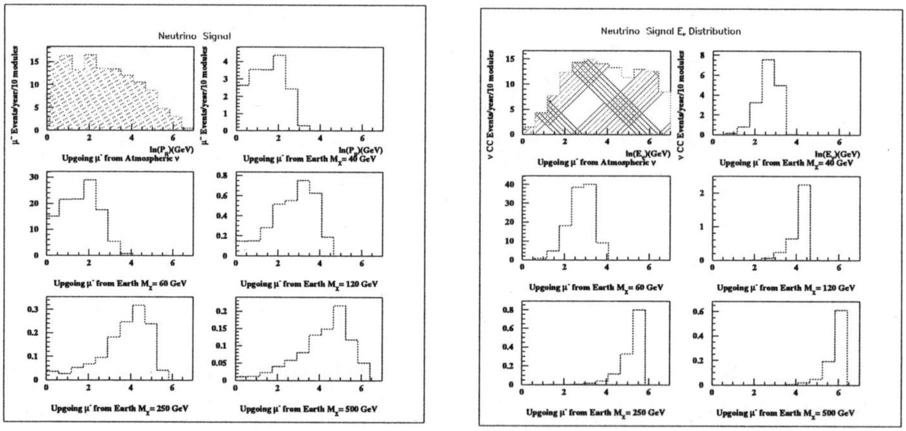

Figure 3: Upward going μ^- and its parent neutrino energy distributions from neutrinos from WIMPs trapped in the Earth for several neutralino masses. The distributions from atmospheric neutrinos are also shown.

The neutrino energy distribution strongly depends on the neutralino mass. In Figure 3 we show that the mean neutrino energy and mean muon energy increase with increasing neutralino mass, and we also show neutrino energy and muon energy distributions for the atmospheric neutrinos.

We can clearly see that the energy distribution of upward going muons will be significantly different from the energy distribution of large angle muons, if there is a statistically significant number of neutralino events. A fundamental question is, "Can we see the neutralino signal?" The signal has two important characteristics. One is that the sources of these neutrinos from inside the Earth and the Sun are almost pointlike [5]. Therefore the angular distribution of the muons from neutralinos should be sharply peaked in the opposite direction away from the sources. The other characteristic is that the muon energy increases as the neutralino mass increases, and shows a strong dependence on the neutralino mass as shown in Figure 3.

As a sample set of parameters of the Minimal Supersymmetric extension of the Standard Model (MSSM) used to obtain neutrino fluxes from the the Earth and the Sun, the case of a neutralino mass of 60 GeV from the Earth (with the other parameters fixed as in the paper by Fornengo *et al.* [5]) would have a more than adequate number of events above background; however, the current data from MACRO already rule this possibility out. For a small signal, which we expect based on the MACRO data, we need to use both the angular and momentum distribution of muons to suppress the background. A two dimensional analysis of muon energy and muon angle should reveal if there are neutralinos in our Galaxy. If there are, we should be able to determine the mass of the neutralino.

6 Summary

A measurement of muon angle and energy distributions should help us pin down atmospheric neutrino fluxes. Once these fluxes are well determined experimentally, we should be able to look for new physics such as neutrinos from extra-galactic sources and neutralinos from the Sun and the Earth, by comparing muon momentum distributions at various angles.

Acknowledgments

HANUL has just been approved as one of the Long Range Basic Research Projects of KOSEF (Korea Science and Engineering Foundation). I would like to thank prof. W. Lee, prof. C.W. Kim, Dr. N. Fornengo, and my colleagues in the HANUL collaboration for many valuable discussions. I also want to thank members of the 3 divisions, particle physics, nuclear physics, and astrophysics of Korean Physical Society for their support for this project.

References

1. O.G. Ryazhskaya, *Proceedings of the 24th International cosmic ray conference*, 655 (1995).
2. T.K. Gaisser *et al.*, *Phys. Rev.* D **54**, 5578 (1996).
3. T.K. Gaisser *et al.*, *Phys. Rev.* D **53**, 1314 (1996).
4. Particle Data Group, *Phys. Rev.* D **50**, 1242 (1994).
5. A. Bottino *et al.*, *Astroparticle Physics*, 65 (1995).

ELECTROWEAK PHASE TRANSITIONS

ZOLTAN FODOR
KEK Theory Division
Tsukuba, 1-1 Oho, Ibaraki, 305, JAPAN[a]

Abstract

There is clear evidence that the world is made exclusively of matter, at least on the $10^{13} M_{\odot}$ scale. The observed baryon asymmetry of the universe has finally been determined at the finite temperature electroweak phase transition. Some features of this phase transition are summarized. Particular attention is paid to analytical and lattice estimates of the sphaleron transition and to the nature of the electroweak phase transition in the standard model and in the minimal supersymmetric standard model. In the standard model the electroweak baryogenesis scenario can be excluded; hoverer, there is a small region in the parameter space of the minimal supersymmetric standard model which might explain the observed baryon asymmetry.

1 Introduction

At high temperatures (e.g. in the early universe) the electroweak gauge symmetry is restored. As the universe expands and supercools there is a phase transition between the high temperature "symmetric" and low temperature "broken" phases. The characteristics of this phase transition (critical temperature: T_c, surface tension: σ, etc.) are clearly of interest. An attractive proposal is to understand the observed baryon asymmetry as produced at the finite temperature electroweak phase transition [1]. In order to provide the observed baryon asymmetry three conditions must be satisfied.

a. existence of baryon number violating processes (in the standard model this can be ensured by sphaleron transitions)
b. C and CP violation
c. departure from thermal equilibrium (the basic question is the strength of the phase transition).

Section 2 will discuss the first (a) Section 3 the last (c) condition. Section 4 contains the conclusion.

2 Sphaleron rate at finite temperature

The basic source of baryon number violating processes in the standard model is the electroweak anomaly for baryon number, which demands

[a] on leave from Inst. Theor. Phys., Eötvös Univ., Budapest, H-1088 Hungary

that

$$\Delta B \sim g^2 \int dt d^3x \mathrm{tr} F\tilde{F}, \tag{1}$$

where g is the weak coupling and F is the field strength of the weak gauge field. This anomaly equation relates the produced baryon number to topological transitions of the weak gauge fields. The vacua with different topological numbers (N_{CS}) are separated by a potential barrier. Note that the anomaly equation holds not only for baryons but for leptons, too. Thus, baryon and lepton numbers are violated simultaneously ($B+L$ violation).

At zero temperature the transition between the vacua is a tunneling event with an unobservably small probability $\approx 10^{-170}$. The transition at high temperatures, but below T_c is a thermal jump. The system jumps up to the top of the barrier (a saddle point: sphaleron) and rolls down to the neighbouring vacuum. In the meantime baryon and lepton numbers are violated. The jump to the barrier, therefore the transition rate is suppressed by the Boltzmann factor

$$\Gamma \propto \exp\left[-E_{sph}/T\right], \tag{2}$$

where the sphaleron energy is proportional [2] to the mass of the W-boson, $E_{sph} \propto m_W/\alpha_W$. Rapid baryon violation processes in the broken phase can wash out the asymmetry generated before. Therefore a "minimal" condition for a successful baryogenesis is that these processes should be slower than the expansion rate of the universe. Comparing (2) with the Hubble constant the minimal condition can be written as $\varphi_b/T_b > 1$, where φ_b is the value of the Higgs field at the transition temperature T_b.

Above T_c the exponential suppression of the transition rate is absent. Naive power counting suggest

$$\Gamma = \kappa \cdot (\alpha_w T)^4, \tag{3}$$

where the constant κ is of order 1. Recently it has been argued [3] that the assumptions used to derive (3) are not valid. The real dependence on α_w is a different one:

$$\Gamma = \kappa' \cdot \alpha_w^5 T^4. \tag{4}$$

The reason is a damping effect for the gauge field. The origin of this damping is the fact that nearly static magnetic fields can be absorbed by the system, which results in a loss of energy for the magnetic fields of interest.

In recent years there has been considerable activity in order to determine the sphaleron transition rate at both sides of the phase transition. No successful numerical method is known for a full Minkowskian theory;

however, important results have been obtained by real time simulations in the classical approximation for the finite temperature theory. The used procedure contains several steps. The classical theory is formulated in terms of fields and their conjugate momenta on a spatial lattice. The phase-space of the system is then sampled with the statistical weight $\exp(-H/T)$. Starting with some initial configuration from the thermal sample the classical canonical equations give the real-time evolution of the fields. As a function of time one can determine different observables and their averages (such as changes of the topological charge). There are fewer physical degrees of freedom in the theory than the number of phase space variables used to formulate it. It is a highly non-trivial task to find effective thermalization algorithms for constrained systems. At present there are two good solutions to this problem [4].

In the broken phase, last year numerical simulations [5] indicated an extreme difference between the lattice results and (2). The observed sphaleron transition rate was a bit smaller than in the symmetric phase; however, no Boltzmann suppression has been seen. Going down into the broken phase the difference between the analytical and lattice results is several orders of magnitude. It is argued [3] that the reason is the problematic definition of the topological charge on a lattice, which give a systematic error in the rate.

A large scale numerical study for the sphaleron rate in the symmetric phase [6] favoured the naive $\Gamma = \kappa(\alpha_w T)^4$ law. The classical motion of the Chern-Simons number consists of two pieces: The thermal fluctuation, which linearly diverges with the lattice cutoff, and a random walk between different vacua. A detailed finite size and finite scaling analysis suggested that the above naive formula is correct with a coefficient of $\kappa = 1.09 \pm 0.05$. According to [3] the observed rate might be only a lattice artifact. The authors reanalysed their data with an improved technique (cooling of lattice field configurations to determine the time evolution of the topological charge) [7]. The new results has shown some deviation from the older one; however, the $\Gamma = \kappa' \alpha_w^5 T^4$ law of the symmetric phase can not be confirmed. A definitely positive outcome of this cooling technique is the new simulation in the broken phase. With the cooling technique no transition between the different vacua has been observed. This result resolves the discrepancy between [5] and eq. (2).

3 The nature of the phase transition

Perturbative studies show that in the realistic Higgs mass range ($m_H >$ 77 GeV) the perturbative approach breaks down [9], it predicts $\mathcal{O}(100\%)$ corrections [8] for relevant quantities (e.g. interface tension, latent heat or correction to the course of the phase transition). A nice illustration (c.f. Figure 1) for that is the interface tension as a function of the Higgs-boson mass in different orders of perturbation theory.

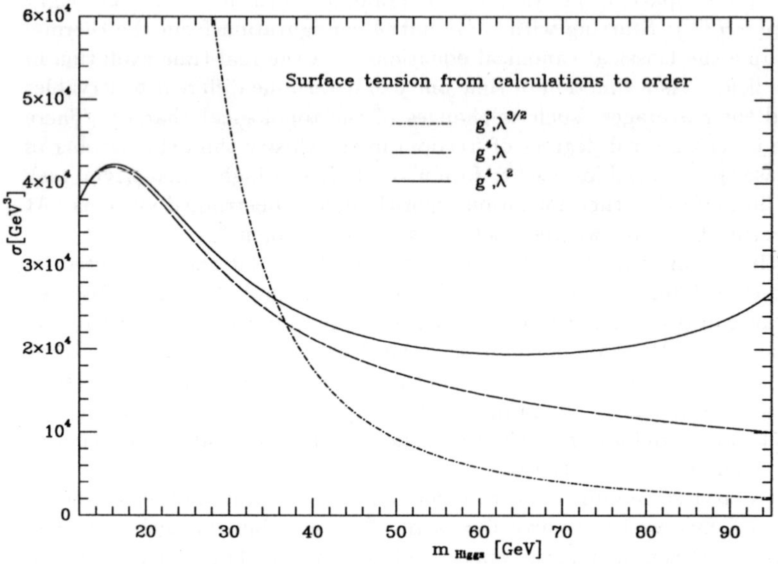

Figure 1: Surface tension from the different potentials as a function of m_H.

A popular way to study the problem is to perform a dimensional reduction in perturbation theory. One starts with the original theory (e.g. Standard Model) and integrates out the heavy degrees of freedom perturbatively. The obtained theory is a three-dimensional bosonic one. The temperature dependent parameters of this theory are determined by matching conditions between the full theory and the reduced one. Static thermodynamical properties, the mass spectrum and other related features have been studied both analytically and numerically for these three-dimensional models (a recent summary is [10]; see also references therein). According to these results the electroweak phase transition is of first order for small Higgs-boson masses; however, it turns out to be an analytic cross-over above $m_H \approx 67$ GeV (critical Higgs-boson mass value for the SU(2)-Higgs model) [11]. Due to the large mass of the top-quark the phase transition can not fulfil the φ_b/T_b condition for any choice of the Higgs-boson mass.

Another possibility is to study the full four-dimensional theory by lattice simulations. Since fermions always have nonzero Matsubara frequencies, the perturbative treatment of these, at high temperatures very massive, modes could be satisfactory. Thus, the starting point of the lattice analyses is the $SU(2)$-Higgs model, which contains the essential features of the standard model of electroweak interactions. In the last three years our group (DESY-Electroweak collaboration) presented a series of papers (see e.g. [12]) in order to clarify the details of the phase transition on four-dimensional lattices. Our work has been done on computers at HLRZ Jülich (CRAY-T90) and DESY-Ifh, Zeuthen (APE-Quadrics). The simulations have been performed for four sets of parameters (m_W = 80 GeV): $m_H \approx$ 18 GeV, 35 GeV, 49 GeV, 80 GeV. For the first three cases the usual lattice formulation can be applied, thus identical lattice spacings in the spatial and temporal directions. In these cases a fairly good agreement is found between the lattice results and perturbation theory [14]. The $m_H \approx 80$ GeV case is much more difficult. The phase transition gets weaker, the lowest excitations have masses small compared to the temperature, T. From this feature one expects that a finite temperature simulation on isotropic lattice would need several hundred lattice points in the spatial directions even for $L_t = 2$ temporal extension. In order to solve this problem we have used the simple idea that finite temperature field theory can be conveniently studied on asymmetric lattices, i.e. lattices with different spacings in temporal and spatial directions [13]. The resulting action contains anisotropies in the couplings. This action has been studied in perturbation theory and on the lattice. The values of the thermodynamical quantities (σ/T_c^3 and $\Delta\epsilon/T_c^4$) for $m_H \approx 80$ GeV are substantially smaller than their perturbative values. They are even consistent with a no first order phase transition scenario on the approximately 1-σ level. These results can be interpreted as a sign for the endpoint for the finite temperature electroweak phase transition.

Similar to the standard model case we have now perturbative two-loop results for the minimal supersymmetric standard model phase transition [15]. An interesting feature of the result is that it opens the baryogenesis window for light stop. Setting sun QCD diagrams (stop-gluon) can give large logarithmic contributions which increase the strength of the phase transition. Nevertheless the allowed parameter space is very constrained. In particular $m_H < 85$ GeV will soon be tested at LEP. Since the setting-sun diagrams have bad infrared behaviour a lattice Monte-Carlo approach is probably needed in order to answer the question in this case, too.

4 Conclusions

Analytical and real time numerical results for sphaleron processes are still not in complete agreement. Above T_c new analytical estimates suggest a $\Gamma \propto \alpha_w^5 T^4$ behaviour for the transition rate, which is not seen in numerical studies. The present numerical techniques are not sensitive enough to determine the sphaleron transition rate below T_c. Perturbative studies show that the electroweak phase transition can not be described above $m_H = 70$ GeV by perturbative methods. Lattice results exclude successful baryogenesis in the minimal standard model. The present bound on the Higgs-boson mass most probably results in a cross-over (no phase transition scenario). The minimal supersymmetric standard model still has some constrained parameter region which perturbatively predicts a phase transition, strong enough for baryogenesis.

References

1. V. A. Kuzmin, V. A. Rubakov and M. E. Shaposhnikov, Phys. Lett. B155 (1985) 36.
2. P. Arnold, L. McLerran, Phys. Rev. D36 (1987) 581.
3. P. Arnold, D. Son, L. Yaffe, Phys. Rev. D55 (1997) 6264.
4. A. Krasnitz, Nucl. Phys. B455 (1995) 320; G.D. Moore Nucl. Phys. B480 (1996) 657.
5. J. Smit, W.H. Tang, Nucl. Phys. B482 (1996) 265.
6. J. Ambjorn, A. Krasnitz, Phys. Lett. B362 (1995) 97.
7. J. Ambjorn, A. Krasnitz, Nucl. Phys. B506 (1997) 387.
8. Z. Fodor, A. Hebecker, Nucl. Phys. B432 (1994) 127.
9. W. Buchmüller, Z. Fodor, T. Helbig and D. Walliser, Ann. Phys. (NY) 234 (1994) 260.
10. M. Laine, hep-ph/9707415.
11. F. Karsch, T. Neuhaus, A. Patkós, J. Rank, Nucl. Phys. Proc.Suppl. 53 (1997) 623; M. Gurtler, E.M. Ilgenfritz, A. Schiller, Phys. Rev. D56 (1997) 3888.
12. Z. Fodor et al. Phys. Lett. B334 (1994) 405; Nucl. Phys. B439 (1995) 147; F. Csikor et al. Nucl. Phys. B474 (1996) 421.
13. F. Csikor, Z. Fodor, Phys. Lett. B380 (1996) 113.
14. W. Buchmüller, Z. Fodor and A. Hebecker, Nucl. Phys. B447 (1995) 317.
15. B. de Carlos, J.R. Espinosa, Phys. Lett. B407 (1997) 12.

LEPTON CP ASYMMETRY IN B FACTORIES: ϵ_B AND NEW PHYSICS

ZHI-ZHONG XING

Physics Department, Nagoya University, Chikusa-ku, Nagoya 464-01, Japan

In this talk, we first make some brief comments on the phase convention dependence of the CP-violating parameter ϵ_B. A simple framework is then presented for analyzing new physics contributions to B^0-$\bar{B}^0$ mixing and their effects on the lepton CP asymmetry, a promising signal to be further searched for in the first–round experiments at B factories.

1 Introduction

The study of various possible $CP-$violating phenomena in weak B meson decays, starting from the end of the 1970's, becomes more intensive today – on the eve of KEK and SLAC B factories, which will provide a unique opportunity to test the Kobayashi–Maskawa (KM) mechanism of CP violation in the standard model (SM) and to discover possible new physics (NP) beyond the SM.

Phenomenologically three different types of $CP-$violating signals are expected in neutral B meson transitions: (a) CP violation from $B^0 - \bar{B}^0$ mixing; (b) CP violation from direct b quark decay; and (c) CP violation from the interplay of decay and mixing. Type (a) can be detected most appropriately from the decay rate asymmetry between two semi–leptonic channels $B^0 \rightarrow l^+\nu_l X^-$ and $\bar{B}^0 \rightarrow l^-\bar{\nu}_l X^+$, the so-called "lepton CP asymmetry" (denoted as $\mathcal{A}_{\rm SL}$ below). Type (b) may involve large uncertainties associated with the evaluation of hadronic matrix elements and/or penguin contributions. Type (c) should play a crucial role in determining the KM $CP-$violating phase and/or the NP phase(s), as it is independent of hadron or penguin pollution to a good degree of accuracy in some decay modes [1].

Experimentally, however, only the lepton CP asymmetry $\mathcal{A}_{\rm SL}$ has been searched for to date. An upper bound on this asymmetry was first set by the CLEO time–integrated measurement: $|\mathcal{A}_{\rm SL}| < 0.18$ or ${\rm Re}\epsilon_B < 0.045$ at the 90% confidence level [2,3], where ϵ_B is a $CP-$violating parameter of $B^0 - \bar{B}^0$ mixing defined like ϵ_K in the neutral kaon system. The OPAL time–dependent measurement has recently given a more accurate result [4]: ${\rm Re}\epsilon_B = 0.002 \pm 0.007 \pm 0.003$ in a special convention used for ϵ_B, which is equivalent to $\mathcal{A}_{\rm SL} = 0.008 \pm 0.028 \pm 0.012$. Both experiments are consistent with the standard model prediction, i.e., $|\mathcal{A}_{\rm SL}| \sim 10^{-3}$ or smaller. It should be noted, nevertheless, that the existence of NP in $B^0 - \bar{B}^0$ mixing might enhance $\mathcal{A}_{\rm SL}$ up to the percent level [5], observable in the first–round experiments of a B meson factory with about 10^8 events of $B^0\bar{B}^0$ mesons [6].

The purposes of this talk are: (1) to comment on the phase convention dependence of the parameter ϵ_B; and (2) to present a simple framework for analyzing the NP effect on $\mathcal{A}_{\rm SL}$. We expect that further measurements of $\mathcal{A}_{\rm SL}$ and its correlation with other CP asymmetries may serve as sensitive probes for NP in $B^0 - \bar{B}^0$ mixing.

2 Comments on ϵ_B

In some literature, the CP–violating parameter ϵ_B is defined to relate the flavor eigenstates $|B^0\rangle$ and $|\bar{B}^0\rangle$ to the mass eigenstates $|B_1\rangle$ and $|B_2\rangle$:

$$|B_{1,2}\rangle \;=\; \frac{1}{\sqrt{2(1+|\epsilon_B|^2)}}\left[(1+\epsilon_B)\,|B^0\rangle \;\pm\; (1-\epsilon_B)\,|\bar{B}^0\rangle\right] \,. \qquad (1)$$

Certainly ϵ_B has the phase freedom arising from the bound states $|B^0\rangle$ and $|\bar{B}^0\rangle$. The reason is simply that $|B^0\rangle$ and $|\bar{B}^0\rangle$ are defined by strong interactions only, leaving the relative phase between them undetermined. The dependence of ϵ_B on the bound state phase convention can be removed, if one adopts the following definition to link the CP eigenstates $|B_+\rangle$ (CP-even) and $|B_-\rangle$ (CP-odd) with the mass eigenstates $|B_1\rangle$ and $|B_2\rangle$:

$$|B_{1,2}\rangle \;=\; \frac{1}{\sqrt{1+|\varepsilon_B|^2}}\left(|B_\pm\rangle \;+\; \varepsilon_B|B_\mp\rangle\right) \,. \qquad (2)$$

Note that ε_B and ϵ_B are not identical[7], unless we fix the CP transformation between $|B^0\rangle$ and $|\bar{B}^0\rangle$ as $(CP)|B^0\rangle = \pm|\bar{B}^0\rangle$. Therefore it is better to adopt the phase convention independent parameter ε_B, rather than ϵ_B, in phenomenological applications [8]. In the analyses of CLEO and OPAL measurements [2,4], however, the phase convention dependent parameter ϵ_B has been used.

The more important point is that both ε_B and ϵ_B depend on the phase convention of the KM matrix for quark flavor mixings, although ε_B itself is independent of the bound state phases of $|B^0\rangle$ and $|\bar{B}^0\rangle$. The KM phase convention comes from the freedom in defining the quark field phases, thus different parametrizations of the KM matrix may result in different values for ε_B and ϵ_B. In this sense, even ε_B is *not* a physically well–defined parameter.

In the CLEO and OPAL analyses [2,4], ${\rm Im}\epsilon_B = 0$ or $|{\rm Im}\epsilon_B| \ll 1$ has been assumed implicitly [9,10]. That is why these analyses can yield an upper bound on ${\rm Re}\epsilon_B$ solely from the measurement of the lepton CP asymmetry. To see this point more clearly, let us consider the CLEO constraint on $\mathrm{Re}\epsilon_B$. The dilepton CP asymmetry measured on the $\Upsilon(4S)$ resonance reads

$$\mathcal{A}_{\rm SL}(\varepsilon_B) = \frac{4\mathrm{Re}\varepsilon_B\,(1+|\varepsilon_B|^2)}{(1+|\varepsilon_B|^2)^2 + 4(\mathrm{Re}\varepsilon_B)^2}\,, \tag{3}$$

and $\mathcal{A}_{\rm SL}(\epsilon_B) = \mathcal{A}_{\rm SL}(\varepsilon_B)$ holds. Only under the assumption $|\mathrm{Im}\epsilon_B| \ll 1$ (and $|\mathrm{Re}\epsilon_B| \ll 1$), one can get $\mathcal{A}_{\rm SL}(\epsilon_B) \approx 4\mathrm{Re}\epsilon_B$. Then $|\mathcal{A}_{\rm SL}(\epsilon_B)| < 0.18$ leads to $|\mathrm{Re}\epsilon_B| < 0.045$. If one takes $|\mathrm{Im}\epsilon_B| \sim 1$, however, $\mathcal{A}_{\rm SL}(\epsilon_B) \approx 2\mathrm{Re}\epsilon_B$ appears [11] and a different upper bound on $\mathrm{Re}\epsilon_B$ follows from the same measurement of $\mathcal{A}_{\rm SL}(\epsilon_B)$. Indeed the condition $\mathrm{Im}\epsilon_B = 0$ corresponds to a special parametrization of the KM matrix which has $\mathrm{Im}V_{td} = \mathrm{Im}V_{tb} = 0$ (e.g., the one proposed recently by Fritzsch and the author [12]). In contrast, the condition $|\mathrm{Im}\epsilon_B| \sim O(1)$ may be satisfied in the "standard" parametrization or the Wolfenstein form [10].

3 NP effect on $\mathcal{A}_{\rm SL}$

Now we present a simple framework to analyze NP contributions to $\mathcal{A}_{\rm SL}$ and other CP asymmetries via $B^0 - \bar{B}^0$ mixing. In terms of the off–diagonal elements of the 2×2 $B^0 - \bar{B}^0$ mixing Hamiltonian $\mathbf{M} - i\mathbf{\Gamma}/\mathbf{2}$, $\mathcal{A}_{\rm SL}$ can be written as

$$\mathcal{A}_{\rm SL} = \mathrm{Im}\left(\frac{\Gamma_{12}}{M_{12}}\right). \tag{4}$$

In most extensions of the SM, NP can significantly contribute to M_{12}. However, NP is not expected to significantly affect direct B meson decays via tree–level W mediated channels. Thus $\Gamma_{12} = \Gamma_{12}^{\rm SM}$ holds as a good approximation, where [5]

$$\Gamma_{12}^{\rm SM} = -\frac{G_{\rm F}^2\, B_B\, f_B^2\, M_B\, m_b^2}{8\pi}\left[(\xi_u^*)^2 T_u + (\xi_c^*)^2 T_c + (\xi_t^*)^2 T_t\right] \tag{5}$$

with KM factors $\xi_i \equiv V_{ib}^* V_{id}$ ($i = u, c, t$) for three quark families and QCD correction factors $T_u \sim -T_c \sim 0.1$ and $T_t \sim 1$. In the presence of NP, M_{12} can be written as

$$M_{12} = M_{12}^{\rm SM} + M_{12}^{\rm NP}\,. \tag{6}$$

Rescaling three complex quantities in Eq. (6) by $|M_{12}| = \Delta M/2$, where $\Delta M = (0.464 \pm 0.018)\ \mathrm{ps}^{-1}$ has been measured [3], we obtain a parametrized triangle in the complex plane (see Fig. 1).

The correlation between the SM and NP parameters reads

$$R_{\rm NP} = -R_{\rm SM}\,\cos 2(\theta - \phi_1) \pm \sqrt{1 - R_{\rm SM}^2\,\sin^2 2(\theta - \phi_1)}\,, \tag{7}$$

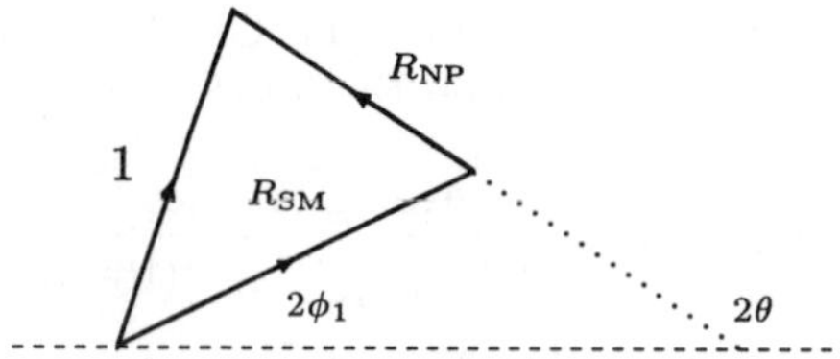

Figure 1: Triangular relation of $M_{12}^{\rm SM}$, $M_{12}^{\rm NP}$ and M_{12} (rescaled by $\Delta M/2$).

where $R_{\rm SM}$ can be calculated in the SM box diagram approximation [5]:

$$R_{\rm SM} \;=\; \frac{G_{\rm F}^2 \; B_B \; f_B^2 \; M_B \; m_t^2}{6\pi^2 \; \Delta M} \; \eta_B \; F\left(\frac{m_t^2}{m_W^2}\right) \; |\xi_t|^2 \tag{8}$$

with $\eta_B \approx F(m_t^2/m_W^2) \approx 0.55$. We see that there exist two solutions for $R_{\rm NP}$.

The lepton CP asymmetry $\mathcal{A}_{\rm SL}$ turns out to be [5]:

$$\begin{aligned} \frac{\mathcal{A}_{\rm SL}}{C_m} \;=\;& R_{\rm SM} R_{\rm NP} \left[{\rm Im}\left(\frac{\xi_u}{|\xi_t|}\right)^2 T_u + {\rm Im}\left(\frac{\xi_c}{|\xi_t|}\right)^2 T_c + {\rm Im}\left(\frac{\xi_t}{|\xi_t|}\right)^2 T_t \right] \cos(2\theta) \\ & + R_{\rm SM}^2 \left[{\rm Im}\left(\frac{\xi_u}{\xi_t}\right)^2 T_u + {\rm Im}\left(\frac{\xi_c}{\xi_t}\right)^2 T_c \right] + \\ & R_{\rm SM} R_{\rm NP} \left[{\rm Re}\left(\frac{\xi_u}{|\xi_t|}\right)^2 T_u + {\rm Re}\left(\frac{\xi_c}{|\xi_t|}\right)^2 T_c + {\rm Re}\left(\frac{\xi_t}{|\xi_t|}\right)^2 T_t \right] \sin(2\theta) \; , \end{aligned} \tag{9}$$

where $C_m \approx 1.3 \times 10^{-2}$. Clearly the second term of $\mathcal{A}_{\rm SL}$ comes purely from $M_{12}^{\rm SM}$ itself and its magnitude is expected to be of $\mathcal{O}(10^{-3})$ due to the absence of the T_t contribution. The first and third terms of $\mathcal{A}_{\rm SL}$ arise from the interference between $M_{12}^{\rm SM}$ and $M_{12}^{\rm NP}$; but they depend on nonvanishing ${\rm Im}(M_{12}^{\rm SM})$ and ${\rm Im}(M_{12}^{\rm NP})$, respectively. For appropriate values of θ and ϕ_1, the magnitudes of both the first and third terms of $\mathcal{A}_{\rm SL}$ may be at the percent level! To obtain $|\mathcal{A}_{\rm SL}| \sim \mathcal{O}(10^{-2})$, however, there should not be a large cancellation between two dominant terms in Eq. (9).

The CP asymmetry in $B_d \rightarrow J/\psi K_S$ can be calculated in the same framework. We obtain

$$\mathcal{A}_{\psi K} \;=\; R_{\rm SM} \sin(2\phi_1) \;+\; R_{\rm NP} \sin(2\theta) \; . \tag{10}$$

As $R_{\rm NP}$, $R_{\rm SM}$ and ϕ_1, θ are dependent on one another through Eq. (7), $|\mathcal{A}_{\psi K}| \leq 1$ is always guaranteed within the allowed parameter space.

Two interesting cases, corresponding to $\mathrm{Im}M_{12}^{\mathrm{SM}} = 0$, $\mathrm{Im}M_{12}^{\mathrm{NP}} \neq 0$ and $\mathrm{Im}M_{12}^{\mathrm{NP}} = 0$, $\mathrm{Im}M_{12}^{\mathrm{SM}} \neq 0$, have been numerically illustrated by Sanda and the author[5]. It is found that $\mathcal{A}_{\mathrm{SL}}$ does have good chances to reach the percent level, and the correlation between $\mathcal{A}_{\mathrm{SL}}$ and $\mathcal{A}_{\psi K}$ (as well as other CP asymmetries) does reflect the information from NP.

4 Summary

We have commented on the dependence of ϵ_B or ε_B on the KM phase convention. We emphasize that neither of them can prove much advantage in describing data of CP violation from $B^0 - \bar{B}^0$ mixing. Also, a simple framework has been presented for the analysis of possible NP contributions to $B^0 - \bar{B}^0$ mixing and of their effects on $\mathcal{A}_{\mathrm{SL}}$ and other CP asymmetries. If we are lucky, we should be able to detect the lepton CP asymmetry at the percent level in the first–round experiments of B factories.

Acknowledgments: I would like to thank Y.Y. Keum for his warm hospitality and the APCTP for their generous support, which made my participation in this wonderful workshop realizable. I am also grateful to A.I. Sanda for sharing physical ideas with me in a recent paper[5], whose results were partially presented here. Finally useful discussions with S. Pakvasa, A.I. Sanda, S.Y. Tsai and D.D. Wu are acknowledged.

References

1. I.I. Bigi and A.I. Sanda, Nucl. Phys. **B193**, 85 (1981); **B281**, 41 (1987).
2. CLEO Collaboration, J. Bartelt *et al.*, Phys. Rev. Lett. **71**, 1680 (1993).
3. Particle Data Group, R.M. Barnett *et al.*, Phys. Rev. **D54**, 1 (1996).
4. OPAL Collaboration, K. Ackerstaff *et al.*, CERN-PPE/97-036 (1997).
5. A.I. Sanda and Z.Z. Xing, Phys. Rev. **D56**, 6866 (1997).
6. H. Yamamoto, Phys. Lett. **B401**, 91 (1997); Y. Sakai, talk given at the *B* Physics Mini-workshop, September, Izu, Shizuoka, Japan (1997).
7. M.C. Bañuls and J. Bernabéu, CERN-TH/97-216, hep-ph/9710348.
8. See, e.g., S.Y. Tsai, Mod. Phys. Lett. **A11**, 2941 (1996).
9. I would like to thank Y. Nir for pointing out this fact to me.
10. Z.Z. Xing, hep-ex/9709005.
11. See, e.g., T. Brown and S. Pakvasa, Phys. Rev. **D31**, 1661 (1985).
12. H. Fritzsch and Z.Z. Xing, Phys. Lett. **B413**, 396 (1997); CERN-TH/97-201, hep-ph/9708366 (Phys. Rev. D in press).

THE PARTON MODEL *vs.* THE ACCMM MODEL

YEONG GYUN KIM
Department of Physics, Korea Advanced Institute of Science and Technology, Taejon 305-701, KOREA

We study the distribution function of the b quark for inclusive semi–leptonic B decays in the framework of the parton model. With the help of the $1/m_b$ expansion we attempt to test the models by extracting two HQET parameters, the effective mass of light degrees of freedom $\bar{\Lambda}$ and the average kinetic energy of the b quark μ_π^2.

1 Introduction

Two models are known which predict the lepton energy spectrum of inclusive decays of B meson (the ACCMM Model[1] and the Parton Model[2,3]). In this paper, we study both models from the viewpoint of the parton model picture. With the help of the $1/m_b$ expansion suggested in Ref. [4] we attempt to test the models by extracting two HQET parameters, the effective mass of light degrees of freedom $\bar{\Lambda}$ and the average kinetic energy of the b quark μ_π^2.

2 The Parton Model

We first review the parton model approach applied to inclusive semi–leptonic B meson decays. The distribution function $f(x)$ is defined as the probability of finding a b quark in a B meson carrying a fraction x of the meson momentum in the infinite momentum frame. We write the Lorentz invariant decay width as follows

$$E_B\ d\Gamma(B \to X_q e\nu) = \int dx\ f(x)\ E_b\ d\Gamma(b \to qe\nu), \tag{1}$$

with the relation $p_b = xp_B$. With properly defined m_b, the parton model is well represented by an $1/m_b$ expansion[4]. We define

$$m_b \equiv \langle x \rangle m_B, \tag{2}$$

where $\langle x \rangle = \int dx\ xf(x)$. If we take the final state quark to be a u quark so that $m_q = 0$ hereafter, the triple differential decay rate is given by

$$\frac{d\Gamma}{dE_l dq^2 dq_0} \approx \frac{G_F^2|V_{ub}|^2}{192\pi^3}\frac{2}{m_B}f(x_+)\frac{24(q_0 - E_l)(2m_bE_l - q^2)}{m_b - q_0} \tag{3}$$

where m_b is as defined in Eq.(2) and $x_+ = \frac{q_0+\sqrt{\mathbf{q}^2+m_q^2}}{m_B}$.

In the HQET framework, the triple differential decay rate for $B \to X_q l\nu$ shows scaling behavior [5,6,7]:

$$\frac{d\Gamma}{dE_l dq^2 dq_0} = \frac{G_F^2 |V_{qb}|^2}{192\pi^3} \frac{2}{\bar{\Lambda}} F(x_B) \frac{24(q_0 - E_l)(2m_b E_l - q^2)}{m_b - q_0}, \tag{4}$$

where $x_B = -\frac{m_b^2+q^2-2m_b q_0}{2\bar{\Lambda}(m_b-q_0)}$. From Eqs.(3) and (4), the line shape function $F(x_B)$ of QCD is related to the distribution function $f(x_+)$ of the parton model after appropriate variable changes

$$\frac{1}{m_B} f(x_+) = \frac{1}{\bar{\Lambda}} F(x_B). \tag{5}$$

In practice, we know only the first few moments of $F(x_B)$

$$a_n = \int dx_B \; x_B^n F(x_B), \qquad n = 0, 1, \ldots \,. \tag{6}$$

In particular, we know that $a_0 = 1$ from b number conservation, and that $a_1 = 0$ up to $\mathcal{O}(1/m_b^2)$, which indicates the lack of $\mathcal{O}(1/m_b)$ corrections. The value of the first moment $a_1 = 0$ is related to the value of the b quark mass m_b or the parameter $\bar{\Lambda}$. The second moment of the spectrum is related to the average kinetic energy of the b quark inside the B meson, usually represented by the parameter μ_π^2, through $a_2 \equiv \frac{\mu_\pi^2}{3\bar{\Lambda}^2}$. After the variable changes of Eq.(5), we can calculate the moments of the distribution function in the framework of the parton model. Under the definition of m_b in Eq.(2), we have $a_1 = 0$. The second moment is derived as

$$a_2 = \int_{-\infty}^{1} dx_B \; x_B^2 F(x_B) = \left(\frac{m_B}{\bar{\Lambda}}\right)^2 \int_0^1 dx \; (x - \frac{m_b}{m_B})^2 f(x). \tag{7}$$

3 The ACCMM Model

The ACCMM model [1] treats the B meson as consisting of a b quark and a spectator quark. The spectator quark is treated as a quasi–free particle having definite mass m_{sp} and momentum $\mathbf{p}$, while the b quark is treated as a virtual particle of invariant mass m_b:

$$m_b^2(p) = m_B^2 + m_{sp}^2 - 2m_B\sqrt{p^2 + m_{sp}^2}, \tag{8}$$

where m_B is the B meson mass and $p \equiv |\mathbf{p}|$ in the B meson rest frame. The b quark energy is given by $E_b(p) = \sqrt{p^2 + m_b^2(p)}$. The distribution of the spectator momentum is usually assumed to be a Gaussian with an adjustable parameter p_F. After convoluting the distribution of the spectator momentum with the subprocess of the virtual b quark decay in a frame where its momentum is $-\mathbf{p}$, we can obtain the lepton energy spectrum at the hadronic level.

In this paper we derive the distribution function $f(x)$ in the IMF from the given Gaussian distribution in the B rest frame, $\Phi(p)$, in order to study the ACCMM model by comparing it with the parton model approach. Let's choose a coordinate system where the B meson is at the origin in the B rest frame. We boost the B meson along the positive z direction with the velocity β. As is well known, for the B meson at rest we have the boosted energy of B meson as $E_B^* = \gamma E_B = \gamma m_B$, where $\gamma = 1/\sqrt{1-\beta^2}$, and that of b quark as $E_b^*(p) = \gamma(E_b(p) - \beta p_z)$, where p is the magnitude of the spatial momentum of the b quark and p_z is the z component of $\mathbf{p}$ in the B rest frame. In the limit of $\beta \to 1$, $\gamma \to \infty$,

$$\frac{E_b^*}{E_B^*} = \frac{(E_b - p_z)}{m_B} \equiv x, \tag{9}$$

which represents the ratio of b quark energy to B meson energy in the IMF. Because no specific direction for $\mathbf{p}$ is preferred in B meson decay, the value of p_z is distributed from $-p$ to $+p$ with equal probability for a fixed value of p. The value of x can thus be from $x_{min}(p)$ to $x_{max}(p)$ with equal weight for a fixed p, where $x_{min}(p) = \frac{1}{m_B}(E_b(p) - p)$, $x_{max}(p) = \frac{1}{m_B}(E_b(p) + p)$, with the weight $m_B/2p$, because $x_{max}(p) - x_{min}(p) = 2p/m_B$. Each value of p in the B meson rest frame corresponds to a distribution of x in the IMF such that the probability of finding a b quark in a B meson carrying a fraction x of the meson momentum in the IMF is obtained by integrating over p with Gaussian weight, $\Phi(p)$,

$$\begin{aligned} f(x) &= \int_0^{p_{max}} dp\, p^2\, \Phi(p) \cdot \frac{m_B}{2p}\, \theta(x - x_{min}(p))\, \theta(x_{max}(p) - x) \\ &= \int_{p_{min}}^{p_{max}} dp\, p^2\, \Phi(p) \cdot \frac{m_B}{2p} \end{aligned} \tag{10}$$

where $p_{min} = \frac{|m_B^2(1-x)^2 - m_{sp}^2|}{2m_B(1-x)}$ is obtained by the conditions $x_{min}(p) < x < x_{max}(p)$. We take the value of p_{max} to be infinity as far as the integration is concerned, because the area of phase space for $p > p_{max}$ is negligible. After the integration over p, we obtain

$$f(x) = \frac{m_B}{\sqrt{\pi} p_F} \exp\left(-\frac{1}{4} \left(\frac{\rho p_F}{m_B(1-x)} - \frac{m_B}{p_F}(1-x) \right)^2 \right) , \tag{11}$$

where $\rho = m_{sp}^2/p_F^2$. Using Eq.(5) and the relation $(1 - x_+) = \frac{\bar{\Lambda}}{m_B}(1 - x_B)$, we find Eq.(11) becomes Eq.(30) of Ref. [6], the light–cone distribution function for the ACCMM model derived by Bigi *et al.*

We thus have derived a distribution function $f(x)$, Eq.(10), for the parton model approach from a Gaussian distribution in the B rest frame. Following the method discussed in section 2, we can calculate the moments of the line shape function and the parameters $\bar{\Lambda}$ and μ_π^2 for given model parameters:

$$\bar{\Lambda} = \int_0^\infty dp\, p^2 \Phi(p) (\sqrt{p^2 + m_{sp}^2}) = \langle \sqrt{p^2 + m_{sp}^2} \rangle_{B-rest}. \tag{12}$$

$$\mu_\pi^2 = 3(\langle E_b^2 \rangle_{B-rest} - \langle E_b \rangle_{B-rest}^2) + \langle p^2 \rangle_{B-rest}. \tag{13}$$

Acknowledgments

The author thanks C.S. Kim and K.Y. Lee for useful discussions. This work was supported by the KRF postdoctoral program.

References

1. G. Altarelli, N. Cabbibo, G. Corbo, L. Maiani and G. Martinelli, *Nucl. Phys.* B **208**, 365 (1982).
2. A. Bareiss and E.A. Paschos, *Nucl. Phys.* B **327**, 353 (1989).
3. C.H. Jin, W.F. Palmer and E.A. Paschos, *Phys. Lett.* B **329**, 364 (1994); C.H. Jin, W.F. Palmer and E.A. Paschos, preprint DO–TH 93/31 (1993).
4. K.Y. Lee and J.K. Kim, *Phys. Lett.* B **377**, 153 (1996).
5. A.V. Manohar and M.B. Wise, *Phys. Rev.* D **49**, 1310 (1994); B. Blok, L. Koyrakh, M. Shifman and A.I. Vainstein, *Phys. Rev.* D **49**, 3356 (1994).
6. I.I. Bigi, M.A. Shifman, N.G. Uraltsev and A.L. Vainstein, *Phys. Lett.* B **328**, 431 (1994).
7. I.I. Bigi, M.A. Shifman, N.G. Uraltsev and A.L. Vainstein, *Int. J. Mod. Phys.* **A 9**, 2467 (1994).

NON-SPECTATOR CONTRIBUTIONS TO $B \to X_s\eta'$ DECAYS

M.R. AHMADY
LINAC Laboratory, The Institute of Physical and Chemical Research (RIKEN), 2-1 Hirosawa, Wako, Saitama 351-01, Japan

E. KOU, A. SUGAMOTO
Department of Physics, Ochanomizu University, 1-1 Otsuka 2, Bunkyo-ku, Tokyo 112, Japan

We propose a nonspectator mechanism in which η' is produced by two gluon fusion as the underlying process to explain the inclusive $B \to X_s\eta'$ as well as the exclusive $B \to K^{(*)}\eta'$ decays.

The CLEO collaboration has recently discovered an unexpectedly large branching ratio for the semi-inclusive hadronic $B \to X_s\eta'$ decay[1]:

$$BR(B \to X_s\eta') = (7.5 \pm 1.5 \pm 1.1) \times 10^{-4} \quad 2.0 \leq p_{\eta'} \leq 2.7 \ GeV \ . \tag{1}$$

The corresponding exclusive decay rate has also been measured:

$$BR(B \to K\eta') = (7.8^{+2.7}_{-2.2} \pm 1.0) \times 10^{-5} \ . \tag{2}$$

Possible mechanisms behind this large production of fast η' meson have been discussed in recent papers[2,3,4,5,6]. In this paper, we investigate the possibility that a somewhat different process might be the underlying mechanism for $B \to X_s\eta'$. We propose a non-spectator process in which η' is produced via fusion of the gluon from QCD penguin $b \to sg^*$ and another one emitted by the light quark inside the B meson. We also calculate the branching ratios $BR(B \to K\eta')$ and $BR(B \to K^*\eta')$ in the context of factorization. The effective Hamiltonian can be written as:

$$H_{eff} = CH(\bar{s}\gamma_\mu(1-\gamma_5)T^a b)(\bar{q}\gamma_\sigma T^a q)\frac{1}{p^2 - M_g^2}\epsilon^{\mu\sigma\alpha\beta}q_\alpha p_\beta \ , \tag{3}$$

where

$$C = \lambda_t \frac{G_F}{\sqrt{2}}\frac{\alpha_s}{2\pi}E_0 \ , \tag{4}$$

and the effective gluon mass M_g is due to bound state effects. Alternatively, one may use the usual gluon propagator (no effective mass) along with a model which incorporates the binding effects and off-shellness of the light quark inside B meson. H is the form factor parametrizing the $gg\eta'$ vertex. In writing (3) we considered only the dominant chromo-electric operator. A re-arrangement of (3) via Fierz transformation and using the definition for the B meson decay constant f_B results in:

$$\langle \eta' X_s | H_{eff} | B \rangle = \frac{C f_B H}{9(p^2 - M_g^2)} \left[-(\bar{s}\gamma_\sigma \gamma_\rho \gamma_\mu (1-\gamma_5) q) p_B^\rho \right. \\ \left. + \left(\frac{M_B^2}{m_q + m_b} \right) (\bar{s}\gamma_\sigma \gamma_\mu (1+\gamma_5) q) \right] \epsilon^{\mu\sigma\alpha\beta} q_\alpha p_\beta \ . \quad (5)$$

Hereafter, we take the light quark mass $m_q = 0$. Using the usual convention for the invariant variables $s = (p_{\eta'} + k')^2$, $t = (p_s + k')^2$ and $u = (p_s + p_{\eta'})^2$,

$$\frac{d\Gamma(B \to X_s \eta')}{dtdu} = \frac{C^2 f_B^2 H^2}{648\pi^3 M_B^3 (p^2 - M_g^2)^2} \times \\ \left[p^2 X \left\{ (W - Y - \frac{p^2}{2})(W - X) - XZ + (X - \frac{s+p^2}{2}) W \right\} \right. \\ - q^2 Z W^2 + XYZW + (s - 2Y - q^2)(X - \frac{s+p^2}{2}) W^2 \\ - \left(\frac{M_B^2}{m_b} \right)^2 \left\{ p^2 (X - \frac{s+p^2}{2})(s - Y - q^2) - p^2 q^2 (W + Z - Y - \frac{p^2}{2}) \right. \\ \left. \left. + Y^2 Z - (s - 2Y - q^2)(W - Y - \frac{p^2}{2}) Y \right\} \right] , \quad (6)$$

where $W = (u - m_b^2)/2$, $X = (m_b^2 - m_s^2 + s)/2$, $Y = (m_{\eta'}^2 - p^2 - q^2)/2$ and $Z = (t - m_s^2)/2$. $m_{X_s}^2 = t$ is the invariant mass of the final state strange hadron. The differential decay rate (6) depends on the virtualities of the internal gluons both explicitly and implicitly through the form factor H. However, $H(p^2, q^2, m_{\eta'}^2)$ is suppressed for large values of p^2 and q^2. Therefore, the dominant contribution to the decay rate is expected to arise from the small virtuality region. On the other hand, in the non-spectator mechanism, due to kinematical freedom, one can impose a constraint such as $p^2 = 0$. Consequently, q^2 can be expressed as:

$$q^2 = m_b^2 + m_{\eta'}^2 - u + (t - m_s^2) \frac{u - m_b^2}{m_B^2 - u} \ . \quad (7)$$

For our numerical evaluations, we have taken $m_b = 4.5$ GeV, $m_s = 0.15$ GeV[a], $M_g \approx \Lambda_{QCD} \approx 0.3$ GeV (see our explanation following eqn. (3)),

[a] For phase space calculation, m_s is taken to be the constituent quark mass $m_s^{\text{constituent}} \approx 0.45$ GeV.

$\alpha_s = 0.2$, $f_B = 0.2$ GeV and $|V_t| = |V_{tb}V_{ts}^*| \approx |V_{cb}| \approx 0.04$. In order to obtain the total branching ratio with the experimental cut $2.0 \le p_{\eta'} \le 2.7$ GeV, the differential decay rate is integrated over the range $m_s^{\rm constituent} = 0.45 \le m_{X_s} \le 2.32$ GeV resulting in:

$$BR(B \to X_s\eta') = 4.7 \times 10^{-3} \quad 2.0 \le p_{\eta'} \le 2.7 \ \text{GeV} \ . \tag{8}$$

The momentum dependence of $H(q^2, 0, m_{\eta'}^2)$ has not been taken into account in the above estimate. However, even if one considers up to an order of magnitude suppression due to this form factor, our result indicates that the non-spectator mechanism is indeed the dominant process making up the bulk of the experimental data (1).

Using eqn. (5) in conjunction with the factorization assumption and the definition of the decay constants f_K and f_{K^*} for K and K^* respectively, we obtain

$$\langle \eta' K | H_{eff} | B \rangle = -i \frac{CHf_Bf_K}{9(p^2 - M_g^2)} \left(p_B \cdot q p_K \cdot p - p_B \cdot p p_K \cdot q\right) \ , \tag{9}$$

leading to the exclusive decay rate

$$\Gamma(B \to K\eta') = \frac{C^2H^2f_B^2f_K^2}{1944\pi M_g^4} |\vec{p}_K|^3 \left(m_{\eta'}^2 + 4|\vec{p}_K|^2\right) p_0^2 \ , \tag{10}$$

and

$$\langle \eta' K^* | H_{eff} | B \rangle = -i \frac{CHf_Bf_{K^*}}{9(p^2 - M_g^2)} \Big[p_B \cdot q \epsilon \cdot p - p_B \cdot p \epsilon \cdot q \tag{11}$$
$$+ \frac{2m_B^2 m_s}{m_b m_{K^*}^2} \left(-i\epsilon^{\mu\sigma\alpha\beta} p_{K^*\mu} \epsilon_\sigma q_\alpha p_\beta + p_{K^*} \cdot q \epsilon \cdot p - p_{K^*} \cdot p \epsilon \cdot q\right)\Big] \ ,$$

resulting in

$$\begin{aligned}
\Gamma(B \to K^*\eta') &= \frac{C^2H^2f_B^2f_{K^*}^2}{1296\pi M_g^4 m_B^2} |\vec{p}_{K^*}| \int_{-1}^{1} F(x)dx \ , \\
F(x) &= (C_1 + C_3C_5)^2 \frac{C_4^2}{m_{K^*}^2} + (C_2 + C_4C_5)^2(-q^2 + \frac{C_3^2}{m_{K^*}^2}) \\
&- 2(C_1 + C_3C_5)(C_2 + C_4C_5)(-C_6 + \frac{C_4C_3}{m_{K^*}^2}) \\
&+ C_5^2 \left[m_{K^*}^2 C_6^2 - C_3C_6C_4 - C_4(C_3C_6 - q^2C_4)\right] \ ,
\end{aligned} \tag{12}$$

where

$$
\begin{aligned}
C_1 &= \frac{m_B^2 + m_{\eta'}^2 - m_{K^*}^2}{2} - C_2 \ , \\
C_2 &= m_B p_0 \ , \\
C_3 &= \frac{m_B^2 - m_{\eta'}^2 - m_{K^*}^2}{2} - C_4 \ , \\
C_4 &= \left[(m_{K^*}^2 + |\vec{p}_{K^*}|^2)^{1/2} - |\vec{p}_{K^*}| x \right] p_0 \ , \\
C_5 &= \frac{2 m_s m_B^2}{m_b m_{K^*}^2} \ , \\
C_6 &= \frac{m_{\eta'}^2 - q^2}{2} \ , \\
|\vec{p}_{K^*}| &= \frac{\left[(m_B^2 - (m_{\eta'} + m_{K^*})^2)(m_B^2 - (m_{\eta'} - m_{K^*})^2) \right]^{1/2}}{2 m_B} \ .
\end{aligned}
$$

Eqns. (10) and (12) are derived by imposing the $p^2 = 0$ constraint. $|\vec{p}_K|$ ($|\vec{p}_{K^*}|$) and p_0 are the three momentum of the $K(K^*)$ meson and the energy change of the light quark in the B meson rest frame, respectively:

$$
|\vec{p}_{K^{(*)}}| = \left[\frac{(m_B^2 + m_{K^{(*)}}^2 - m_{\eta'}^2)^2}{4 m_B^2} - m_{K^{(*)}}^2 \right]^{\frac{1}{2}} , \tag{13}
$$

$$
p_0 = \frac{m_B^2 - m_b^2}{2 m_B} - E_q \ , \tag{14}
$$

where E_q is the energy of the light quark in the $K^{(*)}$ meson. Inserting $E_q = m_K - m_s^{\rm constituent} \approx 0.05$ GeV in eq. (10) results in

$$
BR(B \to K\eta') = 7.0 \times 10^{-5} \ , \tag{15}
$$

and using $E_q \approx m_{K^*} - m_s^{constituent} \approx 0.44\mathrm{GeV}$ in (12) we obtain[b]

$$
BR(B \to K^*\eta') = 3.4 \times 10^{-5} \ . \tag{16}
$$

We note that the results for exclusive decays should not be altered significantly by the momentum dependence of H. This is due to the fact that, unlike the inclusive process, q^2 for these decays is fixed at around $1 - 3$ GeV^2. A

[b] We estimate $f_{K^*} \approx 0.205$ GeV by using the experimental value of the ratio $\Gamma(\tau \to K^{*-}\nu_\tau)/\Gamma(\tau \to K^-\nu_\tau)$ and $f_K = 0.167$ GeV[8].

measurement of K^* mode will be a crucial test for various mechanisms suggested for η' production in hadronic B decays. For example, our prediction is in contrast to $\Gamma(B \to K^*\eta') \approx 2\Gamma(B \to K\eta')$ obtained from the proposed $b \to c\bar{c}s \to \eta' s$ process[3]. We should also emphasize that the ratios of decays $\Gamma(B \to K\eta')/\Gamma(B \to X_s\eta')$ and $\Gamma(B \to K^*\eta')/\Gamma(B \to K\eta')$, calculated in our mechanism, are independent of the input parameters like α_s, f_B and M_g, and therefore, free of the uncertainties associated with these parameters.

References

1. Jim Smith, talk given at the 7th International Symposium on Heavy Flavor Physics, Santa Barbara CA (July 1997); B. Behren, talk given at the Princeton BaBar meeting (March 1997); P. Kim, talk given at FCNC 97, Santa Monica CA (Feb,1997); F. Würthwein, CALT-68-2121, HEP-EX/9706010.
2. D. Atwood and A. Soni, Phys. Lett. **B405** (1997) 150.
3. I. Halperin and A. Zhitnitsky, HEP-PH/9705251; HEP-PH/9704412
4. F. Yuan and K. T. Chao, Phys. Rev. **D56** (1997) 2495.
5. W.-S. Hou and B. Tseng, HEP-PH/9705304.
6. A. L. Kagan and A. A. Petrov, HEP-PH/9707354.
7. T. Inami and C. S. Lim, Progr. Theor. Phys. **65** (1981) 297.
8. Particle Data Group, Phys. Rev. **D54** (1996) 1.

DETERMINATION OF THE CKM MATRIX BY SEMI–LEPTONIC RARE B DECAYS

L.T. HANDOKO[a]

Department of Physics, Hiroshima University
1-3-1 Kagamiyama, Higashi Hiroshima - 739, Japan

The possibility to test the unitarity triangle of the CKM matrix is examined by using the semi-leptonic rare B decays $B \to X_q \ell^+ \ell^-$ with $q = d, s$ and ℓ's are leptons. The discussion emphasizes the CP asymmetry in the framework of standard and beyond the standard models. As an example of the latter, an extension of the standard model containing an additional isosinglet charge -1/3 quark, which leads to a deviation from CKM unitarity, is considered. It is found that in the standard model the CP asymmetry is visible only for $B \to X_d \ell^+ \ell^-$, while in the latter model it is sizeable even for $B \to X_s \ell^+ \ell^-$.

1 Introduction

After the measurement of the radiative $B \to X_s \gamma$ decay, the next targets for experiments are other higher order decays, e.g. $B \to X_s \ell^+ \ell^-$. The CLEO Collaboration expects to be able to measure a few events of this mode in the near future[1].

Theoretically, the decays are very attractive and should be good probes to test the standard model as well as open a window to the theory beyond it. In general rare decays, that means decays which occur only through loop diagrams, are very sensitive to new contributions. Here according to the unitarity triangle of the CKM matrix, new physics can be divided into two types, 1) new physics which conserves unitarity, and 2) new physics which violates it. In this talk, I consider only the standard model and the latter type of new physics that can be realized by introducing an additional isosinglet charge -1/3 quark.

As for measurements that must be examined in the experiments, the interest is focused on the CP asymmetry. Note that in decays which are governed by a single operator like $B \to X_s \gamma$, the CP asymmetry is tiny as indicated, for example, in some previous papers[2]. In other words, $B \to X_s \gamma$ decays are not useful to study CP violation. Then one needs to consider more promising modes. The best candidates are semi-leptonic $B \to X_q \ell^+ \ell^-$ decays with $q = d, s$.

The reason is very clear: $B \to X_q \ell^+ \ell^-$ decays contain at least three

[a]On leave from P3FT-LIPI, Indonesia. E-mail : handoko@theo.phys.sci.hiroshima-u.ac.jp.

operators in its effective Hamiltonian, that is

$$\mathcal{H}_{\rm eff} = \frac{G_F\,\alpha}{\sqrt{2}\,\pi}\,V_{tq}^*\,V_{tb}\,\Big\{C_9{}^{\rm eff}\,[\bar{q}\,\gamma_\mu\,L\,b]\,[\bar{\ell}\,\gamma^\mu\,\ell] + C_{10}{}^{\rm eff}\,[\bar{q}\,\gamma_\mu\,L\,b]\,[\bar{\ell}\,\gamma^\mu\,\gamma_5\,\ell]$$
$$-2\,C_7{}^{\rm eff}\left[\bar{q}\,i\,\sigma_{\mu\nu}\,\frac{\hat{q}^\nu}{\hat{s}}\,(R+\hat{m}_q\,L)\,b\right][\bar{\ell}\,\gamma^\mu\,\ell]\Big\}\,. \quad (1)$$

Here, $C_i^{\rm eff}$ denotes the Wilson coefficient for each operator, q^μ is the four-momentum of the dilepton and $s = q^2$. Therefore, CP violation will occur if there are nontrivial phases in at least one of the Wilson coefficients.

2 Conserved CKM unitarity case

I am not going to give a specific model of new physics that conserves the unitarity of the CKM matrix, but discuss only the case in the framework of the standard model.

In the standard model, some different phases are possibly appearing in the long-distance contributions to the decay. The reason is that m_b is large enough to generate on-shell states of uponium and/or charmonium, i.e. $u^i\bar{u}^i$ with $u^i = u, c$. These $u^i\bar{u}^i$ loops are generated from the $b \to q\,u^i\,\bar{u}^i$ processes that are governed by the following operators,

$$\mathcal{O}_1 = (\bar{q}_\alpha\,\gamma_\mu\,L\,b_\alpha)\sum_{i=1,2}(\bar{u}^i_\beta\,\gamma^\mu\,L\,u^i_\beta)\,, \quad (2)$$

$$\mathcal{O}_2 = (\bar{q}_\alpha\,\gamma_\mu\,L\,b_\beta)\sum_{i=1,2}(\bar{u}^i_\beta\,\gamma^\mu\,L\,u^i_\alpha)\,, \quad (3)$$

after performing a Fierz transformation. Here, $u^1 = u$, $u^2 = c$ and the lower suffixes denote the color. For $q = s$, the $u\bar{u}$ loop contribution can be ignored [4], while for $q = d$ the situation is quite different, since [3]

$$q = s: \left|\frac{V_{u^i s}^*\,V_{u^i b}}{V_{ts}^*\,V_{tb}}\right| \sim \begin{cases} O(\lambda^2)\,, & u^i = u \\ O(1)\,, & u^i = c \end{cases}$$

$$q = d: \left|\frac{V_{u^i d}^*\,V_{u^i b}}{V_{td}^*\,V_{tb}}\right| \sim \begin{cases} O(1)\,, & u^i = u \\ O(1)\,, & u^i = c \end{cases}$$

if we normalize the amplitude with $V_{tq}^*\,V_{tb}$ as usual. λ is a parameter in the Wolfenstein parametrization of the CKM matrix [7]; numerically, $\lambda \sim 0.22$ [8].

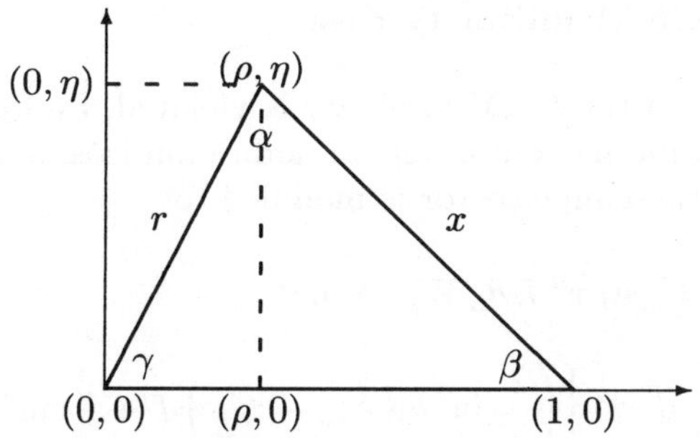

Figure 1: The CKM unitarity triangle on the $\rho - \eta$ plane.

Further, if the triangle is parametrized as in Fig. 1, one finds

$$\frac{V_{uq}^* V_{ub}}{V_{tq}^* V_{tb}} \sim \begin{cases} \dfrac{r\left(e^{-i\gamma} - r\right)}{1 + r^2 - 2r\cos\gamma} & , \quad q = d \\ \lambda^2 r\, e^{-i\gamma} & , \quad q = s \quad (\lambda^2 \sim 5\%) \end{cases} \tag{4}$$

From this result, it is obvious that for $q = d$ the uponium $(u\bar{u})$ contribution is large enough to induce CP violation in the decay. On the other hand, for $q = s$ CP violation will be suppressed by a factor of λ^2. These long-distance effects lead to the Wilson coefficients [3],

$$C_7{}^{\text{eff}} = C_7{}^{\text{NLO}}, \tag{5}$$

$$C_9{}^{\text{eff}} = C_9{}^{\text{NLO}} + \frac{V_{uq}^* V_{ub}}{V_{tq}^* V_{tb}} C_9{}^{\text{CP}}, \tag{6}$$

$$C_{10}{}^{\text{eff}} = C_{10}, \tag{7}$$

where $C_7{}^{\text{eff}}$ and $C_9{}^{\text{eff}}$ have been calculated up to next-to-leading order [5 6], while $C_{10}{}^{\text{eff}}$ obtains no correction at all. $C_9{}^{\text{CP}}$ is a CP violation factor given by:

$$\begin{aligned} C_9{}^{\text{CP}} &= (3C_1 + C_2 + 3C_3 + C_4 + 3C_5 + C_6) \\ &\times \left[\underbrace{g(\hat{m}_c, \hat{s}) - g(\hat{m}_u, \hat{s})}_{\text{continuum}} - \underbrace{\frac{16\pi^2}{9} \left(\sum_{V=\psi,\cdots} F_V(\hat{s}) - \sum_{V=\rho,\omega} F_V(\hat{s}) \right)}_{\text{resonances}} \right] \end{aligned} \tag{8}$$

Expressions for the continuum and resonances contributions can be found in reference [3].

3 Violated CKM unitarity case

For the case when the CKM unitarity is violated, I adopt a definite extension of the standard model containing an additional isosinglet charge -1/3 quark [9]. In this model, the gauge sector is modified to:

$$\mathcal{L}_{W^\pm} = \frac{g}{\sqrt{2}} \underline{V_{i\alpha}} \bar{u}_i \gamma^\mu L \, d_\alpha \, W_\mu^+ + \text{h.c.}, \tag{9}$$

$$\mathcal{L}_Z = \frac{g}{2\cos\theta_W} \bar{d}_\alpha \gamma^\mu \left[\left(\frac{2}{3} \sin^2\theta_W \delta_{\alpha\beta} - \underline{z_{\alpha\beta}} \right) L + \frac{2}{3} \sin^2\theta_W \delta_{\alpha\beta} R \right] d_\beta Z_\mu \tag{10}$$

where $z_{\alpha\beta} \equiv \sum_{i=1}^{3} {U^d}_{\alpha i} {U^d}^*_{i\beta} = \delta_{\alpha\beta} - {U^d}_{\alpha 4} {U^d}^*_{4\beta}$ and the new 3×4 CKM matrix satisfies the relation $\sum V^*_{iq} V_{ib} = z_{qb}$ that shows unitarity violation for $z_{qb} \neq 0$.

As long as one considers only the tree-level contributions of new physics, in this case the $Z-$exchange tree-level diagram, the Wilson coefficients read [10],

$$C_7{}^{\text{eff}} \to C_7{}^{\text{eff}} \tag{11}$$

$$C_9{}^{\text{eff}} \to C_9{}^{\text{eff}} + \left(1 + \frac{\alpha_s(\mu)}{\pi} \omega(\hat{s}) \right) \frac{\pi}{\alpha} \frac{z_{qb}}{V^*_{tq} V_{tb}} \left(4 \sin^2\theta_W - 1 \right) \tag{12}$$

$$C_{10}{}^{\text{eff}} \to C_{10}{}^{\text{eff}} + \frac{\pi}{\alpha} \frac{z_{qb}}{V^*_{tq} V_{tb}} \tag{13}$$

The problem is then how to determine the size of the new mixings $z_{qb}/V^*_{tq} V_{tb}$. Fortunately, in principle upper bounds on these mixings can be extracted from the experimentally known $B \to X_s \gamma$ decay for $q = s$, and from $B^0_d - \bar{B}^0_d$ mixing for $q = d$ [10].

4 Summary

By including long-distance contributions to $B \to X_q \, \ell^+ \, \ell^-$ decays, there will be additional phases in the amplitude. These should induce a CP asymmetry that is defined as,

$$\bar{\mathcal{A}}_{\text{CP}} = \frac{\mathrm{d}\mathcal{B}/\mathrm{d}\hat{s} - \mathrm{d}\bar{\mathcal{B}}/\mathrm{d}\hat{s}}{\mathrm{d}\mathcal{B}/\mathrm{d}\hat{s} + \mathrm{d}\bar{\mathcal{B}}/\mathrm{d}\hat{s}} = \frac{-2\,\mathrm{d}\mathcal{A}_{\text{CP}}/\mathrm{d}\hat{s}}{\mathrm{d}\mathcal{B}/\mathrm{d}\hat{s} + 2\,\mathrm{d}\mathcal{A}_{\text{CP}}/\mathrm{d}\hat{s}}, \tag{14}$$

with $\mathcal{B} = \text{Br}(B \to X_q \, \ell^+ \, \ell^-)$ and $\bar{\mathcal{B}}$ is its complex conjugate.

Lastly, as results in both cases I found :

1. In the standard model where CKM unitarity is conserved, $\bar{A}_{\rm CP}(B \to X_s \ell^+ \ell^-) \sim$ few $\times 10^{-3}$ and $\bar{A}_{\rm CP}(B \to X_d \ell^+ \ell^-) \sim 5\%$.
2. In an extension of the standard model where CKM unitarity is violated, $\bar{A}_{\rm CP}(B \to X_s \ell^+ \ell^-) \sim 2\%$ and $\bar{A}_{\rm CP}(B \to X_d \ell^+ \ell^-) \sim 5\%$, by including only the $Z-$exchange tree diagram.

Therefore, one can say that once a CP asymmetry in $B \to X_s \ell^+ \ell^-$ is detected, it will be a strong proof of physics beyond the standard model.

Acknowledgments

The author would like to thank the Organizing Committee of the Workshop for their warm hospitality during his stay in Seoul. This work is partially supported by the Ministry of Education and Culture of Japan under the Monbusho Fellowship Program.

References

1. Y. J. Kwon (CLEO Collaboration), these Proceedings.
2. G. Bhattacharyya, G. C. Branco and D. Choudhury, *Phys. Lett.* B **336**, 487 (1994). [Err. *Phys. Lett.* B **340**, 266 (1994)];
   L. T. Handoko and T. Morozumi, *Mod. Phys. Lett.* A **10**, 309 (1995).
3. L. T. Handoko, to appear in *Phys. Rev.* **D** (1998) (hep-ph/9707222).
4. B. Grinstein, M. J. Savage and M. B. Wise, *Nucl. Phys.* B **319**, 271 (1989);
   R. Grigjanis, P. J. O'Donnel, M. Sutherland and H. Navelet, *Phys. Lett.* B **223**, 239 (1989).
5. M. Jeżabek and J. H. Kühn, *Nucl. Phys.* B **320**, 20 (1989);
   M. Misiak, *Nucl. Phys.* B **393**, 23 (1993) [Err. *ibid.* **B439**, 461 (1995)];
   A. J. Buras and M. Münz, *Phys. Rev.* D **52**, 186 (1995).
6. A. J. Buras, A. Kwiatkowski and N. Pott, (hep-ph/9710336).
7. L. Wolfenstein, *Phys. Rev. Lett.* **51**, 1945 (1983).
8. Particle Data Group, *Phys. Rev.* D **54**, 1 (1996).
9. G. C. Branco and L. Lavoura, *Nucl. Phys.* B **278**, 738 (1986);
   L. Lavoura and J. P. Silva, *Phys. Rev.* D **47**, 2046 (1993);
   V. Barger, M. S. Berger and R. J. N. Phillips, *Phys. Rev.* D **52**, 1663 (1995).
10. L. T. Handoko, submitted to *Phys. Lett.* **B** (1998) (hep-ph/9708447).

NEW CLASS OF QUARK MASS MATRIX AND THE FLAVOR MIXING MATRIX

KYUNGSIK KANG
Department of Physics, Brown University, Providence, RI 02912, USA

SIN KYU KANG
School of Physics, Korea Institute for Advanced Study, Seoul, Korea

We discuss a new general class of mass matrix ansätze that respect the fermion mass hierarchy and *calculability* of the flavor mixing matrix. This is a generalization of various specific forms of the mass matrix, which is obtained by successive breaking of the maximal permutation symmetry. By confronting the experimental data, a large class of mass matrices is shown to survive, while certain specific cases are phenomenologically ruled out.

(Presented by Sin Kyu Kang)

The flavor mixing and fermion masses and their hierarchical patterns remain one of the basic problems in particle physics. Within the standard model(SM), all masses and flavor mixing angles are free parameters and no relations among them are provided. As an attempt to derive relationships between the quark masses and flavor mixing hierarchies, a mass matrix ansatz was suggested about two decades ago [1]. This in fact reflects the *calculability* [1,2] of the flavor mixing angles in terms of the quark masses. Of several ansätze proposed, the canonical mass matrices of the Fritzsch type [1,3] have been generally assumed to predict the entire Kobayashi-Maskawa (KM) matrix [5] or the Wolfenstein mixing matrix [6]. Though the Fritzsch texture [1,2] is attractive because of its maximal *calculability*, it predicts the top quark mass to be no larger than 100 GeV and thus is ruled out [3]. Thus, the next move is to modify the Fritzsch mass matrix by introducing just one more parameter while maintaining the *calculability* property [4]. In what follows, we will discuss a possible modification of the Fritzsch mass matrix.

A natural choice for the next nonvanishing entry in the Fritzsch matrix is the (2,2) element. Although such type a of mass matrices has been studied in the literature [7,8,10,11], we will show that these can be identified as special cases of the general form in which the (2,2) and (2,3) elements are related in a particular way. The general form can be achieved by breaking the democratic flavor symmetry $S(3)_L \times S(3)_R$ successively down to $S(2)_L \times S(2)_R$ and then to $S(1)_L \times S(1)_R$.

As is well known, the 3×3 "democratic mass matrix ",

$$\frac{c}{3}\begin{pmatrix} 1 & 1 & 1 \\ 1 & 1 & 1 \\ 1 & 1 & 1 \end{pmatrix}, \tag{1}$$

exhibits the maximal $S(3)_L \times S(3)_R$ permutation symmetry. This can be achieved by breaking the chiral symmetry $U(3)_L \times U(3)_R$ to $S(3)_L \times S(3)_R$, where $U(3)$ is the symmetry group connecting the three generations [7,12]. One may say that the scale of this chiral symmetry breaking is the electroweak symmetry breaking scale at which the third generation quarks get masses. Indeed, one can see this by making a unitary transformation of (1) with the help of $U = (u_1^T, u_2^T, u_3^T)$, where $u_1 = (\frac{1}{\sqrt{2}}, \frac{1}{\sqrt{6}}, \frac{1}{\sqrt{3}})$, $u_2 = (-\frac{1}{\sqrt{2}}, \frac{1}{\sqrt{6}}, \frac{1}{\sqrt{3}})$ and $u_3 = (0, -\frac{2}{\sqrt{6}}, \frac{1}{\sqrt{3}})$. In order to account for the hierarchical pattern of the second and first generation quark masses, one has to break the $S(3)_L \times S(3)_R$ symmetry successively in two stages to $S(2)_L \times S(2)_R$ and $S(1)_L \times S(1)_R$. This can be achieved by adding the following two matrices to (1):

$$\begin{pmatrix} 0 & 0 & a \\ 0 & 0 & a \\ a & a & b \end{pmatrix}, \qquad d\begin{pmatrix} 1 & 0 & -1 \\ 0 & -1 & 1 \\ -1 & 1 & 0 \end{pmatrix}, \tag{2}$$

where the parameters (a, b) and d are responsible for the breakdown of $S(3)_L \times S(3)_R$ and $S(2)_L \times S(2)_R$ symmetries, respectively. It is also reasonable to anticipate that this two stage breaking happens at around 1 GeV, the chiral symmetry breaking scale, in view of the proximity of the second and first generation quark masses compared to the third generation quarks. Since the evolution from the electroweak scale to 1 GeV scale cannot alter the "democratic" pattern of the mass matrix, the resulting mass matrix can be regarded as the one at 1 GeV scale. Then the mass matrix in the hierarchical basis reduces after the unitary transformation with U to,

$$M_H = \begin{pmatrix} 0 & A & 0 \\ A & D & B \\ 0 & B & C \end{pmatrix}, \tag{3}$$

where $A = \sqrt{3}d, D = -\frac{2}{3}(2a - b), B = -\frac{\sqrt{2}}{3}(a + b)$ and $C = \frac{1}{3}(4a + b) + c$.

Note that in order to get a hermitian mass matrix instead of (3), one can use the following two matrices in place of (2),

$$\begin{pmatrix} p & p & a+q \\ p & p & a+q \\ a+q^* & a+q^* & b-2p \end{pmatrix}, \qquad d\begin{pmatrix} \cos\sigma & -i\sin\sigma & -e^{-i\sigma} \\ i\sin\sigma & -\cos\sigma & e^{-i\sigma} \\ -e^{i\sigma} & e^{i\sigma} & 0 \end{pmatrix}, \tag{4}$$

where $p = \frac{4}{9}(a+b)\sin^2\frac{\delta}{2}$ and $q = p(1 + i\frac{3}{2}\frac{e^{i\delta/2}}{\sin\frac{\delta}{2}})$. Then, after the unitary transformation with U, the (1,2) and (2,3) elements in M_H become $Ae^{-i\sigma}$ and $Be^{-i\delta}$, respectively. However, since only one phase factor is sufficient to describe CP-violation in the SM containing three family generations of quarks, we may introduce only one phase factor in the hermitian matrix M_H, i.e., only to the (1,2) and (2,1) elements. In this way, a hermitian mass matrix of the type (3), with complex elements at (1,2) and (2,1), can be obtained.

At a glance, the matrix M_H contains four independent parameters even in the case of real parameters so that *calculability* is lost. However, one can make an additional ansatz to relate a to b, so that $a = kb$ in general, with the same ratio parameter k for both the up- and down-quark sectors, so as to maintain *calculability*. Then, the (2,2) element is related to the (2,3) element by $w \equiv B/D = (k+1)/\sqrt{2}(2k-1)$ in the hierarchical mass eigenstates. Moreover, various specific mass matrices proposed by others can be identified as special cases of the new mass matrix i.e., $w = \frac{5}{3}$ $(k = 0.9)$ for Ref. 7, $w = -\frac{1}{\sqrt{2}}$ $(k = 0)$ for Fritzsch *et al.* [8], $w = \pm 2\sqrt{2}$ $(k = \frac{5}{7}$ or $\frac{1}{3})$ for Ref. 10, and $w = \sqrt{2}$ $(k = 1)$ for Ref. 11. The case of $k = \frac{1}{2}$ reduces to the old Fritzsch type with $D = 0$.

The next step is then to constrain k for the general class of mass matrices by confronting them with experiment. The mass matrix M_H of the type (3) can be brought to a diagonal form by a biunitary transformation, $U_L^{(u,d)} M_H^{(u,d)} U_R^{(u,d)\dagger} = diag[m_{u,d}, m_{c,s}, m_{t,b}]$. Since $U_L M_H U_L^\dagger$ and $U_R M_H U_R^\dagger$ are diagonal, $U_L U_R^\dagger \equiv K$ is again diagonal. It turns out in general that, because of the empirical mass hierarchy $m_1 \ll m_2 \ll m_3$, $K = diag[1,-1,1]$ irrespective of the sign of D and $K = diag[-1,1,1]$ only for positive D. This point was not clearly understood in previous works[8,10,11,13]. The parameters A, B, C and D can be expressed in terms of the quark masses. In view of the hierarchical pattern of the quark masses, it is natural to expect that $A \ll |D| \ll C$, and then the case of $K = diag[1,-1,1]$ for positive D can be excluded if the same ratio parameter w is required for both up- and down-quark sectors. Otherwise, the masses of the second family could be unacceptably large.

The Case $K = diag[-1,1,1]$: The hermitian matrix can be written as $M_H^{(u,d)} = P^{(u,d)} M_r^{(u,d)} \tilde{P}^{(u.d)}$, where $P^{(u,d)} = diag[\exp(-i\sigma^{(u,d)}), 1, 1]$, and the real matrix $M_r^{(u,d)}$ can be diagonalized by a real orthogonal matrix $R^{(u,d)}$ so that $R^{(u,d)} M_r^{(u,d)} \tilde{R}^{(u,d)} = diag[-m_{(u,d)}, m_{(c,s)}, m_{(t,b)}]$. Then the flavor mixing matrix is given by $V = U_L^{(u)} U_L^{(d)\dagger} = \tilde{R}^{(u)} P R^{(d)}$ where $P = diag[e^{i\sigma}, 1, 1]$ with $\sigma = \sigma^{(u)} - \sigma^{(d)}$.

From the characteristic equation for M_r, this matrix can be written as

$$M_r = \begin{pmatrix} 0 & \sqrt{\frac{m_1 m_2}{1-\frac{\epsilon}{m_3}}} & 0 \\ \sqrt{\frac{m_1 m_2}{1-\frac{\epsilon}{m_3}}} & m_2 - m_1 + \epsilon & w(m_2 - m_1 + \epsilon) \\ 0 & w(m_2 - m_1 + \epsilon) & m_3 - \epsilon \end{pmatrix} \tag{5}$$

in which the small parameter ϵ is related to w, i.e., $w \simeq \pm \frac{\sqrt{\epsilon m_3}}{m_2}\left(1 + \frac{m_1}{m_2} - \frac{m_2}{2m_3}\right)$, whose range is to be determined from experiment. We now can obtain analytic expressions for the flavor mixing matrix V which gives in the leading approximation:

$$|V_{us}| \simeq \left|\sqrt{m_d/m_s}\exp(i\sigma) - \sqrt{m_u/m_c}\right|, \tag{6}$$

$$|V_{cb}| \simeq |w\,(m_s/m_b - m_c/m_t)|, \tag{7}$$

$$|V_{ub}|/|V_{cb}| \simeq \sqrt{m_u/m_c}, \quad |V_{td}|/|V_{ts}| \simeq \sqrt{m_d/m_s}. \tag{8}$$

Since the second term of $|V_{cb}|$ is negligible compared to the first one, it is easy to examine the range of w for which $|V_{cb}|$ is compatible with experiments. Using the quark masses given in Ref. 14 and the experimental value $|V_{cb}| = 0.036 - 0.046$ [9], (7) leads to $1.01 \leq |w| \leq 2.02$ so that $0.82 \leq k \leq 1.31$ if $w > 0$ and $0.11 \leq k \leq 0.28$ if $w < 0$ in the leading approximation, which is close to the exact result $0.97 \leq |w| \leq 1.87$ so that $0.85 \leq k \leq 1.36$ if $w > 0$ and $0.10 \leq k \leq 0.26$ if $w < 0$. Note that $\epsilon \simeq O(m_1)$ for the allowed range of k and w.

Next, we examine if this range of w is consistent with experimental determinations of the other KM elements. Since several KM elements depend on the phase factor σ, we have to determine the allowed range of the phase factor first. We see from (6) that $|V_{us}|$ depends on the phase factor σ, while it is independent of w. Using the experimental value $|V_{us}| \simeq 0.219 - 0.224$ [9] the allowed range of σ turns out to be $26° - 111°$. The exact numerical result gives $39° \leq \sigma \leq 117°$. In addition we find that all other KM elements are in good agreement with experiments for the above ranges of w and σ.

The Case $K = diag[1, -1, 1]$: For negative D, the real symmetric matrix $M_r^{(u,d)}$ can be diagonalized as $R^{(u,d)} M_r^{(u,d)} \tilde{R}^{(u,d)} = diag[m_{(u,d)}, -m_{(c,s)}, m_{(t,b)}]$, thus reversing the signs of m_1 and m_2 in (5). Following a similar analysis as in the previous case, we get $1.14 \leq |w| \leq 2.76$ so that $0.72 \leq k \leq 1.17$ if $w > 0$ and $0.14 \leq k \leq 0.33$ if $w < 0$, and the same range of σ as in the previous case in the exact numerical calculation, while we find the same result of w and σ as in the previous case in the leading approximation. Consequently the ansatz adopted by Fritzsch *et al.* [8], corresponding to $k = 0$, is not consistent with experimental data of V_{cb} and the ansatz adopted by Ref. 10, corresponding to $w^2 = 8$, is slightly beyond the upper bound of w. Finally, we note that the predicted ratio $|V_{ub}|/|V_{cb}|$ (≤ 0.07) tends to be on the low side of (but consistent with) the present experimental range, $|V_{ub}|/|V_{cb}| = 0.08 \pm 0.02$ [9] or 0.08 ± 0.016 [15].

References

1. A. C. Rothman and K. Kang, *Phys. Rev. Lett.* **43**, 1548 (1979); *Phys. Rev.* D **24**, 167 (1981); H. Fritzsch, *Phys. Lett.* B **73**, 317 (1978); *Nucl. Phys.* B **155**, 189 (1979).
2. A. C. Rothman and K. Kang (Ref. 2); K. Kang in *Particles and Fields 2*, ed. A. Z. Capri and A. N. Kamal (Plenum Pub. Co., 1983); K. Kang and S. Hadjitheodoridis, in Proc. of the 4th Symp. Theor. Phys., ed. H. -S. Song (Mt. Sorak, Korea, 1985).
3. K. Kang and S. Hadjitheodoridis, *Phys. Lett.* B **193**, 504 (1987); K. Kang*et al.*, *Z. Phys.* C **55**, 75 (1992);H. Harari and and Y. Nir, *Phys. Lett.* B **195**, 586 (1987).
4. K. Kang and S. K. Kang, *Phys. Rev.* D **56**, 1511 (1997), hep-ph/9704253.
5. M. Kobayashi and T. Maskawa, *Prog. Theor. Phys.* **49**, 625 (1973).
6. L. Wolfenstein, *Phys. Rev. Lett.* **51**, 1945 (1983).
7. P. Kaus and S. Meshkov, *Phys. Rev.* D **42**, 1863 (1990).
8. H. Fritzsch and D. Holtmannspötter,*Phys. Lett.* B **338**, 290 (1994); H. Fritzsch and Z. Xing, *Phys. Lett.* B **353**, 114 (1995), see also H. Fritzsch, hep-ph/9709300.
9. Particle Data Group, *Phys. Rev.* D **54**, 1 (1996).
10. H. Lehmann, C. Newton, and T. T. Wu, *Phys. Lett.* B **384**, 249 (1996).
11. Z. Xing, hep-ph/9609204.
12. H. Fritzsch, *Phys. Lett.* B **184**, 391 (1987); H. Fritzsch and J. Plankl, *Phys. Lett.* B **237**, 451 (1990).
13. S. N. Gupta and J. M. Johnson, *Phys. Rev.* D **44**, 2110 (1991); S. Rajpoot, *Mod. Phys.Lett* A **7**, 309 (1992); D. Du and Z. Xing, *Phys. Rev.* D **48**, 2349 (1993); P. S. Gill and M. Gupta, hep-ph/9707445.
14. See also H. Leutwyler, *Phys. Lett.* B **378**, 313 (1996); hep-ph/9609467; J. Gasser and H. Leutwyler, *Phys. Rep.* **87**, 77 (1982); Y. Koide, hep-ph/9410270.
15. L. Gibbons (CLEO Collaboration), invited talk at the Int. Conf. on High Energy Physics, Warsaw, Poland July 25-31, 1996.

EVIDENCE FOR THE DECAY $K^+ \to \pi^+\nu\bar{\nu}$

E787 COLLABORATION[a]
Presented by M. Kuriki

An event consistent with the signature expected for the rare kaon decay $K^+ \to \pi^+\nu\bar{\nu}$ has been observed. In the pion momentum region examined, $211 < P < 230 MeV/c$, the backgrounds are estimated to contribute 0.08 ± 0.03 events. If the event is due to $K^+ \to \pi^+\nu\bar{\nu}$, the branching ratio is $4.2^{+9.7}_{-3.5} \times 10^{-10}$

1 Introduction

1.1 Physics motivation

Charged current weak interactions are characterized by the Cabbibo-Kobayashi-Maskawa, CKM, matrix in the Standard Model. The CKM matrix is a 3×3 matrix which has nine elements, but these elements are determined by four parameters as follows:

$$V_{CKM} = \begin{pmatrix} V_{ud} & V_{us} & V_{ub} \\ V_{cd} & V_{cs} & V_{cb} \\ V_{td} & V_{ts} & V_{tb}, \end{pmatrix}$$
$$\simeq \begin{pmatrix} 1-\lambda^2/2 & \lambda & A\lambda^3(\rho - i\eta) \\ -\lambda & 1-\lambda^2/2 & A\lambda^2 \\ A\lambda^3(1-\rho-i\eta) & -A\lambda^2 & 1 \end{pmatrix},$$

where $\lambda = \sin\theta_c$, A, ρ, and η are the four Wolfenstein parameters.

[a]E787 collaboration is:

Brookhaven National Laboratory : S. Adler, M.S. Atiya, I-H. Chiang, M Diwan, J.S. Frank, J.S. Haggerty, S.H. Kettell, T.F. Kycia, K.K. Li, L.S. Littenberg, C. Ng, A. Sambamurti, A.J. Stevens, R.C. Strand, and C.H. Witzig

KEK Tanashi-branch : T. Komatsubara, M. Kuriki, N. Muramatsu, and S. Sugimoto

KEK : T. Inagaki, S. Kabe, M. Kobayashi, Y. Kuno, T. Sato, T. Shinkawa, and Y. Yoshimura

Osaka University : Y. Kishi and T. Nakano

Princeton University : M. Ardebili, A.O. Bazarko, M .R. Convery, M.M. Ito, D.R. Marlow, R.A. McPherson, P.D. Meyers, F.C. Shoemaker, A.J.S. Smith, and J.R. Stone

TRIUMF : M. Aoki, P. Bergbusch, E.W. Blackmore, D.A. Bryman, A. Konaka, J.A. Macdonald, J. Mildenberger, T. Numao, P. Padley, J-M. Poutissou, R. Poutissou, G. Redlinger, J. Roy, and A.S. Turcot

University of Alberta : P. Kitching and R. Soluk

This CKM matrix is explaining CP violation in the Standard Model. The confirmation of this theory is essential to understand Nature.

Because two of these parameters, ρ and η, always appear with a factor λ^3, the elements V_{ub} and V_{td} which are only the CKM elements containing ρ and η are expected to be small. As a result, ρ and η are hard to determine experimentally.

Flavour Changing Neutral Currents (FCNC) are completely suppressed at tree level by the GIM mechanism in the Standard Model. In second or higher order interactions, this suppression is not complete due to the different quark masses. One such FCNC decay, $K^+ \to \pi^+ \nu\bar{\nu}$ is induced by loop effects as shown in Fig. 1.

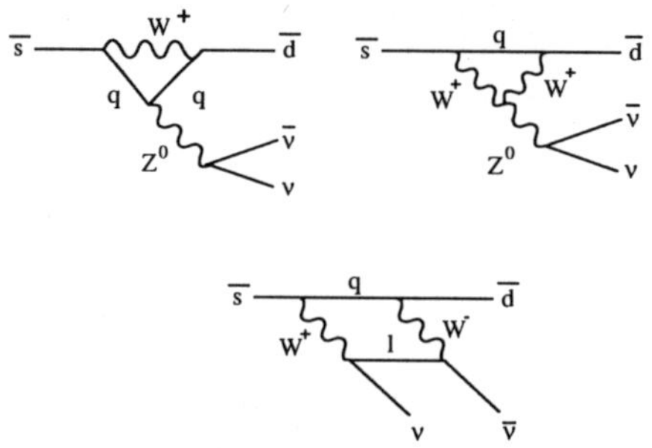

Figure 1: Diagrams of FCNC

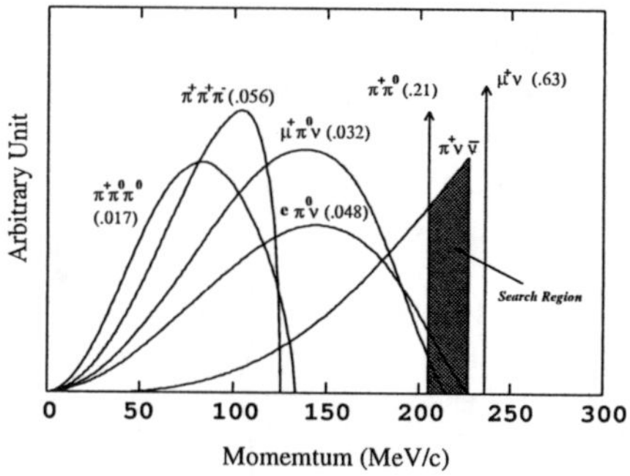

Figure 2: Momentum of charged track from K decay

Because the top quark is much heavier than the up or charm quark, the interaction with the top quark dominates this process. The theoretical uncertainty of the branching ratio of this decay mode is only 7 % from an ambiguity of a charm quark contribution.

Finally, this branching ratio is expressed to be ;

$$BR(K^+ \to \pi^+ \nu\bar{\nu}) \sim 1.97 \times 10^{-11} A^4 x_t^{1.18} [(\rho_0 - \rho)^2 + \eta^2] \quad (1)$$

From this equation, we can determine ρ and η by measuring this branching ratio. $K^+ \to \pi^+ \nu\bar{\nu}$ is a good channel to determine V_{td}.

Based on current knowledge of Standard Model parameters, this branching ratio is expected to be $(.6 - 1.5) \times 10^{-10}$

1.2 BNL E787 experiment

Fig. 2 shows the momentum spectra of charged particles produced in K^+ decay. There is no significant decay modes with π^+ in the region of momentum more than 205 MeV/c other than $K^+ \to \pi^+ \nu\bar{\nu}$. We can therefore search for $K^+ \to \pi^+ \nu\bar{\nu}$ decay in this region.

The $K^+ \to \pi^+ \nu\bar{\nu}$ decay is a three body decay. Because two particles are invisible, there is no kinematic constraint. The signal is identified only as "Not backgrounds". Background suppression is essential.

E787 did its first phase experiment from 1989 to 1991. For 28 weeks data taking, no event was found. It gave an upper limit of 2.4×10^{-9} with 90% confidence level.

After the upgrade of the beam line and the detector, a one month commissioning run was done in 1994. The first physics run was done in 1995 for 18 weeks. In 1996 and 1997, data were taken for 13 and 6 weeks respectively. In this talk, I will give the result for the 1995 data.

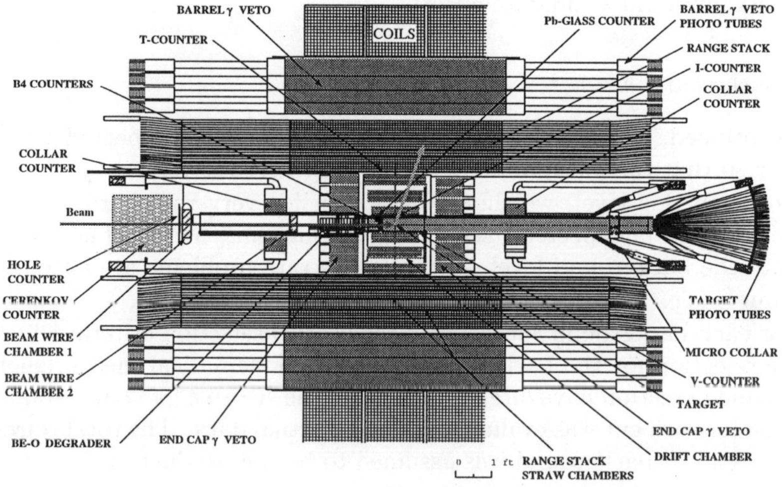

Figure 3: Cross sectional view of E787 detector

Fig. 3 shows the cross sectional view of the E787 detector system. 790 MeV/c K^+ are degraded by BeO and stopped in the active target which was composed of scintillation fibers. 1 Tesla magnetic field along the beam axis is induced by the solenoidal magnet. The momentum of charged particles from K^+ decays was analyzed by the drift chamber, followed by 21 layers of scintillation counters which measured the range and the energy of the particles.

The signal from the scintillation counters was digitized and recorded with a 500 MHz sampling rate, so that π^+ from K^+ decay was identified by tagging $\pi^+ \to \mu^+ \to e^+$ decays.

The barrel EM calorimeter, the end cap CsI counter, the collar and micro collar counters cover 4π solid angle and detect photons to reject backgrounds with gammas.

In the 1995 run, the number of K^+ stopped in the target was 1.49×10^{12}.

2 Analysis

The analysis to search for $K^+ \to \pi^+ \nu\overline{\nu}$ events was done in the following four steps:

1. Background estimates
2. Cross checks of the background estimates
3. Investigating the signal region
4. Examine the candidate event(s)

2.1 Background estimates

As I mentioned, background suppression is essential. The expected signal level is 10^{-10} so that the background level has to be even lower.

To reach this level, we should care even for very rare hardware or software errors and unexpected background mechanisms. We used actual data to estimate the background level, so that these rare problems are automatically picked up.

For each background source, we prepare two selection criteria which have a large rejection for the background. E.g. the kinematical measurement and the gamma detection have large rejection for the $K^+ \to \pi^+ \pi^0$ background. The rejection for each cut was evaluated by using actual data. The total rejection of the $K^+ \to \pi^+ \pi^0$ background was assumed to be the product of these rejection factors.

For each background source, the expected number of background events in the signal region was calculated as shown in Tab. 1.

2.2 Cross check of the background estimates

A cross check is needed to confirm that all types of background are accounted for, and that there is no correlation between the two cuts used to get the total rejection for a background source.

Table 1: The expected background

Background source	# of expected BG
$K_{\pi 2}$	0.03 ± 0.02
$K_{\mu 2}$	0.02 ± 0.02
Single beam	0.012 ± 0.009
Double beam	0.007 ± 0.007
Charge Exchange	0.01 ± 0.01
Total background	0.08 ± 0.03

For this purpose, we investigated the area adjacent to the signal region. We loosened the final selection criteria simultaneously and we compared the expected background with the number of background events which were actually discovered in this area.

In this study, the observed number of background events was consistent with our expectation. All events were consistent with known background sources, i.e. there was no unknown background source. From these results, we concluded that there were no loopholes in our background analysis.

2.3 Investigating the signal region

In Fig. 4, the events that passed the final selection criteria except those on range and energy are plotted. The signal region is indicated by the box. One event was found in the signal region.

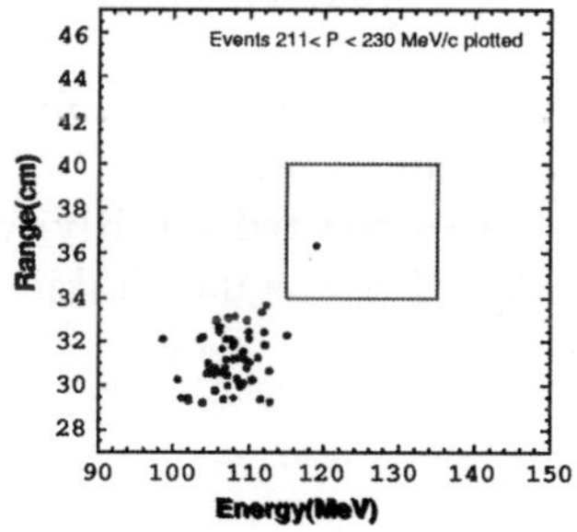

Figure 4: Range x Energy plot of the final events

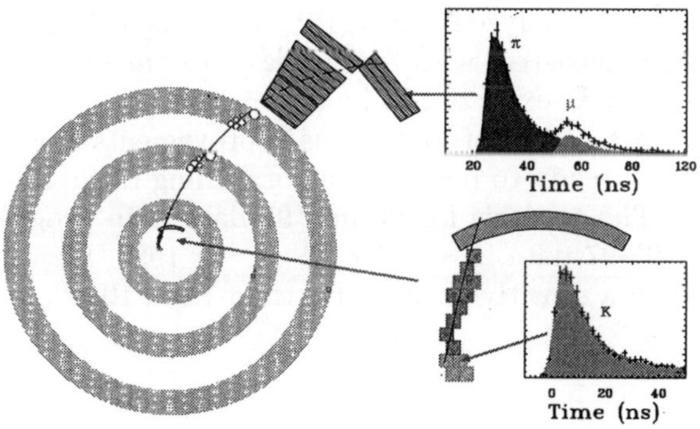

Figure 5: Display of the event

Fig. 5 shows the display of this event. The signal from a target scintillation fiber and the scintillation counter in which the π^+ stopped are also plotted. The charged track was well reconstructed.

A clean signal showing $\pi^+ \to \mu^+$ decay in the scintillation counter was observed. There was no extra activity by gammas. In addition, the kinematics of the event are not consistent with $K^+ \to \pi^+\pi^0$ or $K^+ \to \mu^+\nu_\mu$ backgrounds, so that the event is unlikely to come from those sources.

There was no extra activity in the target, and the π^+ went out from the K^+ vertex after a delay of more than 20 ns, so that the event is also far from any type of beam background.

Because the event is so clean and is not consistent with any type of background, the event survives even after the pre-defined selection criteria tighter than the regular cuts. The event was in the region containing 55% of the signals, but only 10% of the background.

The acceptance for the $K^+ \to \pi^+\nu\overline{\nu}$ signal is $0.16 \pm 0.01(stat.) \pm 0.02(sys)\%$, so that the branching ratio was evaluated to be

$$BR(K^+ \to \pi^+\nu\overline{\nu}) = (4.2 + 9.7 - 3.5) \times 10^{-10}(68\%C.L.) \tag{2}$$

which was consistent with the Standard Model expectation. Without any extra assumption, the CKM matrix element of V_{td} has been determined to be

$$6 \times 10^{-3} < |V_{td}| < 6 \times 10^{-2} \tag{3}$$

3 Conclusion

We have analyzed 1.49×10^{12} stopping K^+ taken in 1995 and have found an event consistent with being a $K^+ \to \pi^+\nu\overline{\nu}$ event. The background in the signal region was expected to be 0.08 ± 0.03.

After years of continuous improvements/upgrades, the E787 detector/analysis is now ready to measure the branching ratio, if we get enough beam.

The analysis for 96 and 97 data is in progress. The expected sensitivity for 95-97 data is $\sim 1.7 \times 10^{-10}$. In 1998, we will have 24 weeks data taking giving a sensitivity of better than 1.0×10^{-10}.

EXCITED HEAVY MESONS FROM QCD SUM RULES

Chun Liu
Center For Theoretical Physics, Seoul National University, Seoul 151-742, Korea

Orbitally excited $L = 1$ charm mesons are studied by QCD sum rules in the framework of heavy quark effective theory. The meson masses and their strong decay widths are obtained. This talk is based on our works of refs. 1 and 2, in collaboration with Y.B. Dai, C.S. Huang, M.Q. Huang and H.Y. Jin.

The study of excited heavy mesons is interesting for the following reasons[3]. First, it may help us understand nonperturbative aspects of QCD more deeply. Second, it is an alternative application of the heavy quark effective theory (HQET), in addition to that of ground state heavy mesons. Finally, they are particularly useful for tagging in CP experiments [4].

In this talk, we focus on the $L = 1$ excited charm mesons. They can heuristically be described as follows (note that this description is not accurate, since L is not a conserved quantity). For the ground state ($L = 0$) mesons which we are familiar with, there are two states with the J^P quantum numbers $D(0^-)$ and $D^*(1^-)$. Their difference only lies in the heavy quark spin direction. For the $L = 1$ case, there are four states: $D_0^*(0^+)$, $D_1'(1^+)$, $D_1(1^+)$ and $D_2^*(2^+)$. In the constituent picture, D_0^* contains one heavy quark and one light quark with same spin direction which is opposite to that of the orbital angular momentum. D_1' is the same as D_0^* up to a heavy quark spin flip. D_1 differs from D_0^* by a light quark spin flip. And D_2^* is the same as D_1 except for the heavy quark spin direction.

These kinds of heavy mesons can be systematically studied using HQET[5]. In the heavy quark limit, the heavy quark SU(2) spin symmetry implies that the four excited heavy mesons can be grouped into two doublets (D_0^*, D_1') and (D_1, D_2^*) [6]. It is interesting to note that although D_1' and D_1 have the same quantum numbers, they are clearly separated in the heavy quark limit. Their mixing is at the order of $1/m_c$, and can be studied in HQET. In the following, we present HQET sum rule calculations for masses and strong decays of the excited charm mesons in leading order.

The QCD sum rule[7] is a nonperturbative method rooted in QCD itself. First, he relevant interpolating current should be fixed. For the excited heavy mesons, we write the currents as[1],

$$J = \frac{1}{\sqrt{2}}\bar{h}_v \Gamma D_\rho q \ , \qquad (1)$$

with Γ denoting some γ matrices. h_v is the heavy quark field with velocity v in HQET. We find that

$$
\begin{aligned}
\Gamma &= -\gamma_t^\rho && \text{for } D_0^* , \\
\Gamma &= \gamma_5\gamma_t^\mu\gamma_t^\rho && \text{for } D_1' , \\
\Gamma &= -\sqrt{\frac{3}{2}}\gamma_5(g_t^{\mu\rho} - \frac{1}{3}\gamma_t^\mu\gamma_t^\rho) && \text{for } D_1 , \\
\Gamma &= \frac{1}{2}(\gamma_t^\mu g_t^{\nu\rho} + \gamma_t^\nu g_t^{\mu\rho}) - \frac{1}{3}g_t^{\mu\nu}\gamma_t^\rho && \text{for } D_2^* ,
\end{aligned}
\tag{2}
$$

where $\gamma_t^\mu \equiv \gamma^\mu - v^\mu \not{v}$ (In the rest frame of the meson, $\gamma_t^\mu = (0, \vec{\gamma})$) and $g_t^{\mu\nu} \equiv g^{\mu\nu} - v^\mu v^\nu$. We have shown that the current for one of the two 1^+ state does not couple to the other in the limit $m_Q \to \infty$. The related decay constant f is defined as

$$\langle 0|J^\dagger|D^{**}(v,\eta)\rangle \equiv f\eta , \tag{3}$$

where D^{**} stands for any of the four excited mesons with polarization vector η, which is the ground state of J (For more details, see ref. 1).

1 Masses

In HQET, the meson masses are expanded as

$$M = m_Q + \bar{\Lambda} + O(\frac{1}{m_Q}) , \tag{4}$$

where $\bar{\Lambda}$ determines the meson masses defined in HQET. They are independent of heavy quark flavor. They can be calculated using the following Green's function:

$$\Gamma(\omega) = i\int d^4x e^{ik\cdot x}\langle 0|\mathrm{T}J^\dagger(x)J(0)|0\rangle , \quad \omega = 2k\cdot v . \tag{5}$$

This can be expressed in the hadronic language,

$$\Gamma(\omega) = \frac{2f^2}{2\bar{\Lambda} - \omega} + \text{resonance} . \tag{6}$$

The corresponding Feynman diagrams can be found in ref. 1. We include condensates up to dimension five. Here the usual duality hypothesis is used for the resonance contribution. After Borel transformation, the sum rules for $\bar{\Lambda}$'s are obtained as

$$\bar{\Lambda}_{\left(\begin{array}{c}0^+\\1^+\end{array}\right)} = \frac{\frac{3}{8\pi^2}\int_0^{\omega_c} d\nu\nu^5 e^{-\nu/T}}{\frac{3}{4\pi^2}\int_0^{\omega_c} d\nu\nu^4 e^{-\nu/T} - m_0^2\langle\bar{q}q\rangle},$$
$$\bar{\Lambda}_{\left(\begin{array}{c}1^+\\2^+\end{array}\right)} = \frac{\frac{1}{32\pi^2}\int_0^{\omega_c} d\nu\nu^5 e^{-\nu/T} - \frac{1}{16\pi^2}\langle\alpha_s GG\rangle T^2}{\frac{1}{16\pi^2}\int_0^{\omega_c} d\nu\nu^4 e^{-\nu/T} - \frac{m_0^2}{3}\langle\bar{q}q\rangle - \frac{T}{8\pi}\langle\alpha_s GG\rangle}. \tag{7}$$

Imposing usual the criterion for the upper and lower bounds on the Borel parameter T, we obtain the numerical results,

$$\bar{\Lambda}_{\left(\begin{array}{c}0^+\\1^+\end{array}\right)} = 0.90 \pm 0.10 \text{ GeV}, \quad \bar{\Lambda}_{\left(\begin{array}{c}1^+\\2^+\end{array}\right)} = 0.95 \pm 0.10 \text{ GeV}, \tag{8}$$

where the error comes from the uncertainty of $\omega_c \sim 0.5$ GeV.

2 Strong Decays

The pionic decays of heavy hadrons can be studied by combining heavy quark symmetry and chiral symmetry [8]. Excited heavy meson decays have been studied in this framework [9, 10]. We will not use the soft pion approximation, however, because the pion energy is about 500 MeV in this case. The decay widths are given by, *e.g.*,

$$\Gamma(D_0^* \to D\pi) = \frac{3}{8\pi} g'^2 |\vec{p_\pi}|,$$
$$\Gamma(D_2^* \to D\pi) = \frac{1}{20\pi} g^2 |\vec{p_\pi}|^5, \tag{9}$$

where a sum over charged and neutral pion final states has been implied, and g' and g are universal quantities describing the decay amplitudes which we are going to calculate by QCD sum rules.

For this purpose, the following Green's function is constructed:

$$\tilde{\Gamma} = i \int d^4x e^{-ik\cdot x} \langle \pi(q)|\mathrm{T} J_{D^{(*)}}(x) J^\dagger(0)|0\rangle, \tag{10}$$

with $J_{D^{(*)}}$ being the currents for $D^{(*)}$ mesons. The hadronic expression is

$$\tilde{\Gamma} = \frac{g^{(\prime)} f_{D^{(*)}} f_{D^{**}}}{(2\bar{\Lambda}_{D^{(*)}} - 2v\cdot k)(2\bar{\Lambda}_{D^{**}} - 2v\cdot k')} + \frac{c}{2\bar{\Lambda}_{D^{(*)}} - 2v\cdot k} + \frac{c'}{2\bar{\Lambda}_{D^{**}} - 2v\cdot k'} + \text{res.}\,, \tag{11}$$

where $k' = k - q$. Instead of taking the soft pion limit, we put $2v\cdot(k-k') = 2(\bar{\Lambda}_{D^{**}} - \bar{\Lambda}_{D^{(*)}})$. In the operator product expansion of HQET,

$$\tilde{\Gamma} = i\int_0^\infty dt e^{\frac{\omega t}{2}} \frac{1}{\sqrt{2}} \langle \pi(q)|\mathrm{T}\bar{q}(vt)\Gamma_{D^{(*)}} \frac{1+\not v}{2} \Gamma D_\rho q(0)|0\rangle\ . \tag{12}$$

We need $\langle \pi|\bar{q}^a_\alpha(x) D_\rho q^b_\beta|0\rangle$, which can be found in ref. 2. With

$$g_2 = \frac{f_\pi}{4}\ , h_1 = 0\ , h_2 = -\frac{1}{12 f_\pi}\langle\bar{q}q\rangle\ , a_1 = -\frac{m_0^2}{8 f_\pi}\langle\bar{q}q\rangle\ , c_1 = \frac{1}{36} f_\pi m_1^2\ ,$$
$$c_2 = -\frac{1}{24 f_\pi}\langle\bar{q}q\rangle\ , b_1 = \frac{a_1}{3}\ , d_1 = 0\ , d_2 = -\frac{e_2}{3} + \frac{1}{12 f_\pi}\langle\bar{q}q\rangle\ , m_1^2 \simeq 0.2\mathrm{GeV}^2\ , \tag{13}$$

and e_2, which is defined as

$$e_2 (n\cdot q)^3 = \frac{i}{2}\langle \pi^i(q)|\bar{q}\frac{\tau_i}{2}\gamma_5 \not n (n\cdot D)^2 q|0\rangle\ , \quad n^2 = 0\ , \tag{14}$$

and is fixed from QCD sum rules [2]: $e_2 \simeq -0.015 \pm 0.002$ GeV, the final sum rules for g' and g are

$$\begin{aligned} g' f_{D_0^*} f_D &= 2\{\bar{\Lambda}_{D_0^*}(3h_1 + 3h_2\Delta - g_2\Delta^2) - [-3b_1 + 3\Delta(c_1 + d_1) \\ &\quad + \Delta^2(4c_2 + 2d_2 - e_2)](1 + \frac{2\bar{\Lambda}_{D_0^*}}{T})\} e^{\frac{2\bar{\Lambda}_{D_0^*}}{T}}\ , \\ g f_{D_2} f_D &= 2\left[\bar{\Lambda}_{D_2} g_2 + (c_2 - e_2\Delta - d_2)(1 + \frac{2\bar{\Lambda}_{D_2}}{T})\right] e^{\frac{2\bar{\Lambda}_{D_2}}{T}}\ , \end{aligned} \tag{15}$$

where $\Delta = \bar{\Lambda}_{D^{**}} - \bar{\Lambda}_D$. The range of the Borel parameter is fixed by requiring higher $\frac{1}{T}$ term to be small in the OPE for the original sum rules, and T not to be much larger than $2(\bar{\Lambda}^1 - \bar{\Lambda})$ where $\bar{\Lambda}^1$ is the mass of first radial excitation in HQET. We obtain the numerical results

$$\begin{aligned} g' &\simeq 1.7 \pm 0.5\ , \\ g &\simeq 4.6 \pm 1.0 \quad \mathrm{GeV}^{-2}\ . \end{aligned} \tag{16}$$

Note that our method improves the calculations using sum rules in full QCD[11], since these have problems with contaminations of other 1^+ states, and use the soft pion approximation. $1/m$ corrections[12] and radiative corrections can be included systematically. The experimental data on the D_2^* width[13] can be explained by assuming the pionic decay mode to dominate. However, that of D_1 cannot be understood in this way. It may be due to the mixing of two 1^+ states at the order of $1/m$ as well as due to $D_1 \to D^{(*)}\rho$ decays having relatively large branching ratios.

Acknowledgments

I am grateful to Professors Yuanben Dai and Chaoshang Huang and Drs Mingqiu Huang and Hongying Jin for an instructive and enjoyable collaboration. This work was supported in part by KOSEF (SRC).

References

1. Y.B. Dai, C.S. Huang, M.Q. Huang and C. Liu, *Phys. Lett.* B **390**, 350 (1997).
2. Y.B. Dai, C.S. Huang, M.Q. Huang, H.Y. Jin and C. Liu, ASITP-97-12, BIHEP-TH-97-005, SNUTP 97-049, hep-ph/9705223.
3. For a review, see A. Falk in *20th Johns Hopkins Workshop on Current Problems in Particle Theory*, ed. M. Jamin *et al* (World Scientific, Singapore, 1997).
4. E. Eichten, C.T. Hill and C. Quigg, Fermilab-Conf-94/118-T.
5. For a review, see N. Isgur and M.B. Wise in *B Decays* ed. S. Stone, 2nd edi. (World Scientific, Singapore, 1994).
6. N. Isgur and M.B. Wise, *Phys. Rev. Lett.* **66**, 1130 (1991).
7. M.A. Shifman, A.I. Vainshtein and V.I. Zakharov, *Nucl. Phys.* B **147**, 358, 448 (1979).
8. For a review, see M.B. Wise in *CCAST Symposium on Particle Physics at Fermi Scale*, ed. Y. Pang *et al* (Gordon and Breach, 1995).
9. A. Falk and M. Luke, *Phys. Lett.* B **292**, 119 (1992).
10. U. Kilian, J.G. Körner, and D. Pirjol, *Phys. Lett.* B **288**, 360 (1992).
11. P. Colangelo *et al*, *Phys. Lett.* B **269**, 204 (1991), B **293**, 207 (1992), *Phys. Rev.* D **52**, 6422 (1995).
12. Y.B. Dai, C.S. Huang and M.Q. Huang, *Phys. Rev.* D **55**, 5719 (1997).
13. Particle Data Group, *Phys. Rev.* D **54**, 1 (1996).

SOLUTION OF THE a_2/a_1 SIGN PROBLEM IN HEAVY MESON DECAYS: NONFACTORIZABLE SOFT GLUONS IN NONLEPTONIC HEAVY MESON DECAYS

B.TSENG and HSIANG-NAN LI

Department of Physics, National Cheng-Kung University, Tainan, Taiwan

For the first time, a solution of the a_2/a_1 problem in heavy meson decays is proposed, based on PQCD factorization theorems. The point is to include QCD evolution effects among various characteristic scales involved in decay processes.

1 Introduction

Nonleptonic weak decays are notoriously difficult to handle. Approximations are needed to simplify the analysis. The most widely adopted approach is the Bauer-Stech-Wirbel (BSW) model [1] based on the factorization hypothesis, in which decay rates are expressed in terms of various hadronic transition form factors multiplied by some coefficients. The coefficients corresponding to external and internal W boson emission are $a_1 = c_1 + c_2/N_c$ and $a_2 = c_2 + c_1/N_c$, respectively, c_i being Wilson coefficients. All nonfactorizable contributions which cannot be expressed in terms of hadronic form factors, and nonspectator contributions from W boson exchange are neglected. In this way the BSW method avoids complicated QCD dynamics. In this approach the Fierz–transformation term is treated as a free parameter ξ $(=1/N_c)$, and it is found that over 20 decay modes can be successfully explained if ξ=0 which means that N_c goes to infinity and the large N_c approach is confirmed[2]. To understand the successes of this large N_c approach, nonfactorizable contributions χ_i are included, and the effective coefficients a_1 and a_2 become

$$a_1^{\text{eff}} = c_1 + c_2\left(\frac{1}{N_c} + \chi_1\right) , \quad a_2^{\text{eff}} = c_2 + c_1\left(\frac{1}{N_c} + \chi_2\right) . \tag{1}$$

Using QCD sum rules[3], it has been found that the parameters χ_i are negative for charm decays, canceling the color–suppressed term $1/N_c$. Hence, the famous "rule of discarding the $1/N_c$ corrections" [3] found its dynamical origin in QCD.

Performing a similar BSW analysis for B meson decays[4], $a2/a1$ is found to be positive, in contrast to the negative a_2/a_1 for the charmed meson. The values $c_1(M_b) \approx 1.12$ and $c_2(M_b) \approx -0.26$ imply that a naive extension of the large N_c approach to bottom decays[5] will lead to a negative a_2/a_1, which is in conflict to experimental data. If nonfactorizable contributions are included into a_2, we will need positive χ to explain the data. Unfortunally, all the existing calculations give negative nonfactorizable contributions[6]. Up to now, it is still unclear what is wrong with these calculations and with the QCD sum rule method. This a_2/a_1 sign problem provides a challenge for the QCD sum rule method.

2 PQCD formalism for nonleptonic heavy meson decays

From the viewpoint of PQCD theory, nonleptonic weak decays are a three scale problem. The relevant scales are M_W, the W boson mass; t, the typical scale of the meson decay; and Λ_{QCD}, which characterizes the hadronic size. Combining effective field theory and the PQCD factorization theorem, a three scale factorization formula has been proposed and applied to the $B \to J/\Psi K^*$ decay with success [8]. We assume that a parton carries a small amount of transverse momentum k_T, which serves as an infrared regulator for radiative corrections. The factorization formula for the decay amplitude is written as

$$\mathcal{M} = H_r(M_W, \mu) \otimes H(t, \mu) \otimes \phi(b, \mu) \otimes U(b, \mu) \ , \tag{2}$$

in the impact parameter b space conjugate to k_T. Here, $\otimes$ represents a convolution, since variables t and b will be integrated out in the end. According to the virtuality of gluon momenta, QCD dynamics is separated into a "harder" function H_r, a hard decay amplitude H and soft parts ϕ and U. Reducible soft gluons are absorbed into $\phi(b, \mu)$, while the irreducible ones go into $U(b, \mu)$.

Let γ_{H_r}, γ_ϕ, and γ_U be the anomalous dimensions of H_r, ϕ, and U, respectively. The anomalous dimension of H is then $\gamma_H = -(\gamma_{H_r} + \gamma_\phi + \gamma_U)$, because a decay amplitude is ultraviolet finite. A RG treatment of Eq. (2) leads to

$$\begin{aligned} \mathcal{M} &= H_r(M_W, M_W) \otimes H(t, t) \otimes \phi(b, 1/b) \otimes U(b, 1/b) \\ &\quad \otimes \exp\left[\int_t^{M_W} \frac{d\bar{\mu}}{\bar{\mu}} \gamma_{H_r}(\alpha_s(\bar{\mu}))\right] \\ &\quad \otimes \exp\left[-\int_{1/b}^{t} \frac{d\bar{\mu}}{\bar{\mu}} \left[\gamma_\phi(\alpha_s(\bar{\mu})) + \gamma_U(\alpha_s(\bar{\mu}))\right]\right] \ , \end{aligned} \tag{3}$$

which becomes explicitly μ-independent. The two exponentials, describing the two evolutions from M_W to t and from t to $1/b$, are the consequence of the summation of large $\ln(M_W/t)$ and $\ln(tb)$. The first exponential can be easily identified as the Wilson coefficient $c(t)$.

Equation (3) sums only single logarithms. In fact, there exist double logarithms $\ln^2(P^+b)$ in the meson wave function ϕ, which arise from the overlap of collinear and soft divergences, P^+ being the largest light-cone component of the meson momentum. Hence, $\phi(b,\mu)$ in Eq. (2) should be replaced by

$$\phi(P^+,b,\mu) = \phi(b,\mu)\exp[-s(P^+,b)] , \tag{4}$$

where the exponential e^{-s}, the so–called Sudakov form factor, comes from the resummation of the double logarithms[9,10]. We shall approximate $H_r(M_W, M_W)$ by its lowest order expression $H_r^{(0)} = 1$, and evaluate $H(t,t)$ perturbatively, since all the large logarithms have been grouped into the exponents. For simplicity, we set the nonperturbative initial condition $U(b, 1/b)$ of the RG evolution to unity due to the strong suppression from the Sudakov form factor at large b, and neglect the b dependence in another nonperturbative initial condition $\phi(b, 1/b)$.

We need to calculate γ_U, which is complicated due to the nonabelian nature of color exchanges. The color structure of soft gluon exchanges can be analyzed as follows. If a soft gluon crosses the hard gluon vertex, the color structure is given by

$$\begin{aligned}&(T^a)_{b_1b_1'}(T^H)_{b_1'b_2}(T^a)_{a_2a_2'}(T^H)_{a_2'a_1}\\&=\frac{1}{2}(T^H)_{a_2b_2}(T^H)_{b_1a_1}-\frac{1}{2N_c}(T^H)_{b_1b_2}(T^H)_{a_2a_1} ,\end{aligned} \tag{5}$$

where T^a (T^H) is the color matrix associated with the soft (hard) gluon. Contracted with $\delta_{a_1b_1}\delta_{a_2b_2}$ from the color singlet initial and final state mesons, the first term, denoting an octet contribution, vanishes. The second term leads to a factor $-1/(2N_c)$ without changing the original tree-level color flow. If a soft gluon does not cross the hard gluon vertex, it introduces a factor $T^aT^a = C_F = 4/3$.

3 $B^- \to D^0\pi^-$ and $D^+ \to \bar{K}^0\pi^+$: solution of the a_2/a_1 sign problem

Now we are ready to see how to solve this a_2/a_1 sign problem within our PQCD approach. Since the analysis of $D^+ \to \bar{K}^0\pi^+$ is similar to that of $B^- \to D^0\pi^-$ (except for kinematics), we will focus on $B^- \to D^0\pi^-$. The decay amplitude is given by

$$\mathcal{M} = f_\pi[(1+r)\xi_+ - (1-r)\xi_-] + f_D\xi_i + \mathcal{M}_e + \mathcal{M}_i , \tag{6}$$

f_D and f_π being the D meson and pion decay constant, respectively. The form factors $\xi_\pm$ come from the factorizable external W contributions and ξ_i from the internal W contributions. The amplitudes M_e and M_i denote the corresponding nonfactorizable contributions. After resumming soft gluons, ξ becomes nonfactorizable. The nonfactorizable contribution can be extracted by subtracting its (lowest order) factorizable part. This is one of the key points to the solution of the a_2/a_1 sign problem. We refer readers to[7] for more details.

After including the soft gluons, the factorizable external W emission contribution $\chi_e^{(f)}$ and the factorizable internal W emission contribution $\chi_i^{(f)}$ to the decay amplitude $\mathcal{M}$ in Eq. (6) should be identified as $\chi_e^{(f)} = f_\pi[(1+r)\xi_+ - (1-r)\xi_-]$, $\chi_i^{(f)} = f_{D(K)}\xi_i|_{S_U=0}$, where $\xi_i|_{S_U=0}$ denotes the lowest order internal W-emission. The nonfactorizable external W emission contribution $\chi_e^{(nf)}$ and the nonfactorizable internal Wd -emission contribution $\chi_i^{(nf)}$ are then: $\chi_e^{(nf)} = \mathcal{M}_e$, $\chi_i^{(nf)} = f_{D(K)}(\xi_i - \xi_i|_{S_U=0}) + \mathcal{M}_i$.

Results for various form factors and amplitudes for the decays $B^- \to D^0\pi^-$ and $D^+ \to \bar{K}^0\pi^+$ are exhibited in the following Table (unit = 10^{-3} GeV).

$B \to D\pi$	external W (factorizable)	$f_D\xi_i$	$\mathcal{M}_e$	$\mathcal{M}_i$
$S_U = 0$	106.5	2.5	$-5.8 + 19.8i$	$18.5 - 12.5i$
$S_U \neq 0$	108.5	2.6	$-5.8 + 20.0i$	$18.8 - 11.0i$
$D \to \bar{K}\pi$	external W (factorizable)	$f_K\xi_i$	$\mathcal{M}_e$	$\mathcal{M}_i$
$S_U = 0$	267.0	-21.3	$-18.8 + 19.0i$	$37.8 - 36.7i$
$S_U \neq 0$	1075.0	-529.3	$-18.5 + 13.6i$	$36.5 - 30.3i$
	$\chi_e^{(f)}$	$\chi_i^{(f)}$	$\chi_e^{(nf)}$	$\chi_i^{(nf)}$
$B \to D\pi$	108.5	2.5	$-5.8 + 20.0i$	$18.9 - 11.0i$
$D \to \bar{K}\pi$	1075.0	-21.3	$-18.5 + 13.6i$	$-471.5 - 30.3i$

The rows labelled $S_U \neq 0$, whose values match the data, are derived with the soft corrections taken into account. Those labelled $S_U = 0$ only help to extract the nonfactorizable contributions $\chi^{(nf)}$, and to investigate the importance of the soft corrections. The exponentials e^{-S_U}, which basically act as enhancement factors, then amplify contributions from the region of smaller t, where the Wilson coefficients are larger. Therefore, the soft gluon effects are more important in charm decays, as shown in the Table.

The factorizable external W emission contributions are positive in both B and D meson decays, and their magnitudes increase, after including the soft corrections with $\gamma_U < 0$. ξ_i changes sign, since the Wilson coefficient c_2 becomes so negative, when evolving from the characteristic scale of the B meson decay to that of the D meson decay, that it overcomes the positive c_1/N_c. Their magnitudes also increase because of $\gamma_U < 0$. The real parts of the nonfactorizable amplitudes $\mathcal{M}_e$ ($\mathcal{M}_i$) are always negative (positive) due to the negative c_2 (positive c_1/N_c) in the B and D meson decays. From the table, we have $Re(\chi_i^{(nf)}) = +0.0189$ GeV for B meson decay and $Re(\chi_i^{(nf)}) = -0.4715$ GeV for D meson decay. It is easy to attribute the sign change of $\chi_i^{(nf)}$ to the stronger enhancement of ξ_i by soft corrections in charm decays.

4 Summary

In this talk, we described a solution to the a_2/a_1 sign problem. Nonfactorizable contributions to nonleptonic heavy meson decays are carefully identified in our formalism, and found to be positive for bottom decays and negative for charm decays with the inclusion of soft gluons.

5 Acknowledgments

B. Tseng would like to thank Profs. K.Hagiwara and A. Ali for valuable discussions during the workshop. This work was supported in part by the National Science Council of ROC under Contract Nos. NSC87-2112-M-006-018.

References

1. M. Bauer, B. Stech and M. Wirbel, Z. Phys. C **29**, 637 (1985); **34**, 103 (1987).
2. A.J. Buras and J.-M. Gerard, Nucl. Phys. **B264**, 371 (1986).
3. B. Blok and M. Shifman, Sov. J. Nucl. Phys. **45**, 135, 301, 522, (1987).
4. H.Y. Cheng and B. Tseng, Phys. Rev. D **51**, 6259 (1995) and references therein.
5. A different large N_c approach, based on additional analysis of the Wilson coefficients, is given by M. Neubert and B. Stech, hep-ph/9705292.
6. B. Blok and M. Shifman, Nucl. Phys. **B389**, 534 (1993), I. Halperin, Phys. Lett. B **349**, 548 (1995), A. Khodjamirian and R. Rückl, hep-ph/9503321.
7. H-n. Li and B. Tseng, Phys. Rev. D **56**,(1997) and references therein.
8. C.H. Chang and H-n. Li, Phys. Rev. D **55**, 5577 (1997); T.W. Yeh and H-n. Li, Phys. Rev. D **56** (1997).
9. H-n. Li and H.L. Yu, Phys. Rev. Lett. **74**, 4388 (1995); Phys. Rev. D **53**, 4970 (1995); Phys. Lett. B **353**, 301 (1995).
10. H-n. Li, Phys. Rev. D **52**, 3958 (1995); C.Y. Wu, T.W. Yeh, and H-n. Li, Phys. Rev. D **53**, 4982 (1996).

A PQCD FORMALISM FOR EXCLUSIVE NON–LEPTONIC HEAVY MESON DECAYS

TSUNG–WEN YEH

Institute of Physics, Academia Sinica, Taipei, Taiwan, R.O.C.

We investigate exclusive non-leptonic heavy meson decays by using a modified PQCD formalism which is built on three scale factorization theorems. This formalism, embodying the effective field theory, the renormalization group, and the Sudakov resummation, appears scale invariant explicitly. There are three basic and exigent problems existing in the BSW approach, the scale setting problem, the problem of matching R_L and R simultaneously in $B \to J/\psi K^{(*)}$ decays, and the sign changing problem of nonfactorizable contributions between B and D decays. These problems are resolvable under this modified PQCD formalism.

1 INTRODUCTION

The most widely adopted approach to exclusive non-leptonic heavy meson decays is the Bauer–Stech–Wirbel (BSW) model [1] based on the factorization hypothesis. It has been known that there are three basic and exigent problems in the BSW approach: (1) The scale setting problem: The amplitudes are expressed as products of Wilson coefficients and factorized matrix elements. The Wilson coefficients are scale dependent, while the factorized matrix elements are scale independent since conserved currents are unrenormalized. In this way the amplitudes are no longer scale independent and suffer from an ambiguity of the choice of the scale. (2) The problem of matching R_L and R simultaneously in $B \to J/\psi K^{(*)}$ decays: $B \to J/\psi K^{(*)}$ decays are class II decay modes wherein the nonfactorizable contributions become important. (3) The sign changing problem of nonfactorizable contributions between B and D decays: The different interference patterns between B and D decays result from the different sign of their nonfactorizable contributions.

The solution to the above three problems is much more explicit and natural from the point of view of PQCD. First, nonfactorizable contributions can not be ignored as indicated by recent experimental data, and are calculable in the PQCD method. Second, the scale setting problem can be resolved once the factorization theorems could be proven. This is because the amplitudes expressed in the PQCD formalism are scale invariant. Third, the sign changing problem is the most difficult and technically involved one. This will also be attacked by introducing effective field theory methods into the PQCD formalism. The Wilson coefficients in the effective Hamiltonian are recovered from the factorization theorem with the help of the renormalization group (RG). After including radiative corrections, the anomalous dimensions of the convolution factors appearing in the factorization theorem could be identified. Further, by solving the RG equations, the evolution factors $a_{1,2}(t)$ with the scale range from M_W, the W-boson mass, to t, the relevant hard subprocess scale, play the role of the Wilson coefficients. As t is running to below M_b and M_c, the constructive and destructive interferences for B and D decays appear naturally. I will explain these facts in more detail in the following.

2 The PQCD formalism

Under this modified PQCD formalism with the three scale factorization theorems included, decay amplitudes can be expressed as

$$A = H_r(M_W, \mu_h) \otimes H(\mu_h, t, \mu_l) \otimes \phi(P, \mu_l, b) \otimes U(\mu_l, b) \qquad (1)$$

where μ_h and μ_l are the high and low factorization scales, and M_W, t and $\Lambda_{\rm QCD}$ are relevant physical scales involved in non–leptonic heavy meson decays. The convolution factors are the "harder" W-emission function, H_r, the hard b quark decay sub-amplitude, H, and the nonperturbative meson wave function, ϕ, which are characterized by M_W, t and $\Lambda_{\rm QCD}$, respectively. The U function denotes the nonfactorizable soft radiative corrections resulting from the above factorizations. Radiative corrections then produce two types of large logarithms $\ln(M_W/t)$ and $\ln(t/\Lambda_{\rm QCD})$. In this three–scale factorization theorem $\ln(M_W/t)$ are summed to give the evolution from M_W down to t described by the Wilson coefficients $a_{1,2}(t)$; and $\ln(t/\Lambda_{\rm QCD})$ are summed into a Sudakov factor $S(t)$ and a soft evolution factor $U(t)$ [5], which describes the evolution from t to the lower hadronic scale. The former has been derived in effective field theory, and the latter has been implemented by the resummation technique [3]. In such a way, we arrive at

$$
\begin{aligned}
A &= H_r(M_W, M_W) \otimes H(t,t,t) \otimes \phi(1/b, b) \otimes U_0(1/b, b) \\
&\quad \otimes C(M_W, t) \otimes S(P, t, b) \otimes U(t, b)
\end{aligned} \tag{2}
$$

This modified PQCD formalism is μ–independent, *i.e.* RG–invariant, and thus the scale setting ambiguity existing in conventional effective field theory is moderated [4,9]. As the variable t runs to below M_b and M_c, the constructive and destructive interference involved in bottom and charm decays, respectively, appear naturally. Furthermore, not only factorizable, but also nonfactorizable and non-spectator contributions are taken into account and evaluated in a systematic way. With the inclusion of the nonfactorizable contributions, we find that $a_{1,2}$ restore their original role as Wilson coefficients, instead of being treated as free parameters of the BSW model. The branching ratios of various decay modes $B \to D^{(*)}\pi(\rho)$ and $D \to K^{(*)}\pi$, and the ratios R and R_L associated with the $B \to J/\psi K^{(*)}$ decays can all be well explained by our formalism.

3 Results and Conclusion

I show below the results obtained by using the PQCD formalism developed above. Table I shows the predictions for the branching ratios of $B \to D^{(*)}\pi(\rho)$ and the CLEO data [2]

modes		data(%)
$B^- \to D^0\pi^-$	0.50	0.534 ± 0.025
$\bar{B}^0 \to D^+\pi^-$	0.33	0.308 ± 0.026
$B^- \to D^{*0}\pi^-$	0.48	0.497 ± 0.044
$\bar{B}^0 \to D^{*+}\pi^-$	0.32	0.304 ± 0.024
$B^- \to D^0\rho^-$	1.21	1.022 ± 0.067
$\bar{B}^0 \to D^+\rho^-$	0.68	0.861 ± 0.078
$B^- \to D^{*0}\rho^-$	1.62	1.444 ± 0.134
$\bar{B}^0 \to D^{*+}\rho^-$	0.83	0.844 ± 0.071

Table II shows predictions and data[7] for the ratios $R_1 = Br(D^+ \to \bar{K}^{*0}\pi^+)/Br(D^0 \to K^{*-}\pi^+)$ and $R_1 = Br(D^+ \to \bar{K}^0\pi^+)/Br(D^0 \to K^-\pi^+)$ associated with $D \to K^{(*)}\pi$ decays.

modes		data
$R1$	0.74	0.72
$R2$	0.35	0.38

Table III gives predictions for the branching ratios of $B \to J/\psi K^{(*)}$ decays. CLEO and CDF data [8] are also shown.

modes		data(%)
$B^- \to J/\psi K^-$	0.11	0.102 ± 0.014
$\bar{B}^0 \to J/\psi K^0$	0.10	0.115 ± 0.023
$B^- \to J/\psi K^{*-}$	0.15	0.158 ± 0.047
$\bar{B}^0 \to J/\psi K^{*0}$	0.14	0.136 ± 0.027
R	1.47	$1.36 \pm 0.17 \pm 0.04$
		$1.32 \pm 0.23 \pm 0.16$(CDF)
R_L	0.56	$0.52 \pm 0.07 \pm 0.04$
		$0.65 \pm 0.10 \pm 0.04$(CDF)

Acknowledgements

I thank H-n. Li, H.Y. Cheng, and B. Tseng for useful discussions. This work was supported by the National Science Council of R.O.C. under the Grant No. NSC-87-2811-M-001-0031.

1. M. Bauer, B. Stech and M. Wirbel, Z. Phys. **C29**, 637 (1985); **34**, 103 (1987).
2. J.L. Rodriguez, hep-ph/9604011; CLEO Collaboration, M.S. Alam *et al.*, Phys. Rev. **D50**, 43 (1994).
3. H-n. Li and H.L. Yu, Phys. Rev. Lett. **74**, 4388 (1995); Phys. Lett. **B353**, 301 (1995); Phys. Rev. **D53**, 2480 (1996).
4. C.H. Chang and H-n. Li, hep-ph/9607214.
5. J.C. Collins and D.E. Soper, Nucl. Phys. **B193**, 381 (1981).
6. CLEO Collaboration, B. Barish *et al.*, Phys. Rev. **D51**, 1014 (1995).
7. Particle Data Group, E.J. Weinberg *et al.*, Phys. Rev. **D54**, 1(1996).
8. T. Miao, talk given at the annual meeting of APS, DPF96 (unpublished).
9. T.W. Yeh and H-n Li, Phys. Rev. **D56**, 1615 (1997).

CHIRAL SYMMETRY BREAKING UNDER AN EXTERNAL FIELD

DEOG KI HONG
Department of Physics, Pusan National University, Pusan 609-735, Korea

In this talk I discuss how an external field catalyzes dynamical symmetry breaking (DSB) by performing both a gap equation analysis and a renormalization group analysis in the NJL model, QED, QCD, and 3D QCD.

1 Introduction

Recently, the effect of an external field on DSB has been studied intensely, since a constant external field is found to catalyze symmetry breaking even at a weak field [1,2,3]. It is then generalized to a nonabelian external field [4,5].

For several reasons, it is important to understand the effect of external fields on dynamical symmetry breaking: (1) In quantum chromodynamics (QCD) under an external gluon field A_μ^a the quark condensate is proportional to the density of eigenvalues of the Dirac operator $i\not{D}(A)$ at zero eigenvalue [6],

$$\langle 0|\,\bar{q}q\,|0\rangle^A = -\pi\rho_A(0). \tag{1}$$

In $N=2$ supersymmetric QCD, it is found that a magnetic monopole condensate is formed and breaks chiral symmetry [7]. (2) In a neutron star the magnetic field is as strong as $10^{15}\ G$, which could lead to a significant effect on its cooling process or on the elementary processes that occur in the neutron star. It is also argued that a primordial magnetic field can be generated in association with a phase transition in the early universe [8] or due to a dynamical dilaton background in quantum cosmology [9]. (3) In a planar condensed system, an external magnetic field has a dramatic effect on the electric conductivity [10].

2 Magnetic Catalysis

By solving the Schwinger-Dyson equations for the fermion two-point function or the Bethe-Salpeter equations for the fermion-antifermion bound state, Gusynin, Miransky and Shovkovy found that, using a constant mass function, fermions get mass dynamically under an external magnetic field, given by

$$m_{\rm dyn}^2 \simeq |eB| \exp\left[-\frac{\pi}{2}\left(\frac{\pi}{2\alpha}\right)^{1/2}\right] \tag{2}$$

for any value of α in QED if electrons are massless, and

$$m_{\text{dyn}}^2 \simeq \frac{|eB|}{\pi} \exp\left[-\frac{4\pi^2}{|eB|\,N_c G}\right] \tag{3}$$

for any $G > 0$ in the Nambu-Jona-Lassinio (NJL) model. The constant-mass-function approximation was improved later and the critical coupling was confirmed to be zero in QED [2].

The physical reason for magnetic catalysis in dynamical symmetry breaking is that the energy spectrum of fermions changes due to the external magnetic field. At low energy the fermions are effectively one-dimensional, moving along the direction of the magnetic field.

Consider a free massless fermion under constant magnetic field $\mathbf{B} = B\hat{z}$. In the symmetric gauge the energy eigenvalue equation is

$$[\vec{\alpha} \cdot (\mathbf{p} - e\mathbf{A})]\Psi = E\Psi \tag{4}$$

with eigenvalues $E_A = \alpha\sqrt{k_z^2 + 2\,|eB|\,n}$. Here, $A = (\alpha, \beta, n, k_z)$ denotes the collective index of the energy eigenvalue and $\alpha = \pm$ the sign of the energy, $\beta = \pm\frac{1}{2}$ the spin component along the magnetic field, and the quantum number n is given by

$$2n = 2n_r + 1 + |m_L| - \text{sign}(eB)(m_L + 2\beta). \tag{5}$$

n is a nonnegative integer that labels the Landau level. Here n_r is the number of nodes of the radial eigenfunction, and m_L is the angular momentum.

We see that for energy much less than the Landau gap $E \ll \sqrt{|eB|}$ the excitations of fermions are effectively one-dimensional since their synchrotron radius is smaller than their de Broglie wavelength. This effective dimensional reduction can also be seen by a renormalization group (RG) analysis [2]. In the basis of the eigenfunctions $U_A(x)$, the fermion fields are

$$\Psi(x,t) = \int\!\!\sum_A \psi_A(t) U_A(x). \tag{6}$$

The kinetic term then becomes

$$S_0 = \int d^4x \bar{\Psi} i \not{D} \Psi = \int\!\!\sum_{A,\omega} \psi_A^\dagger(\omega)(\omega - E_A)\psi_A(\omega), \tag{7}$$

where $\psi_A(\omega)$ is the Fourier transform of $\psi_A(t)$ and it is summed and integrated over A and ω. Under a scaling transformation,

$$k_z \to s k_z, \omega \to s\omega, \quad \text{with} \quad s < 1 \tag{8}$$

we find that $\psi_0(k_z,\omega) \mapsto s^{-3/2}\psi_0(k_z,\omega)$, while $\psi_{n>0}(k_z,\omega) \mapsto s^{-1}\psi_{n>0}(k_z,\omega)$. The $n=0$ mode is more relevant than the other modes. Fermions in the lowest Landau level (LLL) are effectively one-dimensional and the four-Fermi interaction of LLL fermions is marginal. If we introduce a dimensionless coupling $g \equiv G\,|eB|$, we find that at one-loop

$$\beta(g) = -\frac{N_c}{4\pi^2}g^2, \tag{9}$$

which shows that the four-Fermi interaction is relevant and becomes strong at low energy to form fermion-antifermion condensates, breaking chiral symmetry. The dynamical mass for fermion, being a RG invariant scale, agrees with the one obtained by Gusynin *et al.*

3 Effective Field Theory of QED under a constant external B-field

We have seen that at low energy $E < |eB|^{1/2}$ the modes at higher Landau level ($n = 1, 2, \cdots$) are irrelevant degrees of freedom. To derive the low energy effective action, we integrate out those irrelevant degrees of freedom [11];

$$\exp\left(i\int d^4x\mathcal{L}_{eff}\right) = \int \mathcal{D}\Psi_{n\neq 0}\exp\left(i\int d^4x\left[-\frac{1}{4}F_{\mu\nu}^2 + \bar{\Psi}i\not{D}\Psi\right]\right), \tag{10}$$

with the covariant derivative $D_\mu = \partial_\mu + ieA_\mu + ieA_\mu^{\rm ext}$. In practice, one derives the effective Lagrangian by matching all the "one light particle irreducible" (1LPI) diagrams systematically in a loop expansion.

At tree level, $\Psi_{n\neq 0}$ exchange generates a new photon-electron vertex in the effective theory;

$$\mathcal{L}_{eff}^0 = -\frac{1}{4}F_{\mu\nu}^2 + \bar{\psi}(i\not{\partial} - e\not{A})\psi - \frac{e^2}{2|eB|}\bar{\psi}\gamma^\alpha A_\alpha\tilde{\gamma}^\mu i\overleftrightarrow{\partial}_\mu\gamma^\beta\psi A_\beta + \cdots, \tag{11}$$

where ellipses refer to operators containing more photons and higher powers of derivatives and

$$\tilde{\gamma}^\mu = \begin{cases} i\gamma^\mu\gamma^1\gamma^2\ln 2\ \mathrm{sign}(eB) & \text{if } \mu = 0,3, \\ \gamma^\mu & \text{if } \mu = 1,2. \end{cases} \tag{12}$$

Matching a one-loop four-Fermion amplitude, we find that a four-Fermi operator arises in the effective theory,

$$\mathcal{L}_{\rm eff}^1 \ni \frac{g_1}{|eB|}\left[(\bar{\psi}\psi)^2 + (\bar{\psi}i\gamma_5\psi)^2\right]. \tag{13}$$

Below the cutoff scale, the renormalization group equation for g_1 is:

$$\mu \frac{d}{d\mu} g_1 = -\frac{(\ln 2)^2}{8\pi^2} e^4, \tag{14}$$

which is very small, about 0.5×10^{-4}. Hence, g_1 remains almost constant as the scale changes. Integrating this equation from $\sqrt{|eB|}$ to m_e gives

$$g_1(m_e) = g_1(\sqrt{|eB|}) + \frac{e^4 (\ln 2)^2}{16\pi^2} \ln \frac{|eB|}{m_e^2}. \tag{15}$$

We see that the external magnetic field has to be extremely strong $B > e^{1/\alpha}$ G in order for the effective four-Fermi interaction to break chiral symmetry in QED.

4 QCD under a constant external magnetic field

Since quarks have electric charge, magnetic catalysis will operate for quarks when the external magnetic field is sufficiently strong. We will show that if the magnetic field is stronger than 4×10^{15} G the chiral symmetry will break at scale higher than m_ρ, the chiral symmetry breaking scale of QCD. The derivation of the low energy effective Lagrangian is the same as that of QED except for group theoretical factors. At low energy the relevant degrees of freedom are LLL quarks, gluons, and photons. But, since gluons couple to quarks more strongly than photons do, we will neglect the photon interactions. As in QED, the exchange of quarks at the higher Landau level generates a tree-level interaction of quarks and gluons, which induces a four-quark interaction at one-loop matching:

$$\mathcal{L}_{\rm eff}^1 \ni \frac{g_1^s}{4\,|qB|} \left[\left(\bar{Q}_0 Q_0\right)^2 + \left(\bar{Q}_0 i\gamma_5 Q_0\right)^2 \right], \tag{16}$$

where Q_0 is the LLL quark and q is its electric charge. The one-loop RG equation for the four-quark interaction is given by

$$\mu \frac{d}{d\mu} g_1^s = -\frac{40}{9} \alpha_s^2 (\ln 2)^2, \tag{17}$$

where α_s is the coupling constant of strong interactions. Unlike in QED, the strong coupling constant runs at scale below the Landau gap $\Lambda_L = \sqrt{|qB|}$ as

$$\frac{1}{\alpha_s(\mu)} = \frac{1}{\alpha_s(\Lambda_L)} + \frac{11}{2\pi} \ln \frac{\mu}{\Lambda_L}. \tag{18}$$

Note that quarks do not contribute to this running, since the quark loop is finite due to the dimensional reduction. Combining the above RG equations (17) and (18), we find the scale dependence of the four-quark coupling:

$$g_1^s(\mu) = 1.1424\,(\alpha_s(\mu) - \alpha_s(\Lambda_L)) + g_1^s(\Lambda_L). \tag{19}$$

Since $g_1^s(\Lambda_L)$ is of the order of $\alpha_s(\Lambda_L)^2$, we see that the four-quark coupling becomes stronger than the strong coupling constant at a scale larger than the ρ meson mass $\mu > m_\rho$ if $\Lambda_L > 10 m_\rho$ or $B > 4 \times 10^{15}\ G$. Hence the chiral symmetry of QCD will be broken at scale larger than m_ρ, the usual chiral symmetry breaking scale of QCD, if $B > 4 \times 10^{15}\ G$. This may have significant consequences on dynamics inside neutron stars, where the magnetic field is stronger than $10^{15}\ G$ in the core [12].

5 Nonabelian External Fields

As extension to a nonabelian field, let us consider a (2+1)D NJL model and 3D QCD with gauge group $SU(2)$ under a nonabelian field which mimics a magnetic field and has the largest global symmetry [5], given by

$$A_\mu = \frac{1}{2} A_\mu^a \tau^a, \quad \text{with } A_1^1 = A_2^2 = \sqrt{H/g}, \text{ and others} = 0, \tag{20}$$

where τ^a are the Pauli matrices ($a = 1, 2, 3$), g is the gauge coupling constant, and H is a constant. Due to the external field the fermion spectrum changes drastically at low energy,

$$E \simeq \pm \frac{\vec{p}^2}{2h}, \tag{21}$$

where $h = \sqrt{gH}/2$. Because of this change in spectrum the scaling dimension of fermions is (1+1)-dimensional and the four-Fermi interaction becomes relevant at one-loop. It gets strong and generates a dynamical mass at low energy $E < h$, $m_{\rm dyn} \simeq h \exp[-\pi/hG)]$, if $G > 0$, which is of order of the RG invariant mass of the running four-Fermi coupling, $\lambda(\equiv hG)$:

$$\beta(\lambda) = s\frac{\partial \lambda}{\partial s} = -\frac{1}{\pi}\lambda^2. \tag{22}$$

This result can also be derived from the gap equation, obtained by minimizing the vacuum energy of the auxiliary fields π, σ and τ corresponding to fermion bilinears in the four-Fermi interactions of the NJL model. The result is as the same as for weak field $\rho(= \sqrt{\pi^2 + \sigma^2 + \tau^2}) \ll h$:

$$V(\rho) = Nh\left[\frac{2h^2}{3\pi} + \frac{\rho^2}{2\lambda} + \frac{1}{4\pi}\rho^2\left(\ln\frac{\rho^2}{16h^2} - \frac{4}{\sqrt{\pi}} + 3\right)\right] + O(1), \tag{23}$$

In 3D QCD, the calculation is more involved since the gluon propagator gets also modified under the external field; at low energy ($E < h$),

$$S^{ab}_{\alpha\beta} \simeq \frac{ig_{\alpha\beta}}{8h^2}\left(2\delta^{ab} - \delta^{3a}\delta^{3b}\right) + \frac{iF_{\alpha\beta}}{6h^2H}\varepsilon^{3ab} + \frac{i(F^2)_{\alpha\beta}}{3h^2H^2}\left(\delta^{ab} - \delta^{3a}\delta^{3b}\right). \qquad (24)$$

We see that gluons are massive and thus irrelevant at low energy $E < h$. If we integrate them out, the low energy effective Lagrangian contains a four-Fermi interaction $-\frac{g^2}{48h^2}\left[(\bar\psi\psi)^2 + (\bar\psi i\gamma_5\psi)^2 + (\bar\psi\gamma_3\psi)^2\right]$, which is repulsive and has positive β-function. Therefore, the external nonabelian field does not catalyze dynamical symmetry breaking in 3D QCD. This can also be seen easily by solving the Bethe-Salpeter equation for the quark-antiquark bound state. The homogeneous BS equation for the NG bound state reads:

$$\chi = -4\pi\alpha \int \frac{d^3p}{(2\pi)^3}\frac{\tau^a}{2}\gamma^\mu S(p)\chi S(p)\frac{\tau^b}{2}\gamma^\nu S^{ab}_{\mu\nu}, \qquad (25)$$

where $\alpha = g^2/(4\pi)$. The BS wave function for the bound state is given in general as

$$\chi = \gamma^5\left(A + \tau^3\sigma^{12}B + (\vec\gamma\cdot\vec\tau)C\right). \qquad (26)$$

One can verify that the BS equation does not allow a nontrivial solution [5].

6 Conclusion

We see that catalysis of dynamical symmetry breaking by external fields is due to the change of the scaling dimension of relevant degrees of freedom at low energy. Under external fields, the four-Fermi interaction in the NJL model in (2+1)D and (3+1)D becomes marginal and breaks chiral symmetry at low energy. If irrelevant higher-Landau modes are integrated out, an effective four-Fermi interaction arises at low energy in QED and breaks chiral symmetry if the external magnetic field is extremely strong $B > e^{1/\alpha}\ G$, while in QCD the effective four-quark interaction generated by the exchange of higher Landau modes gets rapidly stronger due to the running strong coupling at low energy and it becomes stronger than the gluon-exchange interaction at a scale larger than the ρ meson mass if $B > 4\times 10^{15}\ G$.

Under an external chromo-magnetic field, the fermion spectrum changes at low energy such that its scaling dimension becomes (1+1). Thus the four-Fermi interaction in (2+1)D NJL model becomes marginal and breaks chiral symmetry even with a weak coupling, whereas in 3D QCD the field does not catalyze DSB.

Acknowledgments

This work is supported in part by the academic research fund of the Ministry of Education, Republic of Korea, Project No. BSRI-97-2413, by the KOSEF through SRC program of SNU-CTP. The author wishes to acknowledge partial financial support of the Korea Research Foundation made in the program year of 1997.

References

1. V. P. Gusynin, V. A. Miransky, and I. A. Shovkovy, *Phys. Rev. Lett.* **73**, 3499 (1994); *Phys. Rev.* D **52**, 4747 (1995); *Nucl. Phys.* B **462**, 249 (1996); V.P. Gusynin and I.A. Shovkovy, *Phys. Rev.* D **56**, 5251 (1997).
2. D. K. Hong, Y. Kim, and S. Sin, *Phys. Rev.* D **54**, 7879 (1996).
3. C. N. Leung, Y. J. Ng and A. W. Ackley, *Phys. Rev.* D **54**, 4181 (1996); D.-S. Lee, C. N. Leung, Y. J. Ng, *Phys. Rev.* D **55**, 6504 (1997).
4. D. Ebert and V. Ch. Zhukovsky, *Mod. Phys. Lett.* A**12**, 2567 (1997). For previous related works, see S. Kawati, G. Konisi, and H. Miyata, *Phys. Rev.* D**28**, 1537 (1983); S.P. Klevansky and R.H. Lemmer, *Phys. Rev.* D**39**, 3478 (1989); I. V. Krive and S. A. Naftulin, *Phys. Rev.* D**46**, 2737 (1992); K. G. Klimenko, *Z. Phys.* C**54**, 323 (1992).
5. V. P. Gusynin, D. K. Hong, and I. A. Shovkovy, "Chiral symmetry breaking by a non-Abelian external field in (2+1)-dimensions", hep-th/9711016, submitted to *Phys. Rev.* D.
6. T. Banks and A. Casher, *Nucl. Phys.* **B169**, 103 (1980); E. Marinari. G. Parisi, and C. Rebbi, *Phys. Rev. Lett.* **47**, 1795 (1981); H. Leutwyler and A. Smilga, *Phys. Rev.* D**46**, 5607 (1992).
7. N. Seiberg and E. Witten, *Nucl. Phys.* B**426**, 19 (1994); B**431**, 484 (1994).
8. T. Vachaspati, *Phys. Lett.* B **265**, 258 (1991); K. Enqvist and P. Olesen, *ibid* **319**, 178 (1993); *329*, 195 (1994); G. Baym *et al.*, *Phys. Rev.* D **53**, 662 (1996); J. M. Corwall, *ibid* **56**, 6146 (1997).
9. M. Gasperini, M. Giovannini. and G. Veneziano, *Phys. Rev. Lett.* **75**, 3796 (1995); M. Giovannini, hep-th/9706201.
10. K. Farakos and N. E. Mavromatos, "Gauge-theory approach to planar doped antiferromagnets and external magnetic fields", cond-mat/9710288.
11. D. K. Hong, "Magnetic Catalysis in Quantum Electrodynamics", hep-ph/9707432, to appear in *Phys. Rev.* D, March 15, 1998.
12. D. K. Hong, work under progress.

TOP QUARK MASS MEASUREMENT AT DØ

EUNIL WON
University of Rochester
Rochester, NY 14627

Seoul National University
Seoul 151-742, Korea

For the DØ Collaboration
We present a measurement of the mass of the top quark and a search for charged Higgs bosons in $p\bar{p}$ collisions at $\sqrt{s} = 1.8$ TeV, performed by the DØ experiment at the Fermilab Tevatron. The report is based on data from an integrated luminosity of approximately 125 pb^{-1} accumulated during the 1992–1996 collider run.

1 Introduction

In $p\bar{p}$ collisions at $\sqrt{s} = 1.8$ TeV, top and anti–top quarks are predominantly pair produced through $q\bar{q}$ annihilation ($\approx 90\%$) or gluon fusion ($\approx 10\%$). In the SM, due to their large mass, nearly all ($\geq 99\%$) decay to a W boson and a b quark. The subsequent W decay determines the major signatures of $t\bar{t}$ production. In the dilepton channel, both W bosons decay either to $e\nu$ or $\mu\nu$. The branching fraction for this channel is rather small (4/81), but it has the advantage of small backgrounds. In the lepton+jets channel, one W boson decays to $e\nu$ or $\mu\nu$ and the other hadronically. The branching fraction is 24/81. In this paper we report a measurement of the top quark mass m_t using events consistent with either the dilepton ($t\bar{t} \to bW^+\bar{b}W^- \to bl^+\bar{\nu}\bar{b}l^-\nu$) or the lepton+jets ($t\bar{t} \to bW^+\bar{b}W^- \to bl^+\bar{\nu}\bar{b}q\bar{q}$ or $bq\bar{q}\bar{b}l^-\nu$) hypothesis.

2 Dilepton Mass Analysis

The selection of dilepton top candidate events requires two leptons of $E_T >$ 15(20) GeV for the $e\mu$ and $\mu\mu$ (ee) channels within $|\eta_e| < 2.5$ and $|\eta_\mu| < 1.7$, and two or more jets with $E_T > 20$ GeV and $|\eta| < 2.5$. In addition, the ee and $e\mu$ channels require significant missing transverse energy to discriminate against background sources, while the $\mu\mu$ analysis reduces the Z boson background by rejecting events for which the χ^2 probability of a fit to the $Z \to \mu\mu$ hypothesis is $> 1\%$. The quantity $H_T^e \equiv \sum_{jets} E_T + E_T(e_1)$ is used to reject much of the remaining background. After applying the constraints $m(l_1\nu_1) = m(l_2\nu_2)$ and $m(l_1\nu_1 b_1) = m(l_2\nu_2 b_2)$, the system remains underconstrained due to the two unmeasured neutrinos. We supply the additional constraints by assuming values for the top quark mass and for two quantities associated with the neutrinos. We then solve for the neutrino momenta up to a fourfold ambiguity and assign a weight to each solution to characterise how likely it is to occur in $t\bar{t}$ production for the assumed top quark mass. We employ two weighting schemes.

The first is called matrix element weighting (MWT) and is an extension of the procedure given in references [1]. In solving the event, we require the sum of the neutrino $\vec{p}_T$'s to equal the measured $\vec{\not{E}}_T$. We assign a weight

$$W_o(m_t) = A(m_t)f(x)f(\bar{x})P(E_{l1}^{CM}|m_t)P(E_{l2}^{CM}|m_t) \qquad (1)$$

where $f(x)$ is the parton distribution function at the proton (antiproton) momentum fraction x ($\bar{x}$) required by the solution, and $P(E_l^{CM}|m_t)$ is the probability density for the lepton energy in the top quark rest frame. The factor $A(m_t)$ normalizes the average weight of accepted events to unity independent of the assumed top quark mass. For each mass we add the weights of all solutions.

The other weighting scheme, called neutrino weighting (νWT), steps through a range of assumed η for each neutrino at each m_t. Each step spans an equal fraction of the neutrino η distribution expected in $t\bar{t}$ production. At each step a weight is assigned based on the extent to which the $\vec{\not{E}}_T$ measured in the event agrees with the sum of the neutrino $\vec{p}_T$'s required by the solution. The weights at all η values are summed to give $W_o(m_t)$.

For both analyses, the maximum likelihood fit proceeds by normalizing the $W(m_t)$ distribution for each event to unity and integrating the fractional weights in m_t. Then the maximum likelihood fit is performed assuming a Gaussian constraint on the background and a Poisson constraint on the total number of events. Applying the fit to the data, we determine the top quark mass $m_t = 168.1 \pm 12.4$ GeV/c^2 (MWT) and $m_t = 169.9 \pm 14.8$ GeV/c^2 (νWT), where the uncertainties are statistical only. Figure 1 shows the result of the maximum likelihood fit. The sources of systematic uncertainty are summarized in Table 1; they are in common for both methods. A major uncertainty arises from the jet energy calibration and less significant sources of uncertainty arise from the models of $t\bar{t}$ and background production, multiple interaction, and from the likelihood fit procedure. We measure the top quark mass in the dilepton channel by combining the results of the MWT and νWT analyses, taking into account the 77 % correlation between them:

$$m_t = 168.4 \pm 12.3 \text{ (stat.)} \pm 3.7 \text{ (syst.) GeV}/c^2.$$

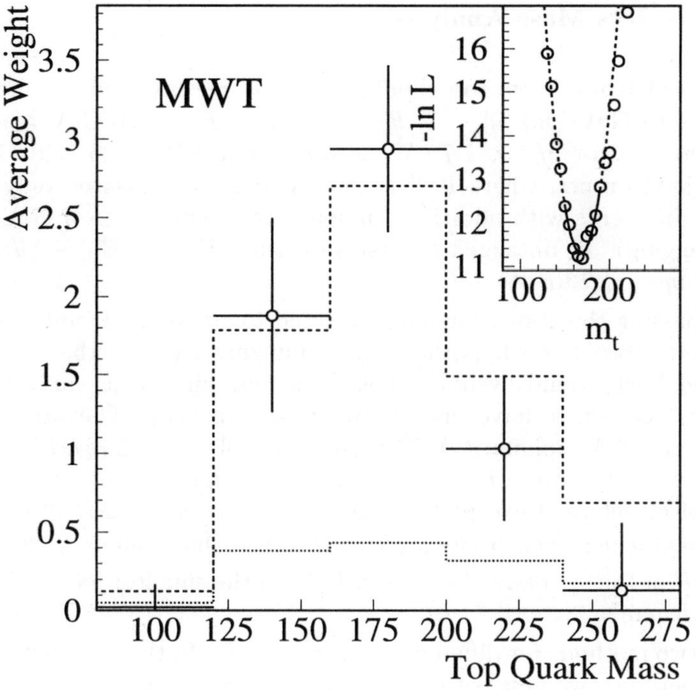

Figure 1: Sum of the normalised candidate weights grouped into five bins. The $-\ln L$ distribution and a quadratic fit are inset at the upper right.

Table 1: Systematic uncertainties in the dilepton channel analysis

Source	Error (GeV/c^2)
Jet energy scale	2.4
Signal model	1.8
Multiple interaction	1.3
Background model	1.2
Likelihood fit	1.3
Total	3.7

3 Lepton + Jets Mass Analysis

The selection of lepton+jets top candidate events requires four or more jets having $E_T > 15$ GeV and $|\eta| < 2$, $\not\!\!E_T > 20$ GeV, $E_T^l > 20$ GeV and pseudorapidity $|\eta_e| < 2$ or $|\eta_\mu| < 1.7$. We also demand $\not\!\!E_T^{cal} > 25$ (20) GeV for e+jets (μ +jets) events, where $\not\!\!E_T^{cal}$ is the missing E_T measured only in the calorimeter. In events with additional muons, we require $p_T^\mu > 4$ GeV/c and $|\eta| < 1.7$ (tagging). In untagged events, we require E_T^L ($\equiv |E_T^l| + |\not\!\!E_T^{cal}|$) $>$ 60 GeV and $|\eta_W| < 2$ for $W \to l\nu$.

Events passing the above cuts are subjected to a two constraint kinematic fit to the $t\bar{t} \to$ lepton + jets hypothesis by minimizing a χ^2. Further separation of signal and background events is based on four kinematic variables $\mathbf{x} \equiv \{x_1, x_2, x_3, x_4\}$ chosen to have small correlation with m_{fit}. The variables are $x_1 = \not\!\!E_T$ and $x_2 =$ A (aplanarity). The third variable $x_3 = H_{T2}/H_z$ measures the event's centrality, where H_z is the sum of $|p_z|$ of l, ν and the jets, and H_{T2} is the sum of all jet $|E_T|$ except the highest. Finally, $x_4 = \Delta R_{jj}^{min} E_T^{min} / E_T^L$ measures the extent to which jets are clustered together, where ΔR_{jj}^{min} is the minimum ΔR of the six pairs of jets, and E_T^{min} is the smallest jet E_T from the minimum ΔR pair.

We use two methods for different definitions of D. In the first method (low bias or LB method), we first parametrize $\mathcal{L}_i(x_i) = s_i(x_i)/b_i(x_i)$, where s_i and b_i are the top signal and background densities in each variable integrating over the others. We form $D_{LB} = \mathcal{L}/(1+\mathcal{L})$ where $\mathcal{L} = \sum_i \omega_i \ln \mathcal{L}_i$. w_i are adjusted slightly away from unity to nullify the average correlation of $\mathcal{L}$ with m_{fit}. In the second method, we use a three layer feed–forward neural network (NN) with 4 input nodes fed by $\mathbf{x}$, 6 hidden nodes, and one output node, trained on samples of top signal and background densities. The NN method is sensitive to correlations (particularly nonlinear) among the x_i as well as to their individual densities. For a given event, the network output ($\equiv D_{NN}$) approximates the ratio $s(x)/(s(x) + b(x))$.

To each m_t for which we have generated MC, we assign a likelihood L which assumes that all samples obey Poisson statistics for both methods. Applying the fit to the data using D_{LB} and using D_{NN}, we determine the top quark mass $m_t = 174.0 \pm 5.6$ and 171.3 ± 6.0 GeV/c^2, respectively (errors are statistical only). Figure 2 shows the result of the fit. The sources of systematic uncertainty are 4.0 (energy scale), 4.1 (generator) and 2.2 (other) GeV/c^2. We measure the top quark mass in the lepton+jets channel by combining the results using D_{LB} and D_{NN}, taking 88 % correlation between them:

$$m_t = 173.3 \pm 5.6 \text{ (stat.)} \pm 6.2 \text{ (syst.) GeV}/c^2.$$

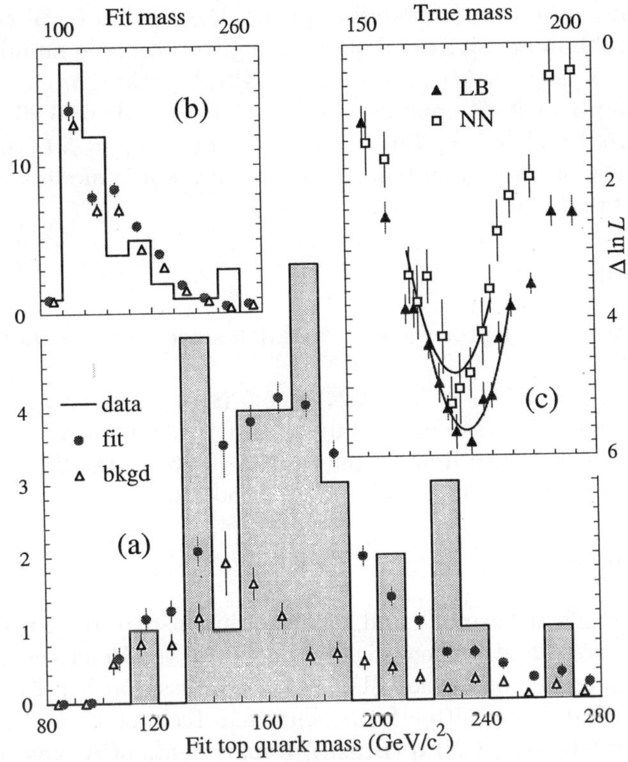

Figure 2: (a-b) Events per bin *vs.* m_{fit} for events (a) passing or (b) failing the LB cutoff. Histograms are data, filled circles are the predicted mixture of top and background, and open triangles are predicted background only. (c) Log of arbitrarily normalized likelihood L *vs.* true top quark mass m_t for the LB (filled triangles) and NN (open squares) fits. The curves are quadratic fits to the lowest point and its 9 nearest neighbors.

4 Search for Charged Higgs

If one assumes the minimal supersymmetry model (MSSM), charged Higgs bosons can couple to the top quark in such a way that one or both of the produced top quarks could decay to $H^{\pm}b$ rather than to $W^{\pm}b$. Within the MSSM, $t \to H^{+}b$ occurs primarily at low and high $\tan\beta$ and it leads to exclusion regions at the $\tan\beta$ extremities of the M_{H+} *vs.* $\tan\beta$ plane, based on the relative likelihood *vs.* $\log\tan\beta$ of obtaining the observed number of lepton + jets events, for a given M_{H+}. Within the MSSM, taking m_t = 175 GeV/c^2 and σ_{tt} = 5.53 pb, at 95 % confidence the data require 0.96 (0.26) $< \tan\beta$ for M_{H+} = 50 (168) GeV/c^2, or $\tan\beta <$ 35 (96) for M_{H+} = 50 (168) GeV/c^2. The dependence on m_t and on renormalization scale μ is modest over most of the M_{H+} *vs.* $\tan\beta$ plane.

5 Conclusions

Using all dilepton and lepton+jets $t\bar{t}$ candidates, we measure the mass of the top quark to be:

$$m_t = 172.0 \pm 5.1 \text{ (stat.) } \pm 5.5 \text{ (syst.) GeV}/c^2,$$

where the correlated errors in both analyses were taken into account. Assuming MSSM, a 95 % confidence limit is set as 0.96 (0.26) $< \tan\beta$ for M_{H+} = 50 (168) GeV/c^2, or $\tan\beta <$ 35 (96) for M_{H+} = 50 (168) GeV/c^2.

Acknowledgments

We thank the staffs at Fermilab and collaborating institutions for their contributions to this work, and acknowledge support from the Department of Energy and National Science Foundation (U.S.A.), Commissariat à L'Energie Atomique (France), State Committee for Science and Technology and Ministry for Atomic Energy (Russia), CNPq (Brazil), Departments of Atomic Energy and Science and Education (India), Colciencias (Colombia), CONACyT (Mexico), Ministry of Education and KOSEF (Korea), and CONICET and UBACyT (Argentina).

References

1. K. Kondo, *Journal of the Physical Society of Japan* **57**, 4126(1988) and **60**, 836 (1991); R.H. Dalitz and G.R. Goldstein, *Phys. Lett.* B **287**, 225 (1992).
2. F.A. Berends *et al, Nucl. Phys.* B **357**, 32(1991), release 3.0.

T-ODD GLUON–TOP QUARK COUPLINGS AT THE LHC

JAKE LEE, S.Y. CHOI and C.S. KIM
Department of Physics, Yonsei University, Seoul 120-749, Korea

The T–odd top quark chromoelectric dipole moment, d_t, is probed through top quark pair production via gluon fusion at the CERN LHC, taking into account the possibility of having polarized proton beams. At 1-σ level, the typical CP-odd lepton energy and tensor correlations enable us to measure $Re(d_t)$ and $Im(d_t)$ up to the order of 10^{-18} ($g_s$cm) in the unpolarized case and the initial CP-odd gluon spin correlation allows us to probe $Im(d_t)$ up to the order of 10^{-20} ($g_s$cm) for $\sqrt{s} = 14$ TeV and the integrated luminosity $\mathcal{L}_{pp} = 10$ fb^{-1}.

The top quark is expected to be sensitive to new physics at the TeV energy scale, not readily observable in lighter quarks, and it is copiously produced at proton-proton collisions at LHC, predominantly through gluon–gluon fusion. Therefore, we probe the possibility of measuring the basic T-odd coupling of gluon and top quark, i.e. the top-quark chromoelectric dipole moment (tCEDM) at LHC [1].

The process $gg \to t\bar{t}$, followed by t and $\bar{t}$ semileptonic decays, has been studied to extract the real part of the T-odd tCEDM form factor by using optimal observables [2]. We extend their work to measuring the imaginary part of tCEDM and extract the real and imaginary parts through typical CP-odd lepton and antilepton correlations [3] in the t and $\bar{t}$ semileptonic decays.

It has been demonstrated [4] that the CP properties of the Higgs boson can be directly probed with the initial CP-odd spin correlations of gluons, whose polarization carries over from polarized proton beams. We use initial gluon-gluon spin correlations to measure the imaginary part of the tCEDM and compare them with the CP-odd lepton correlations from the t and $\bar{t}$ semileptonic decays.

In general the gluon–top quark interaction Lagrangian consists of two Standard Model(SM) dimension-four terms and two dimension-five terms:

$$\mathcal{L}_M = \frac{1}{2}g_s\left(\frac{c_t}{2m_t}\right)\bar{t}\sigma^{\mu\nu}G^a_{\mu\nu}T_a t, \qquad \mathcal{L}_E = \frac{i}{2}g_s\left(\frac{\tilde{c}_t}{2m_t}\right)\bar{t}\sigma^{\mu\nu}\gamma_5 G^a_{\mu\nu}T_a t, \qquad (1)$$

where $\sigma^{\mu\nu} = \frac{i}{2}[\gamma^\mu, \gamma^\nu]$, $G^a_{\mu\nu}$ is the gluon field strength, and $T_a = \frac{1}{2}\lambda_a$ ($a = 1$ to 8). Among the four terms, only $\mathcal{L}_E$ violates T invariance and defines the tCEDM $d_t \equiv g_s(\tilde{c}_t/2m_t)$. In the SM, this tCEDM form factor is extremely small [5] because it arises only at three or more loops. On the other hand, the tCEDM can be much larger in many models of CP violation such as multi-Higgs-doublet models and the Minimal Supersymmetric Standard Model [6]. A

Table 1: Attainable 1-σ limits on $Re(d_t)$ and $Im(d_t)$, through T_{33}, A_E and Q_{33} for the parameter set (4).

observable	Attainable 1-σ limits
T_{33}	$\lvert Re(d_t)\rvert = 0.899 \times 10^{-17} g_s\mathrm{cm}$
A_E	$\lvert Im(d_t)\rvert = 0.858 \times 10^{-18} g_s\mathrm{cm}$
Q_{33}	$\lvert Im(d_t)\rvert = 0.205 \times 10^{-17} g_s\mathrm{cm}$

nonvanishing tCEDM would thus be a strong indication of new physics beyond the SM.

Firstly, let us consider the case that proton beams are unpolarized. In this case the initial unpolarized gluon-gluon system in the process $gg \to t\bar{t}$ is CP invariant and therefore any detection of the tCEDM requires information on the final $t\bar{t}$ spin correlations. These can be indirectly inferred through correlations of the lepton and antilepton produced in semileptonic top and antitop quark decays $t \to bl^+\nu_l$ and $\bar{t} \to bl^-\bar{\nu}_l$ $(l = e, \mu)$. We may employ two typical CP-odd and $CP\tilde{T}$-even correlations[3],

$$A_1 = \hat{p}_g \cdot (\vec{p}_{\bar{l}} \times \vec{p}_l), \qquad T_{33} = 2(\vec{p}_{\bar{l}} - \vec{p}_l)_3(\vec{p}_{\bar{l}} \times \vec{p}_l)_3, \tag{2}$$

and three CP-odd and $CP\tilde{T}$-odd correlations

$$\begin{aligned} A_E &= E_{\bar{l}} - E_l, \qquad A_2 = \hat{p}_g \cdot (\vec{p}_{\bar{l}} + \vec{p}_l), \\ Q^l_{33} &= 2(\vec{p}_{\bar{l}} + \vec{p}_l)_3(\vec{p}_{\bar{l}} - \vec{p}_l)_3 - \frac{2}{3}(\vec{p}_{\bar{l}}^{\,2} - \vec{p}_l^{\,2}). \end{aligned} \tag{3}$$

However, the Bose symmetry of the initial gg system forces the vector correlations A_1 and A_2 to vanish, so that no information on the tCEDM can be extracted from them. We therefore have to use T_{33} as $CP\tilde{T}$-even correlation, which is proportional to the real part of the tCEDM, and A_E and Q_{33} as $CP\tilde{T}$-odd correlations, which are proportional to the imaginary part of the tCEDM.

In our numerical analysis we use the following experimental parameters:

$$\begin{aligned} &\epsilon = 10\%, \qquad B_l = B_{\bar{l}} = 21\% \ \text{ for } \ l = e, \mu, \\ &\sqrt{s} = 14 \ \text{TeV}, \ \ \mathcal{L}_{pp} = 10 \ \text{fb}^{-1}, \ \ m_t = 175 \ \text{GeV}, \end{aligned} \tag{4}$$

where ϵ stands for the detection efficiency. For the unpolarized gluon distribution function, we employ the GRVHO parameterization[7]. Table 1 shows the 1-σ sensitivities of the CP-odd correlations, T_{33}, A_E and Q_{33} to $Re(d_t)$ and

$Im(d_t)$, respectively, for the parameter set (4). Quantitatively, T_{33} and Q_{33} enable us to probe $Re(d_t)$ and $Im(d_t)$ of the order of $10^{-17} g_s$cm, respectively, and A_E allows us to probe $Im(d_t)$ down to the order of $10^{-18} g_s$cm.

Secondly, let us consider the case that proton beams are polarized. It has been claimed in many works [8] that gluons in a polarized proton should be polarized to explain the observed EMC effect. We assume in this work that the polarization transmission from protons to gluons exist and use the gluon polarization to form a CP-odd gluon-gluon spin correlation to measure the tCEDM. This initial CP-odd configuration allows us to consider all detectable $t\bar{t}$ decay modes without any full reconstruction of the decay products. One of the simplest CP-odd asymmetries is the rate asymmetry:

$$A \equiv \frac{\sigma_+ - \sigma_-}{\sigma_+ + \sigma_-}, \tag{5}$$

which has been suggested as a probe of the CP properties of Higgs bosons [4]. Here, $\sigma_\pm$ is the cross section of $t\bar{t}$ production in collision of an unpolarized proton and a proton of helicity $\pm$. It is straightforward to check that the difference and sum of the two cross sections for opposite proton helicities are given by

$$d\sigma_+ - d\sigma_- \sim g_1 \Delta g_2 Im(\tilde{c}_t), \qquad d\sigma_+ + d\sigma_- \sim g_1 g_2, \tag{6}$$

where $g_1/(g_2)$ is the helicity-summed gluon distribution function inside the unpolarized (polarized) proton, $g_{1,2} = g_{1,2+} + g_{1,2-}$, and $\Delta g_2 = g_{2+} - g_{2-}$, where $\pm$ labels the gluon helicity.

The sensitivity of the CP-odd rate asymmetry to $Im(d_t)$ is crucially dependent on the degree of gluon polarization achievable for polarized protons at the CERN LHC. At present the function $\Delta g(x) = g_+(x) - g_-(x)$ indicating the degree of gluon polarization in a polarized proton is not precisely known except for the fact that it should satisfy the asymptotic boundary conditions: $\Delta g(x)/g(x) \to 1$ for $x \to 1$ and $\Delta g(x)/g(x) \propto x$ for $x \to 0$. Several models [8], which satisfy these constraints, suggest that a significant amount of the proton spin is carried by gluons. In our numerical analysis we employ the BQ parameterization [9] as an example of Δg.

Table 2 shows the number of $t\bar{t}$ events, N, and the 1-σ limits on $|Im(d_t)|$ for an integrated luminosity of 10 fb^{-1}. Note that there is no drastic p_T-cut dependence. This feature can be used to make a large p_T cut to reduce backgrounds without spoiling the sensitivity of the rate asymmetry to the tCEDM. Remarkably, $Im(d_t)$ is much more strongly constrained by the rate asymmetry than by the lepton and antilepton energy correlation A_E (see Table 1). Numerically, it is possible to measure $Im(d_t)$ up to the order of 10^{-20} $g_s$cm

Table 2: The number of $t\bar{t}$ events N, and the attainable 1-σ limits on $|\mathrm{Im}(\mathrm{d_t})|$, for various p_T-cut values for $\sqrt{s} = 14$ TeV, $m_t = 175$ GeV, $\mathcal{L} = 10\ \mathrm{fb}^{-1}$, and $\epsilon = 10\%$.

p_T-cuts(GeV)	$N(\times 10^6)$	$\lvert Im(d_t)\rvert(\times 10^{-20} g_s\mathrm{cm})$
0	2.62	0.766
40	2.36	0.847
80	1.74	1.107

at 1-σ level, although precise limits require more precise theoretical estimates and experimental determination of the polarized gluon distribution.

To summarize, we can measure (i) the real and imaginary parts of the T-odd tCEDM up to the order of 10^{-18} $g_s$cm by the CP-odd lepton energy and tensor correlations of the t and $\bar{t}$ semileptonic decays in unpolarized proton-proton collisions, and (ii) the imaginary part of the tCEDM up to the order of 10^{-20} $g_s$cm through the production rate asymmetry for positively versus negatively polarized protons, assuming that the polarization gets transmitted from the proton to gluons. The large enhancement obtained by the use of proton polarization is certainly worth checking experimentally at the CERN LHC.

Acknowledgments

JL thanks APCTP for offering the opportunity to give this talk. The authors wish to acknowledge financial support from the Korean Research Foundation made in the program year of 1997.

References

1. S.Y. Choi, C.S. Kim and Jake Lee, hep-ph/9706379, to appear in PLB.
2. D. Atwood, A. Aeppli and A. Soni, Phys. Rev. Lett. **69** (1992) 2754.
3. W. Bernreuther et al., Nucl. Phys. **388** (1992) 53 and references therein.
4. J.F. Gunion et al., Phys. Rev. Lett. **71** (1993) 488.
5. E.P. Shabalin, Sov. J. Nucl. Phys. **31**, (1980) 864.
6. A. Soni and R.M. Xu, Phys. Rev. Lett. **69** (1992) 33.
7. M. Glück, E. Reya and A. Vogt, Z. Phys. **C53** (1992) 127.
8. S. Brodsky and I.A. Schmidt, Phys. Lett. **B234** (1990) 144.
9. E.L. Berger and J. Qiu, Phys. Rev. **D40** (1989) 778.

PHYSICS WITH HIGH P_T LEPTONS AT CDF: W ASYMMETRY, HIGH MASS DRELL-YAN AND LIMITS ON QUARK-LEPTON COMPOSITENESS

ARIE BODEK

Department of Physics and Astronomy, University of Rochester
Rochester, N.Y. 14627 - Representing the CDF Collaboration

We describe two measurements using 110 ± 7 pb $^{-1}$ of data collected by the CDF detector during 1992-95. The first is a new measurement of the charge asymmetry in the lepton rapidity range $0 < |y_l| < 2.5$ using $W \to \mu\nu$, $e\nu$ events. The asymmetry data constrain the ratio of d and u quark momentum distributions in the proton over the x range of 0.006 to 0.34 at $Q^2 \approx M_W^2$. Although the data in the central rapidity region ($|y_l| < 1.1$) are reproduced by modern parametrizations of parton distributions, there are differences in the high rapidity regions. The second is a measurement of the dilepton mass spectrum in $p\overline{p} \to l^+l^- + X$ using dielectrons (ee) and dimuons ($\mu\mu$). The dilepton data are consistent with standard model predictions. The mass spectrum, being a probe for new physics, is examined for new interactions of quarks and leptons from a common composite structure. Assuming a contact interaction with the conventional coupling $g_0^2/4\pi = 1$, limits on chiral quark-electron and quark-muon compositeness scales in the range of 2.5 to 4.2 TeV are obtained.

1 Measurement of the W Asymmetry

At Tevatron energies, W^+ (W^-) bosons are produced in $p\overline{p}$ collisions primarily by the annihilation of u (d) quarks in the proton and $\overline{d}$ ($\overline{u}$) quarks from the antiproton. Because u quarks carry on average more momentum than d quarks [1 2 3], the W^+'s tend to follow the direction of the incoming proton and the W^-'s that of the antiproton. The charge asymmetry in the production of W's as a function of rapidity (y_W) is related to the u and d quark distributions at $Q^2 \approx M_W^2$, and is roughly proportional [4 5] to the ratio of the difference and the sum of the quantities $d/u(x_1)$ and $d/u(x_2)$, where x_1 and x_2 are the the fraction of nucleon momentum carried by the quarks in the p and $\overline{p}$, respectively.

Previously published W asymmetry results from an analysis of the 1992-93 data [6] collected by the Collider Detector at Fermilab (CDF) demonstrated this sensitivity for the first time at a level better than that of deep inelastic scattering (DIS) experiments within the x range accessible by the W data at the Tevatron. These data, together with those from DIS experiments, are used in a global analysis to extract parametrizations of parton distribution functions (PDFs) in the nucleon [1 2 3]. In hadron colliders, these functions are used in the calculation of all hadronic cross-sections. CDF's previous asymmetry data [6] improved the understanding of the u and d quark momentum distributions in the proton in the x range of W production ($0.007 < x < 0.24$). The use of these previous data resulted in a reduction in the uncertainty from PDFs in the W mass measurement [7] from ≈ 75 to ≈ 50 MeV/c^2. A further reduction of this error to ≈ 20 MeV/c^2 is possible [7] using the new data presented here.

The previous W asymmetry measurements (with an initial sample of ≈ 20 pb^{-1} from 1992-93) are dominated by statistical uncertainties. In this contribution, we describe the asymmetry results at $\sqrt{s} = 1.8$ TeV using data from the entire 1992-95 period. The new data agree with the 1992-93 results, and increase the integrated luminosity to 110 ± 7 pb^{-1}, a 5 fold increase relative to the 1992-93 data. Furthermore, the asymmetry measurement is extended to larger rapidity ($|y_l| < 2.4$) by the introduction of a new charge determination technique [8] for electrons, and also by using data from the forward muon detector ($1.9 < |y_l| < 2.5$). These electron and muon data at large rapidity provide information about PDFs at a larger range ($0.006 < x < 0.34$) than previously available.

Since the W rapidity is experimentally undetermined because of the unknown longitudinal momentum of the neutrino from the W decay, we measure the lepton charge asymmetry which is a convolution of the W production charge asymmetry and the well known asymmetry from the $V - A$ W decay. The two asymmetries are in opposite direction and tend to cancel at large values of rapidity. However, since the $V - A$ asymmetry is well understood, the lepton asymmetry is still sensitive to the parton distributions. The lepton charge asymmetry is defined as:

$$A(y_l) = \frac{d\sigma^+/dy_l - d\sigma^-/dy_l}{d\sigma^+/dy_l + d\sigma^-/dy_l}, \tag{1}$$

where $d\sigma^+$ ($d\sigma^-$) is the cross section for W^+ (W^-) decay leptons as a function of lepton rapidity, with positive rapidity being defined in the proton beam direction. If the detection efficiencies and acceptances for l^+ and l^- are equal, then the uncertainties from acceptance, efficiencies and luminosity cancel in $A(y_l)$. By CP invariance $A(y_l) = -A(-y_l)$ and the two values are combined.

The tracking detectors at CDF are the Vertex Time Projection Chambers (VTX), the Central Tracking Chamber (CTC, $|\eta| < 1.8$), the Silicon Vertex Detector (SVX, $|\eta| < 2.3$), the Central Muon Chambers, and the Forward Muon Detector (FMU, $1.9 < |\eta| < 3.5$).

The data are divided into five samples: central electrons ($|y_l| < 1.1$), central muons ($|y_l| < 1.0$), forward muons($1.9 < |y_l| < 2.5$), plug electrons within the SVX fiducial region (plug-SVX, $1.1 < |y_l| < 2.4$), and plug electrons outside the SVX geometrical acceptance region but with a CTC track (plug-CTC, $1.1 < |y_l| < 1.8$). A reliable charge determination and a good energy measurement are essential for this analysis. Central electrons, central muons, and plug electrons outside the SVX are required to have an associated CTC track.

A new charge determination technique[8] is introduced to identify the charge of plug electrons within the SVX geometrical fiducial region. At CDF, positively (negatively) charged particles are bent in increasing (decreasing) ϕ inside the solenoidal magnetic field. The SVX track stub measures the precise initial track direction in ϕ. The plug EM calorimeter (PEM) shower centroid measurement combined with the location of the vertex position yields another measurement of ϕ, and the sign of the difference of the two ϕ measurements determines the charge.

The transverse energy (E_T) of the lepton and the missing transverse energy ($\not{E}_T$) are required to be greater than 25 GeV. To further reduce the QCD background from dijet events where one jet may be misidentified as a lepton, events with a extra jet with $E_T^J > 30$ GeV are rejected.

Figure 1 shows the fully corrected asymmetry after taking the weighted mean of the various data sets and combining the bins for positive and negative y_l. Also shown are the predictions of Quantum Chromodynamics[9] (QCD) calculated to Next-to-Leading-Order (NLO) using the program DYRAD[9] with several parametrizations of parton distributions as input. Note that these PDFs (MRSA[1],CTEQ3M[2] and GRV94[3]) have been extracted with the inclusion of the 1992-93 asymmetry data in the global fit. As shown in Figure 1, these PDFs are in good agreement with the new data in the central region ($|y_l| < 1.1$). However, at high rapidity, the various predictions are generally higher than the data, indicating that the PDF parametrizations should be modified in the range $0.006 < x < 0.34$.

Although the W decay lepton charge asymmetry is mostly sensitive to the d/u ratio, at large rapidity it is also affected by the W production P_t spectrum. The NLO QCD calculations do not reproduce the production P_t spectrum of W's at very low P_t. Therefore, it is necessary to check how well the NLO QCD calculation compares to an independent calculation. The NLO QCD prediction is compared to a calculation including soft gluon resummation at all orders in perturbation theory implemented in the program ResBos[10]. Both programs yield identical results for the W rapidity distribution, but different results for the W P_t distribution. At Tevatron energies soft gluon radiation is mainly responsible for the P_T of W's and Z's in the range of $P_T < 30$ GeV, and ResBos can reproduce the P_T spectrum of Z's. Figure 1 also shows the comparison between predictions from DYRAD NLO and ResBos for the

CTEQ3M[2] structure functions. The two calculations agree very well in the central region ($|y_l| < 1.1$), which is the region used in the W mass determination[7]. The difference between two calculations is mainly at $|y_l| > 1.7$. The disagreement between our data and the ResBos calculation is even bigger at high lepton rapidity than the disagreement of the data with the NLO DYRAD calculation.

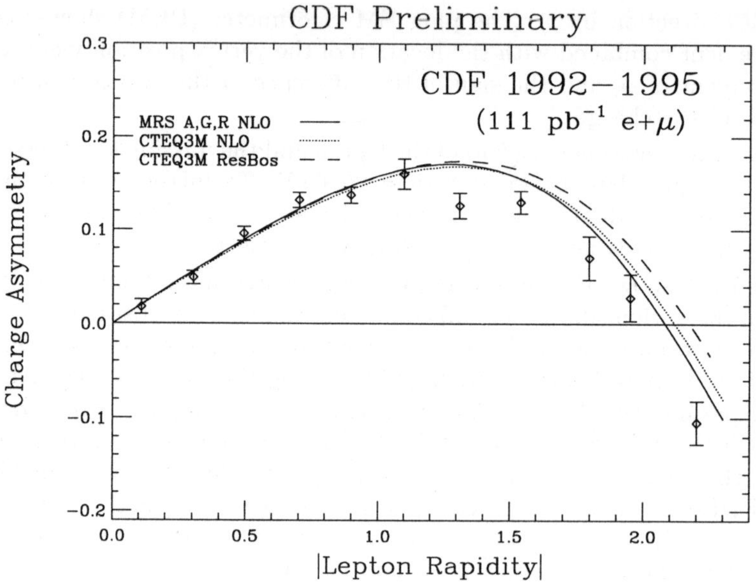

Figure 1: The fully corrected charge asymmetry. Data from all the detectors for positive and negative y_l are combined. The statistical and systematic errors are added in quadrature.

In summary, the significant reduction in the errors of CDF's new W asymmetry data in the central region can be used to reduce the uncertainties from PDFs to measurements of the W mass using collider data[7]. The asymmetry data demonstrate the value of collider data in the measurement of proton structure functions and place the strongest constraint on the d/u ratio of quark momentum distributions in the proton over the range of 0.006 to 0.34 at $Q^2 \approx M_W^2$.

2 High Mass Drell-Yan and Limits on Quark Lepton Compositeness

In hadron-hadron collisions at high energies, massive lepton pairs are produced via the Drell-Yan process, where pointlike quarks and antiquarks annihilate to form the dileptons. The dimuon and dielectron invariant mass spectra of $p\bar{p}$ collisions at a center-of-momentum energy of 1.8 TeV are from 110 pb^{-1} of collisions taken by the Collider Detector at Fermilab (CDF) during 1992-95. They are used to measure the shape of the cross section and to search for continuum anomalies from new constituent interactions of quarks and leptons with a common composite structure.

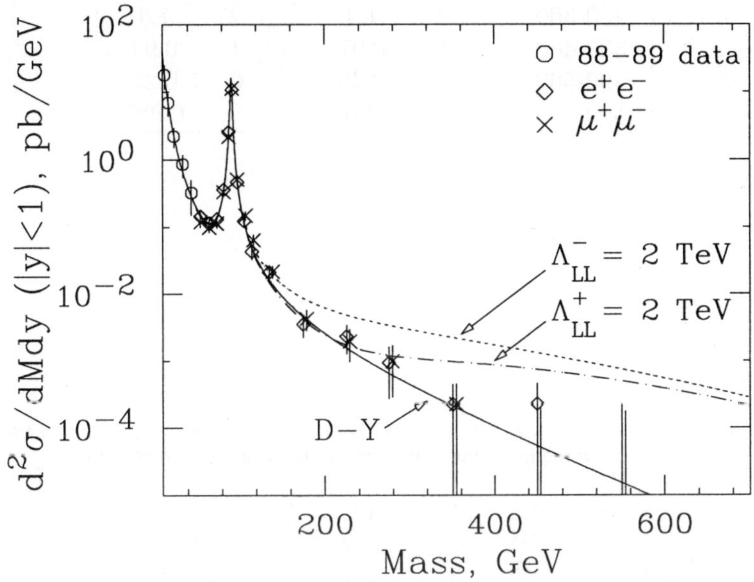

Figure 2: $d^2\sigma/dMdy$, $|y| < 1$ for $p\bar{p} \to l^+l^- + X$. The circles ($M < 50$ GeV) are from earlier CDF data; the diamonds and crosses are the dielectron and dimuon data, repectively, normalized from 50 to 150 GeV to the standard model value. Above 110 GeV, the data symbols are displaced for clarity. The D-Y curve is a standard model NLL calculation. Superimposed are LL model calculations with $\Lambda = 2$ TeV.

At collision energies far below the mass scale of new physics, the new physics can be described by an effective four-fermion contact interaction [11]. To establish compositeness scales, any deviation of the data from a prediction based on just the standard model is assumed to be due to composite fermions. The compositeness scale is defined using $g_0^2/4\pi = 1$. Each channel of $\mathcal{L}_{EQ}$,

Table 1: The dilepton event samples. The SM columns are the standard model event predictions normalized to the data over the range 50-150 GeV. There are no events above 500 GeV.

Mass Bin	Dielectrons		Dimuons	
(GeV)	Data	SM	Data	SM
50-150	2581	$\equiv 2581$	2533	$\equiv 2533$
150-200	8	10.8	9	9.7
200-250	5	3.5	4	3.2
250-300	2	1.4	2	1.3
300-400	1	0.97	1	0.94
400-500	1	0.25	0	0.27
500-600	0	0.069	0	0.087

Table 2: The one-sided 95% confidence level lower limits of the compositeness scales. The last columns give the combined electron and muon results.

Model	Λ^+_{qe}	Λ^-_{qe}	$\Lambda^+_{q\mu}$	$\Lambda^-_{q\mu}$	Λ^+	Λ^-
ij	TeV	TeV	TeV	TeV	TeV	TeV
LL	2.5	3.7	2.9	4.2	3.1	4.3
LR	2.8	3.3	3.1	3.7	3.3	3.9
RL	2.9	3.2	3.2	3.5	3.3	3.7
RR	2.6	3.6	2.9	4.0	3.0	4.2
VV	3.5	5.2	4.2	6.0	5.0	6.3
AA	3.8	4.8	4.2	5.4	4.5	5.6
SC	2.9	2.9	3.1	3.1	3.3	3.3

LL, LR, RL, RR, and SC, is tested one at a time.

Figure 2 shows the measured dilepton cross section compared against calculations. The observed numbers of events for $M > 150$ GeV in Table 1 are compared against model predictions using a binned likelihood technique. The lower limits on the quark-electron (qe), quark-muon ($q\mu$) compositeness scales, and quark-lepton scales assuming lepton universality (combined likelihood) are given in Table 2.

In conclusion, this search for quark-lepton substructure in $p\bar{p} \to e^+e^-, \mu^+\mu^- + X$ interactions finds no significant deviation from the standard model. The contact interaction analysis yields improved limits on the qe and $q\mu$ compositeness scales.

References

1. A.D.Martin *et al.*, Phys. Rev. D **50**, 6734 (1994).
2. H.L.Lai *et al.*, Phys. Rev. D **51**, 4763 (1995).
3. M.Gluck *et al.*, Z. Phys. **C67**, 433-448 (1995).
4. E.L.Berger *et al.*, Phys. Rev. D **40**,83 (1989).
5. A.D.Martin *et al.*, Mod. Phys. Lett. A **4**, 1135 (1989).
6. F. Abe *et al.*, Phys Rev. Lett. **74**, 850 (1995).
7. F. Abe *et al.*, Phys Rev. Lett. **75**, 11 (1995); M. Lancaster, Proceedings of "Les Rencontres de Physique de la Vallee d'Aoste", La Thuile, March 1997, ed. M. Greco.
8. Q.Fan and A.Bodek, Proceedings of VIth International Conference on Calorimetry in High Energy Physics, June 8-14, 1996, Frascati (Rome), Italy, p553.
9. W. Giele *et al.*, Nucl. Phys. **B403**, 633 (1993).
10. C. Balazs *et al.*, Phys.Lett. **B355** 548(1995).
11. E.J. Eichten, K.D. Lane, M.E. Peskin, Phys. Rev. Lett. **50**, 811 (1983); E. Eichten, I. Hinchliffe, K. Lane, C. Quigg, Rev. Mod. Phys. **56**, 579 (1984); T. Lee, Phys. Rev. D **55**, 2591 (1997).

LIGHT SPARTICLE PRODUCTION AS A PROBE OF THE SQUARK MASS SCALE

MIHOKO. M. NOJIRI[a]
YITP, Kyoto University, Kyoto, 606-8502, Japan

Radiative corrections to supersymmetric relations between couplings, and their measurement through the study of sneutrino or chargino production are discussed.

1 Radiative Correction to Supersymmetric Relations

Supersymmetry is an attractive possibility to extend the Standard Model. Because of the relations supersymmetry imposes among dimensionless couplings, quadratic divergences in the Higgs sector are cut off by the superpartner mass scale. These cancellations stabilize the hierarchy between the Planck scale and the weak scale. The minimal supersymmetric standard model (MSSM) is consistent with gauge coupling unification suggested by grand unified theories. Also, it is interesting that one of the hallmarks of supersymmetry, a light Higgs boson ($m_h \lesssim 130$ GeV), is favored by global fits to precision electroweak data.

When a sparticle is found at future experiments such as LHC, a proposed e^+e^- colliders, or a μ colliders, precision measurements of the new particle's mass, event rates, and branching ratios will provide detailed tests of the supersymmetry hypothesis. Recently it has been realized that a hard supersymmetric relation between gaugino and gauge couplings can be measured rather precisely at future e^+e^- colliders.[3,4,1,2] On the other hand, this relation receives corrections proportional to the log of the mass ratio between particles and their superpartners.[5] Measurements of the coupling relation therefore elucidate important features of the scale and pattern of supersymmetry breaking.

For example, (s)quark contributions to the splitting of the $SU(2)$ gaugino/ gauge–(s)lepton couplings grow logarithmically if squarks are much heavier than sleptons. This correction can be estimated by evolving the couplings according to the renormalization group equations (RGE's) of the effective theory[5] below the squark mass threshold. When $M_{\tilde{Q}}/m_{\tilde{\ell}} \simeq 10$ the correction to the $SU(2)$ coupling is about 2%. This gives rise to an enhancement of the square of t-channel amplitudes in slepton or gaugino production of about 8%. Note that large data samples will be available for the processes $e^+e^- \rightarrow \tilde{\nu}\tilde{\nu}^*$ and $\tilde{\chi}_1^+\tilde{\chi}_1^-$. Therefore, if the relevant tree level parameters are constrained and

[a]e-mail: nojiri@yukawa.kyoto-u.ac.jp. This work is based on a collaboration with S. Kiyoura, D. Pierce and Y. Yamada.[1,2] This work is supported in part by a Grant in aid for Science and Culture of Japan (07640428, 09246232)

systematic errors can be controlled, we can constrain the squark mass scale through measurements of these production cross sections.

Notice that it is important to calculate the full one–loop amplitude, because in lowest order the cross section depends on the definition of the tree level parameters. In the $\overline{\rm DR}$ scheme, changing the renormalization scale by a factor of 2 easily leads to a few percent change in the predicted value of the tree level cross section, about the size of the correction of interest to us. Such an ambiguity can be curtailed only by including radiative corrections. Non–logarithmic corrections may be important if squarks are light. We have calculated full one–loop cross sections for slepton and chargino pair production within the MSSM,[1,2] including quark and squark loops. We refer interested readers to those papers for further details.

2 Measurement of Corrections to the SUSY Coupling Relation

In this section, we discuss how accurately we can measure the first generation lepton–slepton–gaugino coupling $g_2^{\tilde\chi} \equiv g_{e\tilde\nu\widetilde W}$ at future e^-e^+ linear colliders through the study of sneutrino or chargino pair productions.[4,1,2]

2.1 Sneutrino Production

$\tilde\nu$ pair production followed by its decay into $\tilde\chi_1^\pm e^\mp$ were studied in MC simulations by Baer *et al.*.[6] The decay modes $\tilde\nu_e\tilde\nu_e{}^* \to e^-e^+\tilde\chi_1^+\tilde\chi_1^- \to e^-e^+\mu 2j$ and e^+e^-4j are (nearly) free from SM and SUSY backgrounds for the parameter point they studied. The chosen parameter set corresponds to $(m_{\tilde\nu_e}, m_{\tilde\chi_1^+}, m_{\tilde\chi_1^0})$ = (207 GeV, 96 GeV, 45 GeV), and the lightest chargino and neutralinos are gaugino–like. Their study suggests that we can take $m_{\tilde\chi_1^+}$, $m_{\tilde\chi_1^0}$, and $m_{\tilde\nu_e}$ as well constrained input parameters, giving a good handle on the tree level parameters needed to calculate the cross section. Therefore we consider their sample point and study how strongly the universal squark mass $M_{\widetilde Q}$ would be constrained through the measurement of $\sigma(e^-e^+ \to \tilde\nu\tilde\nu^*)$, which yields $g_2^{\tilde\chi}$.

We first provide a feel for the sensitivity to the squark mass scale and $\tan\beta$ in the ideal case where we can ignore the slepton and chargino mass uncertainties. In Fig. 1 we show the statistical significance of the loop corrections by plotting contours of constant cross section. Here we fix the sneutrino mass and determine μ, M_1 and M_2 by fixing the one-loop corrected masses $m_{\tilde\chi_1^0}$, $m_{\tilde\chi_3^0}$, and $m_{\tilde\chi_1^+}$. We plot contours corresponding to the number of standard deviations of the fluctuation of the accepted number of events. A 1 σ fluctuation corresponds to $\sqrt{N_{\rm input}}$, where $N_{\rm input} \sim 12800$ is our nominal value of the number of events at $M_{\tilde{Q}} = 1000$ GeV and $\tan\beta(M_Z) = 4$. The accepted number of events N is estimated by multiplying overall constant acceptance to the number of events where both of sneutrino decay into $e^{\mp}\chi_1^{\pm}$.

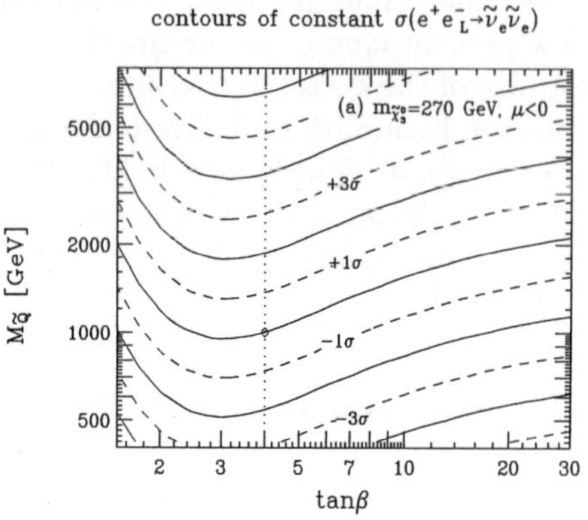

Figure 1: The constraint on $M_{\tilde{Q}}$ and $\tan\beta$ coming from $\sigma(e^-e^+ \to \tilde{\nu}_e\tilde{\nu}_e^* \to e^-e^+\tilde{\chi}_1^+\tilde{\chi}_1^-)$, with $\int dtL = 100\ \mathrm{fb}^{-1}$. The central value is taken as $M_{\tilde{Q}}$=1 TeV and $\tan\beta(M_Z) = 4$. $m_{\tilde{\chi}_3^0} = 270$ GeV, $\mu < 0$.

Here we show contours for $\mu < 0$. If $\tan\beta$ is well measured, $M_{\tilde{Q}}$ is constrained to $M_{\tilde{Q}} = 1000^{+370}_{-280}$ GeV. If instead we assume the constraint $2 < \tan\beta < 8$, the mild $\tan\beta$ dependence yields $700 < M_{\tilde{Q}} < 1900$ GeV.[b] Note that the sneutrino mass uncertainty reduces the sensitivity of the production cross section to $M_{\tilde{Q}}$, because the effect of increasing $M_{\tilde{Q}}$ can be compensated by a small increase in $m_{\tilde{\nu}_e}$. However, even with the expected uncertainty in sneutrino mass, we can reasonably constrain the squark mass scale for this case. At $M_{\tilde{Q}} = 1$ TeV, we constrain $M_{\tilde{Q}}$ to $1^{+1.2}_{-0.5}$ TeV, assuming naive statistical scaling of the mass error given by Baer *et al.*[6] (from 20 fb^{-1} to 100 fb^{-1}). This corresponds to the measurement of the gaugino coupling with an error of 0.6% Notice that the mass error is estimated by using the $e^+e^-\mu 2j$ mode. If we can reduce the mass uncertainties by a factor of 2 by using kinematical distributions of the other decay modes of $\tilde{\nu}\tilde{\nu}^*$, we find the very interesting constraint 600 GeV$< M_{\tilde{Q}} <$ 1500 GeV.

[b]In the case $\mu > 0$, chargino mixing gives rise to a significant $\tan\beta$ dependence.

2.2 Chargino Production

Chen *et al.* considered probing the $SU(2)$ supersymmetric relation of gauge/gaugino couplings $g_2 = g_2^{\tilde{\chi}}$ using chargino production, based on a MC study of the $\ell 2j$ model at the point in parameter space where $(m_{\tilde{\chi}_1^+}, m_{\tilde{\chi}_1^0}) = (172$ GeV, 86 GeV).[4] $\delta g_2^{\tilde{\chi}}/g_2 = 2\%$ was estimated for $(m_{\tilde{\nu}}, \sqrt{s}) = (240$ GeV, 400 GeV) based on MC results at $\sqrt{s} = 500$ GeV.[7] This result is considerably poorer than the result $\delta g_2^{\tilde{\chi}}/g_2 < 0.6\%$, estimated in the previous subsection.

By studying their case carefully, we found the poor result can be traced back to the choice of their (special) parameter point: the mass difference between the parent and the daughter particles $\Delta m = m_{\tilde{\chi}_1^+} - m_{\tilde{\chi}_1^0} - m_W$ is only 6 GeV for the point. When the charginos are produced at $\sqrt{s} = 500$ GeV, W bosons from $\tilde{\chi}_1^{\pm}$ decay are strongly boosted in the $\tilde{\chi}_1^{\pm}$ direction, and the final W-boson pair momenta are roughly balanced. On the other hand, we are relying on a large $\tilde{\chi}W$ angle $\theta_{\tilde{\chi}^+W}$ and large missing transverse momentum to separate the signal from the WW background. As a result, the acceptance η of the signal turns out to be small (11.9%) for the standard set of cuts ($\not{p}_T > 35$ GeV, $\theta_{acop} > 30°$, $m_{\ell\nu_{ISR}} > 120$ GeV), and $S/N = 1$. This must be contrasted with other MC studies which generally claim acceptances of 30 to 50% for SUSY signals.

Moreover, the systematic error on η, when estimated by allowing $m_{\tilde{\chi}_1^+}$, $m_{\tilde{\chi}_2^+}$, and $m_{\tilde{\chi}_1^0}$ to vary within 2 GeV, turns out to be large, $\Delta\eta/\eta = 5\%$. This is comparable to the change in the cross section when $m_{\tilde{q}}$ is changed from 1 TeV to 4.5 TeV at $(m_{\tilde{\nu}}, \sqrt{s})$ =(240 GeV, 400 GeV). This large error is a direct consequence of the small acceptance described above. Notice that the signal event distribution in the ($\not{p}_T$, $\theta_{\rm acop}$, $m_{\ell\nu_{ISR}}$) space sits mostly outside the pre–defined "signal region". When the input parameters are changed slightly within their error, the signal distribution in this space also changes, which can easily change the acceptance by several percent. Similarly, $\theta_{\tilde{\chi}_1^+W}$ is a rather sensitive function of Δm when Δm is small. Note that the above estimate of systematic error is obtained by changing Δm from 2 to 10 GeV!

To illustrate the kinematics, we show the acoplanarity angle distribution of W pairs reconstructed from $e^+e^- \to \tilde{\chi}_1^+\tilde{\chi}_1^- \to W^+W^-\tilde{\chi}_1^0\tilde{\chi}_1^0 \to 4j + \not{p}_T$ events in Fig. 2(left). The distribution shown by points with error bars corresponds to our standard input parameters, ($m_{\tilde{\chi}_1^+}$, $m_{\tilde{\chi}_1^0}$)=(176.6 GeV, 86.9 GeV). On this plot we also show two distributions corresponding to $m_{\tilde{\chi}_1^0} = 92.2$ GeV and $m_{\tilde{\chi}_1^0} = 80.3$ GeV, respectively. The acceptance changes drastically for these different cases, primarily because the distribution is more and more peaked at smaller $\theta_{\rm acop}$ region as Δm is reduced. Implementing $\theta_{\rm acop} > 30°$, we find that

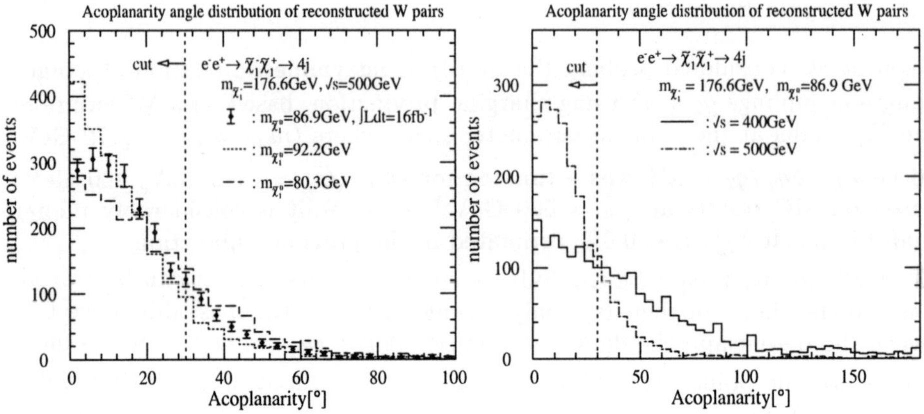

Figure 2: (left): The number of accepted chargino pair events vs. the acoplanarity angle, with $\sqrt{s} = 500$ GeV and $\int \mathcal{L}dt = 16$ fb^{-1}. The points with error bars are for $M_1 = 84.6$ GeV, and the dashed (dot–dashed) line corresponds to $M_1 = 90.6$ (78.6) GeV. See text for other parameters. (right): The acoplanarity angle distribution of W-pair events from chargino production, at $\sqrt{s} = 400$ GeV.

the acceptance varies by a factor of two. For an error in $m_{\tilde{\chi}_1^0}$ of 2%, we expect $\Delta\eta/\eta$ of 10%. If Δm were large, most events would be accepted regardless of a few percent variation in the input parameters. Therefore, the acceptance error is much smaller in a generic region of parameter space. We expect the acceptance uncertainty to scale roughly inversely with the acceptance, for sufficiently large Δm. The variation of Δm is also responsible for the change in the distribution for the $\ell 2j$ mode, although the situation is more complicated because many cuts are involved to reduce the background.

The uncertainty of the acceptance may be reduced by increasing the acceptance itself. In the $4j$ mode, one can increase the acceptance easily by reducing $\sqrt{s}$. In Fig. 2(right) we see the acoplanarity angle distribution of W-pairs is much flatter for $\sqrt{s} = 400$ GeV. We find the acceptance increases to 54.8% from 19.2% at our standard point and the uncertainty in the acceptance from the chargino/neutralino mass errors is reduced by factor of 7 relative to the $\sqrt{s} = 500$ GeV case. However, it should be noted that reducing $\sqrt{s}$ may not result in a larger acceptance for the $\ell 2j$ mode, because the relevant distributions of the signal and the background might have different $\sqrt{s}$ dependencies. Optimization of the cuts requires a careful study of signal and background distributions.

In summary, the previously claimed cross section error due to the uncertainty in the acceptance should not be taken as a generic statement. The point studied has very special kinematic properties making it a very pessimistic case. Because the error in the acceptance scales inversely with the acceptance, the acceptance uncertainty may be minimized by changing the beam energy and the cuts so as to maximize the acceptance. One should always try to find the best possible way to increase the acceptance, not only in order to increase the statistics, but also to reduce the systematic error.

3 Conclusions

The calculation of higher loop effects in the SM, together with the recent precision measurements of electroweak parameters, has given rise to a wealth of information on physics at the weak scale and above. Since particle masses in the SM break the gauge symmetry, those measurements lead to constraints on these masses, in particular on the top and Higgs mass.

Supersymmetry is another beautiful symmetry which might be realized in Nature. This symmetry is broken just like the electroweak gauge symmetry in the SM, and corrections to supersymmetric relations could be sizable. In section 1, we saw as an example that corrections will be large if $m_{\tilde{q}}$ is much larger than other sparticle masses. In section 2, we concentrated on the measurement of $SU(2)$ gauge/gaugino coupling relations through the study of sneutrino and chargino production at future e^+e^- colliders. We found both of the measurements should be very promising if their production cross sections are large. For the case of chargino production, optimizing the beam energy and acceptance cuts might greatly improve the measurements. We also note that information on squark masses from other colliders, such as the LHC, would be very important, because the measurement of the production cross sections at e^+e^- colliders would constrain supersymmetry breaking physics beyond the MSSM if squark masses are known independently.

References

1. M.M. Nojiri, D.M. Pierce and Y. Yamada, Phys. Rev. D**57** 1539 (1998).
2. S .Kiyoura, M.M. Nojiri, D.M. Pierce and Y. Yamada, in preparation.
3. M.M. Nojiri, K. Fujii and T. Tsukamoto, Phys. Rev. D **54**, 6756 (1996).
4. H.–C. Cheng, J.L. Feng and N. Polonsky, Phys. Rev. D **56** 6875 (1997); Phys. Rev. D **57**, 152 (1998).
5. P.H. Chankowski, Phys. Rev. D **41**, 2877 (1990).
6. H. Baer, R. Munroe and X. Tata, Phys. Rev. D **54**, 6735 (1996).
7. J.L. Feng, M.E. Peskin, H. Murayama and X. Tata, Phys. Rev. D **52**, 1418 (1995).

CHARGINO PAIR PRODUCTION AT CERN LEP II

S.Y. CHOI

Department of Physics, Yonsei University, Seoul 120-749, Korea

Charginos are expected to be the lightest observable supersymmetric particles in many supersymmetric models. If lighter charginos are pair–produced at CERN LEP II, we present a straightforward procedure for determining the SUSY parameters, $\tan\beta$, M_2, μ, and the electron sneutrino mass $m_{\tilde{\nu}_e}$ up to a four–fold discrete ambiguity, although a large number of unknown SUSY parameters is involved in chargino decays.

One of the main goals of the LEP II e^+e^- collider at CERN[1] is to search for weak–scale supersymmetry (SUSY)[2]. Among the many SUSY particles, the chargino, a mixture of the W–ino and charged Higgsinos, is of particular interest, since it is expected to be lighter than gluinos and most sfermions. For this reason, chargino searches have been well studied. The current lower bound on the lighter chargino mass is about 90 GeV[3]. Chargino discovery studies have shown that the discovery reach will extend nearly to the LEP II kinematical limit. Moreover, if kinematically accessible, charginos have a large cross section throughout SUSY parameter space and produce a clean signal. As the chargino pair production cross section rises rapidly above threshold, each step in collider energy holds the promise not only of chargino discovery, but also of detailed SUSY studies from chargino events.

The purpose of the present work is to estimate the capability of LEP II to determine the parameters of SUSY in lighter chargino pair production in the Minimal Supersymmetric Standard Model (MSSM). This issue has been studied by Leike[4], Diaz and King[5] and recently by Feng and Strassler[6]. Our approach here is to employ the most dominant chargino decay modes to extract the maximal amount of information on chargino polarizations, while avoiding theoretical assumptions at high energy scales, and exploiting the fact that chargino pair production only depends on a small subset of SUSY parameters. Although chargino decays are very complicated with many diagrams involved, certain final state angular correlations, which are experimentally identifiable, allow us to construct three additional observables besides the production cross section, so that all SUSY parameters relevant to chargino pair production are determined up to a discrete ambiguity.

The MSSM includes the usual matter superfields and two Higgs doublet superfields $\hat{H}_1$ and $\hat{H}_2$, which give masses to the isospin $-\frac{1}{2}$ and $+\frac{1}{2}$ fields, respectively. These two superfields are coupled in the superpotential through the term $-\mu\epsilon_{ij}\hat{H}_1^i\hat{H}_2^j$, where μ is the supersymmetric Higgs boson mass parameter. The ratio of the two Higgs scalar vacuum expectation values is defined to be $\tan\beta \equiv \langle H_2^0\rangle/\langle H_1^0\rangle$. There are two chargino mass eigenstates that result from the mixing of the electroweak gauginos $\tilde{W}^\pm$ with the Higgsinos due to spontaneous electroweak symmetry breaking. The chargino mass term is

$$-\mathcal{L}_m = \left(\tilde{W}_R^- \tilde{H}_{2R}^-\right) \begin{pmatrix} M_2 & \sqrt{2}m_W \cos\beta \\ \sqrt{2}m_W \sin\beta & \mu \end{pmatrix} \begin{pmatrix} \tilde{W}_L^- \\ \tilde{H}_{1L}^- \end{pmatrix} + \text{h.c.}. \tag{1}$$

We assume that the gaugino masses M_i $(i = 1, 2)$ and the parameters μ and $\tan\beta$ are real so that CP violation plays no role in chargino events. Then the 2×2 chargino mass matrix can be diagonalized by two orthogonal matrices O_L and O_R defined by two rotation angles ϕ_L and ϕ_R, respectively, which determine the couplings between the gauge boson Z and the charginos.

In general, studying MSSM is complicated because of the large number of free parameters, so that we make the following assumptions: (i) R parity is conserved, (ii) the lightest supersymmetric particle (LSP) is the lightest neutralino $\tilde{\chi}_1^0$, (iii) sfermions have masses beyond the LEP II kinematical limit, and (iv) intergenerational mixing in the sfermion and quark sectors is small and may be neglected. Under our assumptions, for almost all values of parameters, charginos decay to three–body final states consisting of an LSP and either two quarks or two leptons at LEP II c.m. energy.

One typical observable which is crucial to our analysis is the chargino mass $m_{\tilde{\chi}_1^\pm}$. The mass $m_{\tilde{\chi}_1^\pm}$ along with the LSP mass $m_{\tilde{\chi}_1^0}$ can be measured by the dijet energy distribution in hadronic chargino decays[a]. The end points of the dijet energy E_{jj} and mass spectra are completely determined by $m_{\tilde{\chi}_1^\pm}$ and $m_{\tilde{\chi}_1^0}$, with the maximal and minimal dijet energies, E_{max} and E_{min}, as:

$$m_{\tilde{\chi}_1^\pm} = \frac{\sqrt{E_{max}E_{min}}}{E_{max} + E_{min}}\sqrt{s}, \qquad m_{\tilde{\chi}_1^0} = \sqrt{1 - 2\frac{(E_{max} + E_{min})}{\sqrt{s}}}. \tag{2}$$

The distribution of the invariant mass m_{jj} is between zero and $m_{\tilde{\chi}_1^\pm} - m_{\tilde{\chi}_1^0}$. If at least two of the three end points are sufficiently sharp to be well measured, they can be used to precisely determine $m_{\tilde{\chi}_1^\pm}$ and $m_{\tilde{\chi}_1^0}$.

The amplitude for producing charginos which decay to a certain final state does not factorize into production and decay amplitudes due to the fact that charginos are spin-$\frac{1}{2}$ objects; the angular correlations of chargino decay products are affected by the underlying structure of the production processes. In

[a] An alternate determination of $m_{\tilde{\chi}_1^\pm}$ can be provided by an energy scan at the chargino production threshold[4].

the present work, we will consider the production process $e^+e^- \to \tilde{\chi}_1^-\tilde{\chi}_1^+$ followed by the decays $\tilde{\chi}_1^- \to \tilde{\chi}_1^0 f_1\bar{f}_2$ and $\tilde{\chi}_1^+ \to \tilde{\chi}_1^0 f_3\bar{f}_4$ where the f_i are quarks or leptons. We can then write the amplitude for the sequential process in the narrow width approximation as a product of the production helicity amplitudes $T_{\sigma;\lambda\bar{\lambda}}$ and two decay helicity amplitudes $\mathcal{D}_\lambda$ and $\bar{\mathcal{D}}_{\bar{\lambda}}$ for the negative and positive charginos; here $\sigma = \pm$ is the electron helicity, and $\lambda, \bar{\lambda}$ are the negative and positive chargino helicities. Neglecting fermion masses, $\tilde{\chi}_1^- \to \tilde{\chi}_1^0 f_1\bar{f}_2$ includes two sfermion exchange diagrams and one W−boson exchange. The decay amplitudes only depend on the invariant mass of two final fermions, because sfermions exchange can be well approximated by point propagators[6] under the assumption that sfermions have masses beyond the LEP II kinematical limit.

In order to evaluate the chargino decay helicity amplitudes, we take the chargino rest frames and introduce the angular variables (θ^*, ϕ^*) and $(\bar{\theta}^*, \bar{\phi}^*)$ for the two fermion systems produced in negative and positive chargino decay, respectively. Keeping only the angular dependence and integrating the decay distributions over the invariant masses q_W^2 and $\bar{q}_W^2$, we find that when the $f_3\bar{f}_4$ system is charge–conjugate to the $f_1\bar{f}_2$ system, the decays for polarized negative and positive charginos are described by the decay density matrices

$$\rho_{\lambda\lambda'} = \frac{\mathcal{D}_\lambda \mathcal{D}^*_{\lambda'}}{\sum_\lambda \mathcal{D}_\lambda \mathcal{D}^*_\lambda} = \frac{1}{2}\begin{pmatrix} 1+\kappa\cos\theta^* & \kappa\sin\theta^* e^{i\phi^*} \\ \kappa\sin\theta^* e^{-i\phi^*} & 1-\kappa\cos\theta^* \end{pmatrix},$$
$$\bar{\rho}_{\bar{\lambda}\bar{\lambda}'} = \frac{\bar{\mathcal{D}}_{\bar{\lambda}} \bar{\mathcal{D}}^*_{\bar{\lambda}'}}{\sum_{\bar{\lambda}} \bar{\mathcal{D}}_{\bar{\lambda}} \bar{\mathcal{D}}^*_{\bar{\lambda}}} = \frac{1}{2}\begin{pmatrix} 1-\kappa\cos\bar{\theta}^* & \kappa\sin\bar{\theta}^* e^{i\bar{\phi}^*} \\ \kappa\sin\bar{\theta}^* e^{-i\bar{\phi}^*} & 1-\kappa\cos\bar{\theta}^* \end{pmatrix}, \qquad (3)$$

respectively. The parameter κ, which determines the angular dependence of the decay distributions, is a function of chargino and neutralino mixing parameters, chargino and neutralino masses, and sfermion masses, so that it can take a very wide range of values. Therefore, we consider κ as a phenomenological parameter in our analysis, which is not fixed.

Combining the production and decay amplitudes yields a five–dimensional differential cross section consisting of a kinematical factor and an angle–dependent part $\Sigma(\Theta; \theta^*, \phi^*; \bar{\theta}^*, \bar{\phi}^*)$, where Θ is the $\tilde{\chi}_1^-$ production angle. The angular dependence Σ for the case that the two fermion systems are charge–conjugate to each other is decomposed into eight independent parts:

$$\begin{aligned}\Sigma &= \Sigma_{\rm unpol} + (\cos\theta^* + \cos\bar{\theta}^*)\kappa P + \cos\theta^*\cos\bar{\theta}^*\kappa^2 Q \\ &+ (\sin\bar{\theta}^*\cos\bar{\phi}^* - \sin\theta^*\cos\phi^*)\kappa W \\ &+ (\cos\theta^*\sin\bar{\theta}^*\cos\bar{\phi}^* - \cos\bar{\theta}^*\sin\theta^*\cos\phi^*)\kappa^2 X \\ &+ \sin\theta^*\sin\bar{\theta}^*\cos(\phi^* + \bar{\phi}^*)\kappa^2 Y + \sin\theta^*\sin\bar{\theta}^*\cos(\phi^* - \bar{\phi}^*)\kappa^2 Z, \qquad (4)\end{aligned}$$

where the eight distribution functions describing chargino production are defined in terms of the production helicity amplitudes $T_{\sigma;\lambda\bar{\lambda}}$ (see Ref. 7 for the definition of these eight functions).

Certainly not all of the distribution functions are experimentally measurable. Regarding the reconstruction of chargino production and decay, we have to take into account the following aspects. First of all, because the LSP is not detected, there exists at least a two–fold discrete ambiguity in determining the scattering angle Θ for the hadronic decay modes of the negative and positive charginos. We therefore only consider distributions integrated over the scattering angle Θ. Nevertheless, $\cos\theta^*$, $\cos\bar{\theta}^*$ and $\sin\theta^*\sin\bar{\theta}^*\cos(\phi^* - \bar{\phi}^*)$ are measurable experimentally through the relations

$$\cos\theta^* = \frac{1}{\beta q^*}\left(\frac{E}{\gamma} - E^*\right), \qquad \cos\bar{\theta}^* = \frac{1}{\beta\bar{q}^*}\left(\frac{\bar{E}}{\gamma} - \bar{E}^*\right),$$
$$\sin\theta^*\sin\bar{\theta}^*\cos(\phi^* - \bar{\phi}^*) = \cos\vartheta - \frac{\left(E - \frac{E^*}{\gamma}\right)\left(\bar{E} - \frac{\bar{E}^*}{\gamma}\right)}{\beta^2 q\bar{q}}, \tag{5}$$

where (E, q) and $(\bar{E}, \bar{q})$ are the energy and absolute momentum of the two fermion systems in the laboratory frame, $\gamma = \sqrt{s}/2m_{\tilde{\chi}_1^\pm}$, ϑ is the angle between the two hadronic systems, and the superscript $*$ denotes energies and momenta in the chargino rest frames. Secondly, P is parity–odd and C–odd, so that its determination requires charge identification of charginos. This can be accomplished in the mixed mode of the leptonic and hadronic decays of the negative and positive charginos, where only the lepton charge needs to be identified. On the other hand, the C–even Q and Z do not require charge identification, so that the dominant hadronic decay mode for both negative and positive charginos can be used. Therefore, if the total cross section σ_{tot} and the total leptonic and hadronic branching ratios of the chargino decays are experimentally determined, we can have three additional independent observables integrated over the scattering angle Θ: $\kappa\langle P\rangle$, $\kappa^2\langle Q\rangle$, and $\kappa^2\langle Z\rangle$. However, the parameter κ is not known so that it should be factored out by taking ratios of those observables. Consequently, there are four reconstructible quantities:

$$m_{\tilde{\chi}_1^\pm}, \qquad \sigma_{tot}(e^+e^- \to \tilde{\chi}_1^+\tilde{\chi}_1^-), \qquad \frac{\langle P\rangle^2}{\langle Q\rangle}, \qquad \frac{\langle Z\rangle}{\langle Q\rangle}. \tag{6}$$

Since chargino pair production is described by the four parameters $\cos 2\phi_L$, $\cos 2\phi_R$, $\tan\beta$, and $m_{\tilde{\nu}}$, the four independent observables listed above should allow us to determine the SUSY parameters $\tan\beta$, M_2 and μ (almost) completely. In order to explicitly demonstrate the possibility of measuring these

SUSY parameters, we make a case study with the assumption that the experimentally measured values of the four observables at $\sqrt{s} = 200$ GeV are

$$m_{\tilde{\chi}_1^\pm} = 90 \ \text{GeV}, \quad \sigma_{tot} = 2.5 \ \text{pb}, \quad \frac{\langle P \rangle^2}{\langle Q \rangle} = -0.05 \ \text{pb}, \quad \frac{\langle Z \rangle}{\langle Q \rangle} = 0.1. \qquad (7)$$

For the ideal case with no experimental errors the observed values yield $m_{\tilde{\nu}} =$ 197 GeV, $\cos 2\phi_L = 0.14$ and $\cos 2\phi_R = 0.03$.

However, both systematic and statistical errors should be considered in real experimental situations. Also, the resolution power of the observables constructed from angular correlations depends on the unknown parameter κ rather strongly; a reliable estimate thus requires prior understanding of quantities such as the neutralino mass matrix and sfermion mass spectra. Fig. 1 illustrates how those errors may affect the determination of the cosines of the two mixing angles under the assumption that $\Delta\sigma_{tot} = \pm 0.2$ pb, $\Delta[\langle P \rangle^2/\langle Q \rangle] = \pm 0.02$ pb, and $\Delta[\langle Z \rangle/\langle Q \rangle] = \pm 0.02$ for $\sqrt{s} = 200$ GeV with a precisely measured chargino mass $m_{\tilde{\chi}^\pm} = 90$ GeV.

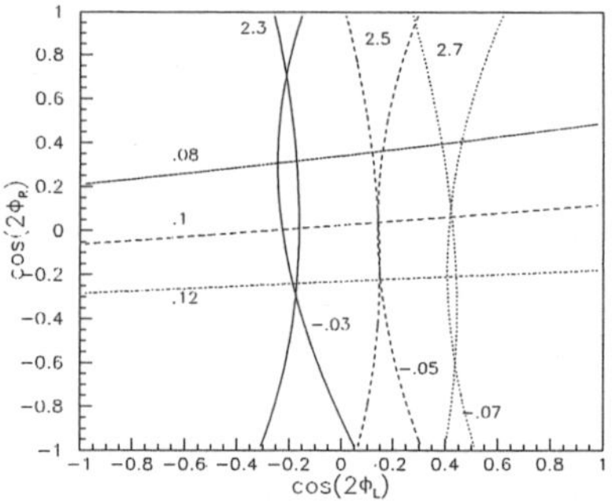

Figure 1: A contour plot for $\sigma_{tot} = 2.5 \pm 0.2$ pb, $\langle P \rangle^2/\langle Q \rangle = -0.05 \pm 0.02$ pb, and $\langle Z \rangle/\langle Q \rangle = 0.1 \pm 0.02$, assuming $m_{\tilde{\chi}_1^\pm} = 90$ GeV and $m_{\tilde{\nu}} = 197$ GeV at $\sqrt{s} = 200$ GeV.

From the measured values of $\cos 2\phi_L$, $\cos 2\phi_R$ and $m_{\tilde{\chi}_1^\pm}$, we can extract the allowed values for $\tan\beta$, M_2 and μ. Defining x and y through $M_2 = m_W(x+y)$ and $\mu = m_W(x-y)$, we find that there are four different solutions of x and y for given $\cos 2\phi_L$ and $\cos 2\phi_R$:

$$(a) \quad \begin{cases} x = \pm\cot(\phi_R - \phi_L)\sin\left(\beta - \frac{\pi}{4}\right), \\ y = \pm\cot(\phi_R + \phi_L)\cos\left(\beta - \frac{\pi}{4}\right), \end{cases}$$

$$(b)\quad \begin{cases} x = \pm\cot(\phi_R + \phi_L)\sin\left(\beta - \frac{\pi}{4}\right), \\ y = \pm\cot(\phi_R - \phi_L)\cos\left(\beta - \frac{\pi}{4}\right), \end{cases} \tag{8}$$

Depending on whether it is larger or smaller than unity, the value of $\tan\beta$ is determined from $m_{\tilde{\chi}^\pm}$, $\cos 2\phi_L$ and $\cos 2\phi_R$, through the relations

$$\tan\beta = \begin{cases} \mp\frac{p^2-q^2\pm 2pq\sqrt{1+\eta}\sqrt{p^2+q^2-(1+\eta)p^2q^2}}{2(1+\eta)p^2q^2-(p-q)^2} & \text{for} \quad \tan\beta \geq 1 \\ \\ \mp\frac{p^2-q^2\pm 2pq\sqrt{1+\eta}\sqrt{p^2+q^2-(1+\eta)p^2q^2}}{2(1+\eta)p^2q^2-(p+q)^2} & \text{for} \quad \tan\beta \leq 1, \end{cases} \tag{9}$$

where the overall $\mp$ is for (a) and (b) in Eq. (8), $\eta = m^2_{\tilde{\chi}^\pm_1}/m^2_W$, $p^2 - q^2 = \sqrt{(1-\cos^2 2\phi_L)(1-\cos^2 2\phi_R)}$, $pq = |\cos 2\phi_L - \cos 2\phi_R|/2$, and $p^2 + q^2 = 1 - \cos 2\phi_L \cos 2\phi_R$. Of course, only the $\tan\beta$ value satisfying each condition should be taken in Eq. (9). To conclude, there exists at most a four-fold discrete ambiguity in determining $\tan\beta$, M_2, and μ in the pair production of lighter charginos at CERN LEP II.

Acknowledgments

The author gratefully acknowledges the collaboration with A. Djouadi, J. Kalinowski and P.M. Zerwas. This work was supported by the KOSEF-DFG large collaboration project, Project No. 96-0702-01-01-2.

References

1. *Physics at LEP2*, Proceedings of the Workshop on Physics at LEP II, edited by G. Altarelli, T. Sjöstrand, and F. Zwirner (CERN Report No. CERN-96-01, Geneva, Switzerland, 1996), Vol. I.
2. H.E. Haber and G.L. Kane, *Phys. Rep.* **117**, 75 (1985); H.P. Nilles, *ibid.* **110**, 1 (1984).
3. ALEPH: *EPS-HEP Conf.*, Jerusalem (1997) Ref. 856; DELPHI: *EPS-HEP Conf.*, Jerusalem (1997) Ref. 858; L3: *EPS-HEP Conf.*, Jerusalem (1997) Ref. 859.
4. A. Leike, *Int. J. Mod. Phys.* A **3**, 2895 (1988).
5. M.A. Diaz and S.F. King, *Phys. Lett.* B **349**, 105 (1995); **373**, 100 (1996).
6. J.L. Feng and M. J. Strassler, *Phys. Rev.* D **51**, 4661 (1995) and **55** (1997) 1326.
7. S.Y. Choi *et al.*, in preparation.

NEW ELECTROWEAK TESTS FOR THE TOPFLAVOUR MODEL

JONG CHUL LEE and KANG YOUNG LEE[a]
Department of Physics, Korea Advanced Institute of Science and Technology
Taejon 305 - 701, Korea

We explore the phenomenology of the topflavour model for the LEP experiments at scale m_Z. Implications of the model for the Z peak data are studied in terms of the precision variables ϵ_i.

1 Introduction

In this work we consider a model with additional SU(2) acting only on the third generation, which has been suggested by several authors [1,2] as a possible solution of the R_b problem. The third generation has different flavour dynamics from the first and second generations and we expect that this type of model would help us to explain the discrepancy of R_b, the measurement of which is about 1.8 (LEP) and 1.5 (LEP+SLC) standard deviations higher than the SM prediction. In analogy to the topcolor model [3], it is called topflavour model [2].

In order to parametrize new physics effects on LEP observables, we calculate the precision variables ϵ_i introduced by Altarelli et al.[4,5] in the framework of the topflavour model, and compare with the new LEP data reported by the LEP Electroweak Working Group[6]. Because there is no evidence for new physics beyond the Standard Model (SM) at LEP until now, any new physics contribution, if it exists, has to be at most comparable with the radiative correction effects of the SM. Hence it will be interesting to study new physics effects in terms of the precision variables. Among the four epsilon variables, we focus on ϵ_b, which has been of particular interest because it encodes the corrections to the $Z \to b\bar{b}$ vertex.

2 The Epsilon Variables and Standard Model Predictions

The precision variables ϵ_1, ϵ_2, ϵ_3 and ϵ_b are defined from basic observables: the mass ratio of W and Z bosons m_W/m_Z, the leptonic width Γ_l, the forward–backward asymmetry for charged leptons A^l_{FB} and the b–quark width Γ_b, which are all defined at the Z–peak. The epsilon variables have the virtue that the most interesting physical results are obtained in a completely model independent manner, without assumptions like the dominance of vacuum polarisation

[a] conference speaker

diagrams. The m_t–dependence of all observables via loops is contained in the ϵ_i.

The $\epsilon_{1,2,3}$ variables are written as linear combinations of the correction terms $\Delta\rho$, Δk and Δr_W in eqs. (9) of ref. [5], which are extracted from the vector and axial–vector couplings of charged leptons, g_{lV}, g_{lA} and the mass ratio of W and Z bosons, m_W/m_Z. The vector and axial–vector couplings are obtained from the observables Γ_l and A^l_{FB}. The formula for ϵ_b is rather complicated. It is given by:

$$g_{bA} = -\frac{1}{2}\left(1+\frac{\Delta\rho}{2}\right)(1+\epsilon_b), \qquad \frac{g_{bV}}{g_{bA}} = \frac{1-\frac{4}{3}(1+\Delta k)s_0^2+\epsilon_b}{1+\epsilon_b} \quad . \tag{1}$$

We obtain the relation between ϵ_b and Γ_b by inserting g_{bV} and g_{bA} into the formula for Γ_b. The SM predictions for the ϵ_i are given by

$$\epsilon_1 = 5.8\times10^{-3}, \quad \epsilon_2 = -7.4\times10^{-3}, \quad \epsilon_3 = 5.0\times10^{-3}, \quad \epsilon_b = -5.3\times10^{-3}, \tag{2}$$

for a Higgs boson mass $m_H = 100$ GeV.

The experimental values of the epsilon variables are obtained using recent LEP data[6]:

$$\begin{aligned} \epsilon_1 &= (2.9 \sim 5.3)\times10^{-3}, \quad & \epsilon_2 &= (-6.6 \sim -3.6)\times10^{-3}, \\ \epsilon_3 &= (1.0 \sim 4.5)\times10^{-3}, \quad & \epsilon_b &= (-4.5 \sim -0.8)\times10^{-3} \; . \end{aligned} \tag{3}$$

Note that lepton universality is assumed for the values of Γ_l and A^l_{FB}.

3 The Topflavour Model

We study the topflavour model with the extended electroweak gauge group $SU(2)_l \times SU(2)_h \times U(1)_Y$, where the first and second generations couple to $SU(2)_l$ and the third generation couples to $SU(2)_h$. The left–handed quarks and leptons in the first and second generations transform as (2,1,1/3), (2,1,-1) under $SU(2)_l\times SU(2)_h\times U(1)_Y$, and those in the third generation as (1,2,1/3), (1,2,-1), while right–handed quarks and leptons transform as (1,1,2Q) where $Q = T_{3l}+T_{3h}+Y$ is the electric charge of a fermion.

The covariant derivative is given by $D^\mu = \partial^\mu + ig_l T^a_l W^\mu_{la} + ig_h T^a_h W^\mu_{ha} + ig'YB^\mu$, where T^a_l and T^a_h denote the $SU(2)_{(l,h)}$ generators and Y is the $U(1)$ hypercharge generator. The corresponding gauge bosons are W^μ_{la}, W^μ_{ha} and B^μ, with coupling constants g_l, g_h and g' respectively. The gauge couplings may be written as $g_l = \frac{e}{\sin\theta\cos\phi}$, $g_h = \frac{e}{\sin\theta\sin\phi}$, $g' = \frac{e}{\cos\theta}$ in terms of the weak mixing angle θ and the new mixing angle ϕ between $SU(2)_l$ and $SU(2)_h$.

Symmetry breaking is accomplished by the vacuum expectation values (VEV) of two scalar fields Σ and Φ: $\langle\Phi\rangle = (0, v/\sqrt{2})^\dagger$, $\langle\Sigma\rangle = uI$ where I is the 2×2 identity matrix. The scalar field Σ transforms as (2,2,0) under $SU(2)_l \times SU(2)_h \times U(1)$ and we choose Φ to transform as (2,1,1), corresponding to the SM Higgs field. In the first stage, the scalar field Σ gets a vacuum expectation value and breaks $SU(2)_l \times SU(2)_h \times U(1)_Y$ down to $SU(2)_{l+h} \times U(1)_Y$ at the scale $\sim u$. The remaining symmetry is broken down to $U(1)_{em}$ by the the VEV of Φ at the electroweak scale. Since the third generation fermions do not couple to Higgs fields with these quantum numbers, they should get masses via higher dimensional operators. We do not study the mass generation problem of this model in detail here. We demand that both $SU(2)$ interactions are perturbative so that the value of the mixing angle $\sin\phi$ is constrained $g^2_{(l,h)}/4\pi < 1$, which results in $0.03 < \sin^2\phi < 0.96$. We also assume that the first symmetry breaking scale is much higher than the electroweak scale, $v^2/u^2 \equiv \lambda \ll 1$.

The basic observables in the topflavour model depend upon the model parameters λ and $\sin^2\phi$ as well as the Higgs mass m_H. Including correction terms, we obtain the epsilon variables in the topflavour model [7]:

$$\begin{aligned}
\epsilon_1 &= \epsilon_1^{SM} - 2\lambda \sin^4\phi \\
\epsilon_2 &= \epsilon_2^{SM} + \lambda \sin^4\phi \left(\frac{s_0^2}{c_0^2 - s_0^2} - 2 \right) \\
\epsilon_3 &= \epsilon_3^{SM} - \lambda \sin^4\phi \\
\epsilon_b &= \epsilon_b^{SM} - \frac{1}{2}\lambda \frac{\sin^4\phi}{g_{lA}^{SM}} (1 + \epsilon_b^{SM} \sin^2\phi) \ .
\end{aligned} \tag{4}$$

We used the program ZFITTER [8] for numerical calculations of the epsilon variables in this model, with $m_t = 175$ GeV as reported by CDF and D0 [9].

In Figs. 1 and 2, the experimental ellipses are shown in the $\epsilon_1 - \epsilon_b$ and $\epsilon_3 - \epsilon_b$ planes, respectively, and compared with predictions of the SM and the topflavour model. We show results for various values of $\sin^2\phi$ and $m_{Z'}$, where

$$m_{Z'}^2 = \frac{m_Z^2}{\lambda \sin^2\phi \cos^2\phi} \left[\cos^2\theta_{SM} + \lambda \sin^4\phi \left(\frac{c_0^2 s_0^2}{c_0^2 - s_0^2} + 2c_0^2 \right) \right] \ , \tag{5}$$

with the Weinberg angle of the SM, θ_{SM}. The ellipses are shifted along the minus direction of ϵ_b compared with those from 1996 data [10,11] and thus come closer to the SM prediction. We find that the SM predictions still lie outside the 90 % C.L. ellipses in both planes. These deviations are caused by the $Zb\bar{b}$ coupling and are related to the R_b discrepancy. The lower limit of the mass

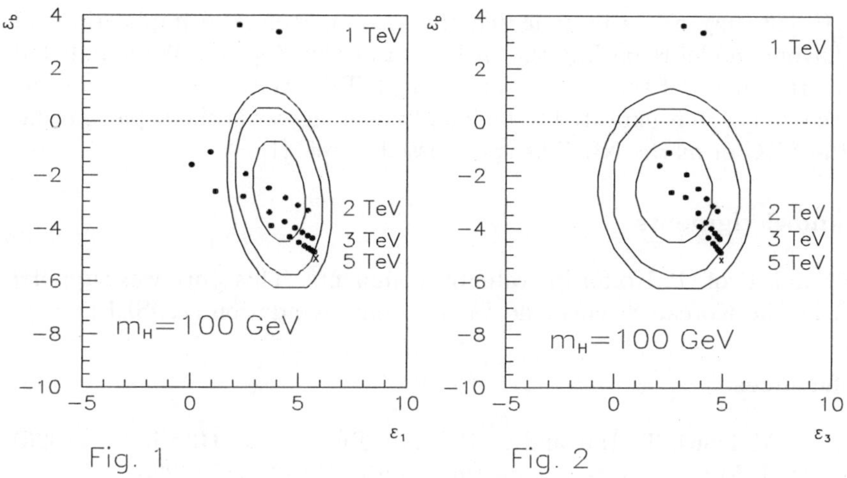

Figure 1: Plots of the model predictions in units of 10^{-3} in the $\epsilon_1 - \epsilon_b$ plane (left) and the $\epsilon_3 - \epsilon_b$ plane (right). The model parameters $sin^2\phi$, $m_{Z'}$ and the Higgs boson mass m_H are varied. The experimental ellipses at 1-σ, 90 % C.L. and 95 % C.L. are given. Points are labelled from $\sin^2\phi = 0.1 \sim 0.9$ from right to left for each value of $m_{Z'}$.

of heavy gauge boson Z' is about 1.2 TeV (1.1 TeV) at 90 (95) % c.l. which agrees with the result of ref. [1].

4 Concluding Remarks

In this work, we explore the phenomenology of the topflavour model, which is an extension of the SM with additional $SU(2)$ symmetry. The third generation is special in this model, and it is expected to explain the hierarchy of the fermion mass spectrum as well as the R_b discrepancy observed by LEP experiments. The experimental values of the ϵ variables are obtained from recent LEP results reported by the LEP Electroweak Working Group. At present the SM predictions are still out of 90 % c.l., ellipses mainly caused by the deviation of ϵ_b originating from the $Zb\bar{b}$ vertex. Since the topflavour model provides different flavour dynamics for the third generations, the value of ϵ_b can shift to the experimental value.

In conclusion, we investigated the phenomenological implications of the topflavour model regarding the LEP data at the Z–peak. We found that the best fitted mass of the Z' boson is about 2 TeV. With this value, a signature for the Z' boson is expected to be found via an excess in the $t\bar{t}$ pair production at the LHC or at the NLC, as is discussed in ref. [1].

Acknowledgments

We thank Prof. P. Ko for his valuable comments. This work was supported in part by the Korean Science and Engineering Foundation (KOSEF).

References

1. E. Malkawi, T. Tait and C.-P. Yuan, Phys. Lett. **B385**, 304 (1996).
2. D. J. Muller and S. Nandi, Phys. Lett. **B383**, 345 (1996).
3. C. T. Hill, Phys. Lett. **B266**, 419 (1991); *ibid.*, 483 (1995); C. T. Hill, Report No. FERMILAB–Conf–97/032–T, invited talk presented at *Strongly Coupled Gauge Theories '96*, Nagoya, (1996).
4. G. Altarelli, R. Barbieri and S. Jadach, Nucl. Phys. **B369**, 3 (1992).
5. G. Altarelli, R. Barbieri and F. Caravaglios, Nucl. Phys. **B405**, 3 (1993); G. Altarelli, R. Barbieri and F. Caravaglios, Phys. Lett. **B349**, 145 (1995).
6. The LEP collaborations, the LEP Electroweak Working Group and the SLD Heavy Flavour Group, Report No. LEPEWWG/97-02.
7. J.C. Lee, K.Y. Lee and J.K. Kim, preprint KAIST-TH-97/19, hep-ph/9711509.
8. D. Bardin et al., Report No. CERN–TH, 6443/92.
9. D.S. Kestenbaum, preprint FERMILAB-CONF-97-016, presented at the 16th International Conference on Physics in Collision, Mexico (1996).
10. K.Y. Lee, preprint hep-ph/9710478, to appear in Phys. Rev. D.
11. G.T. Park and K.Y. Lee, preprint hep-ph/9705420, to appear in Mod. Phys. Lett. A.

ENHANCEMENT OF LOOP–INDUCED $H^{\pm}W^{\mp}Z^0$ VERTEX

SHINYA KANEMURA
Theory Group, KEK
Tsukuba Ibaraki 305, Japan

We discuss non–decoupling effects on the loop-induced $H^{\pm}W^{\mp}Z^0$ vertex in a general two Higgs-doublet model. The decay process $H^{\pm} \to W^{\pm}Z^0$ is analyzed at one loop level and possible enhancement of the decay width is explored. We find that a novel enhancement can be realized by the Higgs non–decoupling effects from a large mass difference between charged and CP–odd Higgs bosons. The branching ratio of the process can be a few % $\sim$ 10%. Therefore the decay mode may be detectable at LHC or future e^+e^- linear colliders (LC's).

1 Introduction

The Higgs sector remains unknown. Although the minimal model (with one Higgs doublet) does not contradict any current experimental data, various purely theoretical arguments often favor extended Higgs sectors [1]. A charged Higgs boson $H^{\pm}$ and a CP–odd Higgs boson A^0 are always introduced in such extensions. Thus the detection of $H^{\pm}$ (or A^0) is very important to confirm the extended Higgs sectors. If $H^{\pm}$ is light, it may be detectable through the modes $H^{\pm} \to \tau\nu$ and cs. However, a $H^{\pm}$ heavy enough to have tb as main decay mode may be elusive because of large QCD backgrounds. In such a case, we have to investigate the possibility of alternative decay modes with sufficiently large branching ratio to yield a substantial number of events. Possible modes may be $H^{\pm} \to \tau\nu$, $h^0W^{\pm}$, $W^{\pm}Z^0$ and $W^{\pm}\gamma$. Unfortunately, it has been known that the latter two modes disappear at tree level in general Higgs models with multi doublet structures, in which they can be induced only at loop levels [2]. Thus these modes have been considered as a clear signal for *exotic* Higgs sectors (for example, including triplets).

In this talk, we discuss the loop–induced $H^{\pm}W^{\mp}Z^0$ vertex in the two Higgs doublet model (2HDM) and MSSM [3]. We examine the possible enhancement of the vertex at one loop level. The decay width of $H^{\pm} \to W^{\pm}Z^0$ is then calculated at one loop level and the branching ratio is estimated in order to see whether it can be sufficiently enhanced to be detectable at the LHC or LC's. We find that a novel enhancement can occur by Higgs non decoupling effects in 2HDM within the constraints from present data. The branching ratio can be a few % $\sim$ 10% for $\tan\beta > 1$ at $m_{H^{\pm}} = 300$ GeV. In the MSSM, the branching ratio is less than 0.01 % for $\tan\beta > 1$. Therefore, $H^{\pm} \to W^{\pm}Z^0$ may be useful as a probe of Higgs sectors between MSSM, 2HDM and *exotic*

Higgs models.

2 Properties of the Model

We here consider the model with a softly broken discrete symmetry under $\Phi_1 \to \Phi_1$, $\Phi_2 \to -\Phi_2$ [3]. This Higgs potential is a popular one and covers the MSSM Higgs potential as a special case [1]. We assume the soft–breaking mass parameter μ_3^2 to be real. This model includes five massive physical scalar bosons; one charged ($H^\pm$), one CP–odd neutral (A^0) and two CP–even neutral ones (h^0 and H^0). Three Nambu–Goldstone bosons $w^\pm$ and z^0 are absorbed into $W_L^\pm$ and Z_L^0.

In 2HDM, whether the internal heavy Higgs bosons are decoupled or not is a model dependent problem. The masses of $H^\pm$, A^0 and H^0 can be expressed as $M_i^2 = \lambda_i v^2 + \mathcal{O}(\mu_3^2)$, where λ_i represent linear combinations of the quartic coupling constants. If large M_i are realized by increasing μ_3^2 while keeping the λ_i small, the effects of heavy Higgs bosons tend to be decoupled from low–energy observables as seen in the MSSM. Alternatively, if M_i are large due to the growing λ_i with a small μ_3^2, the decoupling theorem no longer works and non–decoupling effects appear[3,4].

The Higgs sector, in general, does not have the custodial $SU(2)_V$ symmetry in 2HDM. In terms of 2×2 matrices $\mathcal{M}_i = (i\tau_2\Phi_i^*, \Phi_i)$, $(i = 1, 2)$, the Higgs sector includes the term $\sim \lambda'\{\mathrm{tr}(\mathcal{M}_2\tau_3\mathcal{M}_1^\dagger)\}^2$. Such a term is not invariant under the transformation $\mathcal{M}_i \to g_L^\dagger \mathcal{M}_i g_R$ $(g_{L,R} \in SU(2)_{L,R})$. This term therefore breaks $SU(2)_R$ and thus the custodial $SU(2)_V$ symmetry explicitly. The coupling constant of this term is expressed as $\lambda' = (m_{H^\pm}^2 - m_{A^0}^2)/v^2$. The explicit breaking of $SU(2)_V$ in the Higgs sector is measured by the mass difference between A^0 and $H^\pm$[3].

3 Loop-Induced $H^\pm W^\mp Z^0$ Vertex

We first show the absence of the tree $H^\pm W^\mp Z^0$ coupling in 2HDM (and MSSM). The coupling is expected to be generated in the kinetic part of the Higgs sector: $\mathcal{L}_{\mathrm{THDM}}^{kin} = (D_\mu\Phi)^\dagger D^\mu\Phi + (D_\mu\Psi)^\dagger D^\mu\Psi$, where Φ and Ψ are the Higgs doublets in the basis of the gauge-eigenstates [5];

$$\Phi = \begin{pmatrix} w^+ \\ \frac{1}{\sqrt{2}}(\phi^0 + v + iz^0) \end{pmatrix}, \Psi = \begin{pmatrix} H^+ \\ \frac{1}{\sqrt{2}}(\psi^0 + iA^0) \end{pmatrix}. \tag{1}$$

ϕ^0 and ψ^0 are linear combinations of h^0 and H^0. Since Ψ does not have any vacuum expectation value, we can see the absence of the tree $H^\pm W^\mp Z^0$

coupling. However, it can be induced radiatively through the mixing between Φ and Ψ.

We next discuss the non–decoupling effects of heavy particles on the $H^{\pm}W^{\mp}Z^{0}$ vertex [3]. The effective Lagrangian is [2]

$$\begin{aligned}\mathcal{L}_{\text{eff}} &= f_{H^+W^-Z^0} H^+ W^-_\mu Z^\mu + g_{H^+W^-Z^0} H^+ F_Z^{\mu\nu} F^W_{\mu\nu} \\ &+ h_{H^+W^-Z^0}\; i\epsilon_{\mu\nu\rho\sigma} H^+ F_Z^{\mu\nu} F_W^{\rho\sigma} + \text{h.c.}. \end{aligned} \tag{2}$$

The coefficient $f_{H^+W^-Z^0}$ has mass dimension 1 and the others have dimension -1. The contributions of heavy particles of masses M_i to the coefficients can be estimated at one loop level as

$$\begin{aligned} f_{H^+W^-Z^0} &\sim g \times \frac{g}{\cos\theta_W} \times \frac{M_i^2}{v}(\times \ln M_i) \sim \frac{m_W m_Z}{v^3} \times M_i^2 (\ln M_i), \\ g_{H^+W^-Z^0},\; h_{H^+W^-Z^0} &\sim \frac{m_W m_Z}{v^3} \times \ln M_i . \end{aligned} \tag{3}$$

Therefore the naive power counting shows that there may be substantial non–decoupling effects of the heavy Higgs bosons as well as heavy fermions on the loop induced $H^{\pm}W^{\mp}Z^{0}$ vertex.

The vertex is, however, strongly constrained by the custodial $SU(2)_V$ symmetry. The terms in the Lagrangian (2) comes from the following operators;

$$\text{tr}\left[\tau_3 (D_\mu \mathcal{M})^\dagger (D^\mu \mathcal{N})\right], \text{tr}\left[\tau_3 \mathcal{M}^\dagger \mathcal{N} F_Z^{\mu\nu} F^W_{\mu\nu}\right] \text{ or } i\epsilon_{\mu\nu\rho\sigma} \text{tr}\left[\tau_3 \mathcal{M}^\dagger \mathcal{N} F_Z^{\mu\nu} F_W^{\rho\sigma}\right], \tag{4}$$

where the 2×2 matrices $\mathcal{M}$ and $\mathcal{N}$ are defined by $\mathcal{M} = (i\tau_2\Phi^*, \Phi)$ and $\mathcal{N} = (i\tau_2\Psi^*, \Psi)$. None of the operators (4) is invariant under $SU(2)_R$ and thus $SU(2)_V$, so the vertex is related to $SU(2)_V$ breaking in the model.

4 Enhancement of the Decay Process $H^{\pm} \to W^{\pm}Z^{0}$

We proceed to examine the decay process $H^{\pm} \to W^{\pm}Z^{0}$. The decay width in 2HDM (with Type II Yukawa coupling [1]) with large $\Delta m (= m_{A^0} - m_{H^{\pm}})$ and also that in the MSSM with heavy sparticles are shown in Fig 1. In the MSSM, in addition to the Higgs decoupling property, m_{A^0} is approximately degenerate with $m_{H^{\pm}}$. Thus the Higgs effects are small and the heavy fermion effects are dominant. Since the $H^{\pm}tb$ coupling constant consists of $m_t \cot\beta + m_b \tan\beta$, the top quark contributions are rapidly reduced for larger $\tan\beta$. On the other hand, in 2HDM with large Δm, a novel enhancement of the width is realized for large $\tan\beta$ because of the non–decoupling effects of Higgs bosons [3].

Let us consider the branching ratio of this decay mode in 2HDM next. The other decay modes included here are $H^{\pm} \to tb$, $\tau\nu$ and $h^0 W^{\pm}$. The

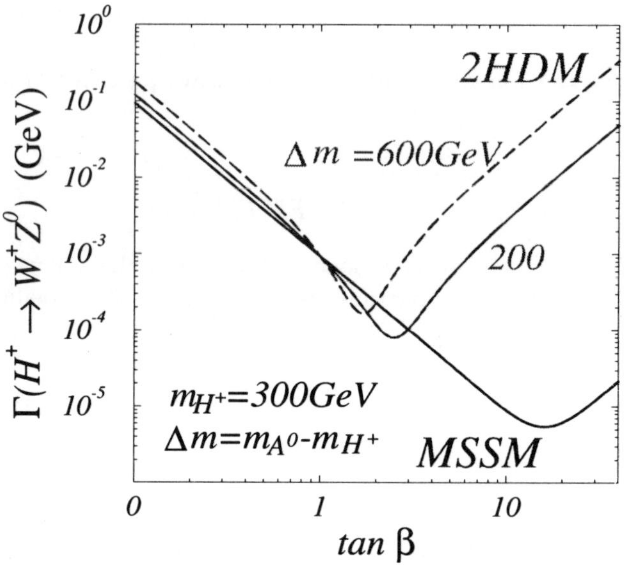

Figure 1: Partial width for $H^{\pm} \to WZ$ as a function of $\tan\beta$.

parameters are fixed as $m_{h^0} = 140$ GeV, $m_{H^0} = 310$ GeV and $\alpha = \beta - \pi/2$. We also assume that $m_t = 175$ GeV and $m_b(m_{H^\pm}) = 3$ GeV. We can see in Fig. 2 that the branching ratio becomes larger than 1% if Δm is greater than 200 GeV for $\tan\beta > 5 \sim 8$. The maximal value can amount to near 10% for very large m_{A^0} and $\tan\beta > 20$. If the Higgs sector is nearly $SU(2)_V$ symmetric ($m_{A^0} \sim m_{H^\pm} = 300$ GeV), the Higgs non–decoupling effects cancel out and only the fermion effects remain, so that the branching ratio becomes less than 0.01%. As mentioned before, a nonzero μ_3^2 reduces the non–decoupling Higgs effects. So far we have tried to maximize the Higgs non–decoupling effects, assuming the soft breaking parameter μ_3^2 to be zero. However, μ_3^2 is often very important in various aspects of physics. The reduction of the branching ratio by $m_3 = \mu_3/\sqrt{\sin\beta\cos\beta}$ is shown in Fig 3.

5 Summary and Discussion

We have discussed the loop induced $H^{\pm}W^{\mp}Z^0$ vertex. The possibility of its enhancement has been explored. The conditions for the enhancement are sum-

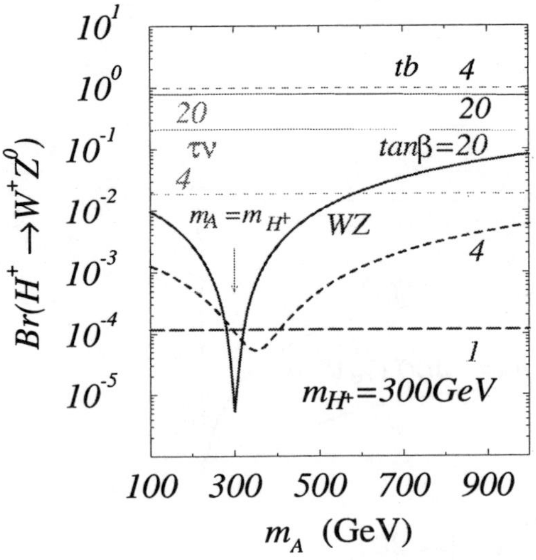

Figure 2: Branching ratios of the charged Higgs boson as a function of m_A.

marized as 1) Non–decoupling properties of Higgs sector with small μ_3^2, and 2) Large $SU(2)_V$ breaking in the Higgs sector by the mass difference between A^0 and $H^\pm$. Although such conditions are not satisfied in the framework of the MSSM, it is possible for 2HDM to satisfy them within the currently experimentally allowed region, while maintaining perturbative unitarity. As a result, the branching ratio of $H^\pm \to W^\pm Z^0$ can be a few % $\sim$ 10%. Such an enhancement may make it possible to detect the decay mode at LHC and also LC's. At the LHC, charged Higgs bosons are mainly produced through the subprocess $gb \to tH^\pm$, where over one hundred $H^\pm \to W^\pm Z^0 \to ll l\nu$ events can be expected per year for $Br(H^+ \to W^+Z^0) > 1\%$. Since the background (mainly $ud \to WZ$) is naively estimated to be such that a few % of the branching ratio are required to see a signal, we can expect to detect the decay mode with the above enhancement. We also expect that it can be detectable at LC's ($\sqrt{s} = 1$ TeV, $L = 160$ fb^{-1}/year), where over a few dozen events of $H^\pm \to W^\pm Z^0 \to lll\nu$, $lljj$, $l\nu jj$ and $\nu\nu jj$ are produced per year for $Br(H^+ \to W^+Z^0) > 1$ % with less background.

Acknowledgments

The author would like to thank A. Miyamoto for much valuable discussion.

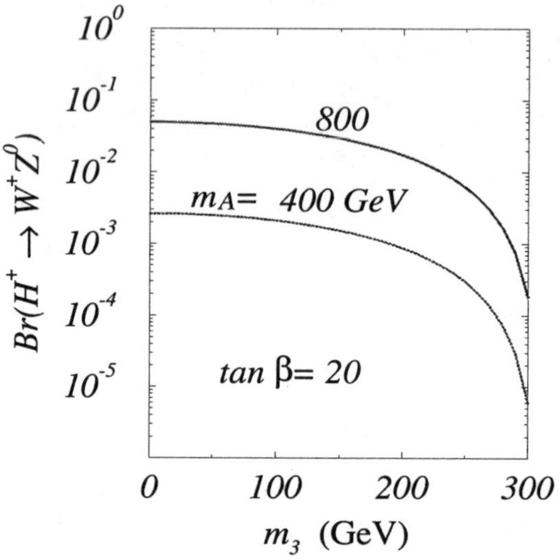

Figure 3: Branching ratio for $H\pm \to WZ$ as a function of $m_3 \equiv \mu_3$.

References

1. J.F. Gunion, H.E. Haber, G. Kane and S. Dawson, *The Higgs Hunter's Guide*, (Addison-Wesley, New York, 1990).
2. A. Méndez and A. Pomarol, *Nucl. Phys.* B **349**, 369 (1991); M. Capdequi Peyranère, H.E. Haber and P. Irulegui, *Phys. Rev.* D **44**, 191 (1991).
3. S. Kanemura, submitted to *Phys. Rev.* **D**, hep-ph/9710237, and references therein.
4. S. Kanemura and H. Tohyama, *Phys. Rev.* **D** in press, hep-ph/9707454.
5. H. Georgi, *Hadr. J. Phys.* **1**, 155 (1978).

POLARIZED J/ψ PRODUCTION AT CLEO

SEUNGWON BAEK, H.S. SONG
Center for Theoretical Physics and Department of Physics, Seoul National University, Seoul 151-742, Korea

P. KO
Department of Physics, KAIST, Taejon 305-701, Korea

JUNGIL LEE
Dept. of Physics, Ohio State Univ., Columbus, OH 43210, USA.

The polarization of prompt J/ψ's produced in e^+e^- annihilation at CLEO energy ($\sqrt{s}$ = 10.58 GeV) is considered in the framework of the NRQCD factorization formalism. We find that the J/ψ polarization strongly depends on the production mechanism. At CLEO energy, the most dominant J/ψ production mechanism in most of phase space is the color–singlet mechanism, $e^+e^- \to J/\psi + gg$, for which J/ψ's are highly longitudinally polarized. On the other hand, the dominant J/ψ production mechanism at the upper end point of the J/ψ energy distribution is the color–octet process, $e^+e^- \to (c\bar{c})^{(8)} + g$, for which J/ψ's are almost unpolarized. Thus, the measurement of the polarization of the end point J/ψ's can give another test of the color–octet mechanism, and constraints on NRQCD matrix elements.

We consider J/ψ polarization in e^+e^- annihilation at the CLEO energy ($\sqrt{s}$ = 10.58 GeV). The longitudinal polarization of the prompt J/ψ is expected to be 42% when the color–octet $c\bar{c}[^1S_0 \text{ or } {}^3P_J]^{(8)}$ contributions are included, whereas the CSM alone predicts 53%. We show that the energy dependence of the longitudinal polarization is sensitive to the J/ψ production mechanism. If the color–octet mechanism dominates at the upper end point of phase space, almost unpolarized J/ψ's are produced. If the color–singlet mechanism dominates, almost longitudinally polarized J/ψ's are produced.

In order to study the polarization of J/ψ's produced in the e^+e^- annihilation, we define an observable $\eta(z)$ as

$$\eta(z) \equiv \frac{d\sigma_L}{dz} \Big/ \frac{d\sigma}{dz}\,, \tag{1}$$

where $z \equiv 2E_{J/\psi}/\sqrt{s}$, $d\sigma_L/dz$ represents the energy spectrum of the longitudinally polarized prompt J/ψ, and $d\sigma/dz$ that of total J/ψ production. The polarization of produced J/ψ's can be obtained by measuring the angular distribution of lepton pairs in the subsequent decay of J/ψ. Specifically, the angular distribution of decay leptons in the process, $J/\psi \to l^+l^-$, is given by

$$\frac{d\Gamma}{d\cos\theta_l^*} \propto \left[1 + \alpha(z)\ \cos^2\theta_l^*\right], \tag{2}$$

with θ_l^* the angle between the lepton three–momentum in the J/ψ rest frame and the J/ψ direction in the lab frame. The relation between α and η is given by the following equation

$$\alpha(z) = \frac{1 - 3\eta(z)}{1 + \eta(z)}. \tag{3}$$

Therefore, unpolarized J/ψ corresponds to $\eta = 1/3$ ($\alpha = 0$), whereas purely transverse polarization corresponds to $\eta = 0$ ($\alpha = 1$).

The dominant color–singlet contributions come from the following two processes:

$$e^+e^- \ \to\ (c\bar{c})[^3S_1]^{(1)} + gg \tag{4}$$
$$e^+e^- \ \to\ (c\bar{c})[^3S_1]^{(1)} + c\bar{c}. \tag{5}$$

Among color–octet contributions, the following two processes are dominant:

$$e^+e^- \ \to\ (c\bar{c})[^1S_0 \text{ or } {}^3P_J]^{(8)} + g \tag{6}$$
$$e^+e^- \ \to\ (c\bar{c})[^3S_1]^{(8)} + q\bar{q}. \tag{7}$$

We show the angular momentum and spin quantum numbers of the $c\bar{c}$ in the spectroscopic notation, and the superscripts (1) and (8) represent its color structures. These $c\bar{c}[^{2S+1}L_J]^{1,8}$ states will eventually evolve into a physical J/ψ by emitting/absorbing soft gluons, the probabilities for which are parametrized in terms of NRQCD matrix elements, $\langle O_{1,8}^{J/\psi}(^{2S+1}L_J)\rangle$ [1].

Virtual Z^0 contributions were neglected, which should be a good approximation if $\sqrt{s}$ is far below M_Z [a]. Our results for the total production cross sections agree with the previous results obtained in Refs. [3] and [4] [b].

For our numerical analyses, we use the following values for the nonperturbative matrix elements that appear in the NRQCD factorization formula for the J/ψ production cross sections :

$$\langle \alpha O_1^{J/\psi}(^3S_1)\rangle \ = \ 0.73 \text{ GeV}^3 \tag{8}$$

[a] See Refs. [2] [5] for $Z^0 \to J/\psi + X$ at LEP, Ref. [6] for the J/ψ polarization therein.
[b] There are some typos in the overall factors of Eqs.(3.8),(A2a) and (A2b) in Ref. [3]. Their results should be multiplied by a factor of 3.

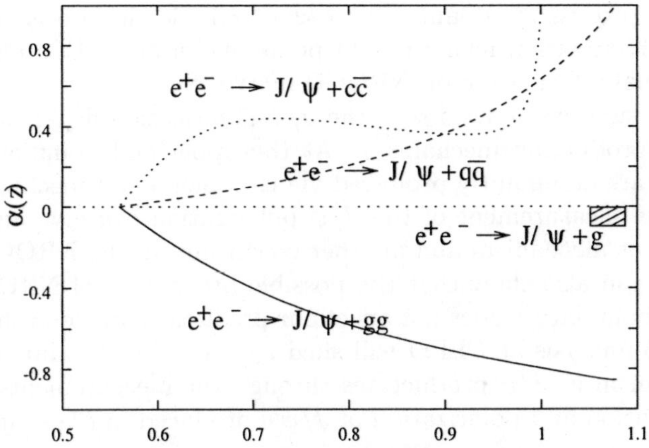

Figure 1: $\alpha(z)$ for each mode.

$$\langle \mathcal{O}_8^{J/\psi}(^3S_1)\rangle = 0.015 \text{ GeV}^3 \quad (9)$$

$$\langle \mathcal{O}_8^{J/\psi}(^1S_0)\rangle \approx 10^{-2} \text{ GeV}^3 \quad (10)$$

$$\langle \mathcal{O}_8^{J/\psi}(^3P_0)\rangle/m_c^2 \approx 10^{-2} \text{ GeV}^3. \quad (11)$$

We set $\alpha_s(2m_c) = 0.28$ with $m_c = 1.48$ GeV.

At low electron beam energies ($E_{\text{beam}} < 10$ GeV) such as CLEO, the color–octet process (6) and the color–singlet process (4) dominate over other mechanisms. In Figure 1, we show the polarizations of J/ψ's for each production mechanism. It turns out that each production mechanism leads to vastly different polarization of J/ψ's. The color–singlet gluon mode (4) makes longitudinally polarized J/ψ's, so that $\alpha(z)$ is negative in all the range of z. Especially the high energy J/ψ's made by (4) are almost completely longitudinally polarized ($\alpha \to -1$). The J/ψ's produced via the color–octet mechanism (6) populate only the end point region of phase space, since it is a two body process at the parton level. And they are almost unpolarized ($\eta = 0.35$, or equivalently, $\alpha = -0.05$). The J/ψ's from the color–octet gluon fragmentation process (7) tend to be transversely polarized over all their energy range, and they become completely transversely polarized at the end point region. However since this process is not dominant at CLEO energy, we do not consider it any further. In Figure 1, we can see that at the upper end point of phase space, if the color–octet process (6) dominates, $\alpha \approx -0.05$, and if the color–singlet process (4) dominates, $\alpha \approx -0.86$. So the precise measurement of the J/ψ polarization near the end point region at CLEO may provide us with additional information on NRQCD matrix elements.

In conclusion, we showed that the J/ψ polarization depends distinctively on the J/ψ production mechanism. At the upper end point of phase space where J/ψ's are dominantly produced via the color–octet mechanism $e^+e^- \to (c\bar{c})_8 + g$, the measurement of the J/ψ polarization can give another test of the color–octet mechanism and another constraint on the NRQCD matrix elements. We can also show that the possible breakdown of NRQCD near the phase space boundary [7] does not affect our predictions made in this work. The ongoing data analysis at CLEO will shed light on the relevance of the color–octet mechanism in J/ψ productions through the measurements of the polar angle distribution and polarization of J/ψ's produced in e^+e^- annihilation at CLEO [8].

Acknowledgements

We are grateful to K.T. Chao, Victor Kim and M.B. Wise for useful discussions. This work is supported in part by KOSEF through CTP, BSRI-97-2418, and Nondirected Research by Ministry of Education (PK), and KOSEF Fellowship (JL).

References

1. G.T. Bodwin, E. Braaten and G.P. Lepage, Phys. Rev. **D51**, 1125 (1995).
2. K. Cheung, W.-Y. Keung and T.C. Yuan, Phys. Rev. Lett. **76**, 877 (1996) ; P. Cho, Phys. Lett. **B368**, 171 (1996) ; S. Baek, P. Ko, Jungil Lee and H.S. Song, Phys. Lett. **B389**, 609 (1996).
3. P. Cho and A.K. Leibovich, Phys. Rev. **D54**, 6690 (1996).
4. F. Yuan , C.-F. Qiao and K.-T. Chao, Phys. Rev. **D56**, 321 (1997).
5. E. Braaten, K. Cheung and T.C. Yuan, Phys. Rev. **D48**, 4230 (1993).
6. S. Baek, P. Ko, Jungil Lee and H.S. Song, Phys. Rev. **D55**, 6839 (1997).
7. M. Beneke, I.Z. Rothstein and M.B. Wise, hep-ph/9705286.
8. A.H. Wolf, private communications.

COLOR-OCTET MECHANISM IN $Z^0 \to J/\psi + X$

JUNGIL LEE
Department of Physics, Ohio State University,
Columbus OH43210, USA.

Recent studies of the process $Z^0 \to J/\psi + X$ in the framework of NRQCD are briefly summarized.

1 Introduction

As an effective field theory of QCD, nonrelativistic QCD (NRQCD)[1] is devised to describe heavy quarkonium physics. Short and long distance effects are well separated in the NRQCD factorization formalism. This enables one to calculate the short distance part with perturbative QCD, and to express the soft process in terms of nonperturbative matrix elements, which are well defined in NRQCD. Before the advent of NRQCD, the color-singlet model (CSM)[2,3] had been used as conventional phenomenological model of heavy quarkonium production and decay. While the CSM was successful in many aspects, an artificial K-factor ($\sim$ 2–5) to compensate for the gap between theoretical (CSM) predictions and experimental data on production rates was needed. The uncertainties related to this large K-factor were attributed to bound state wave function normalization, possible next-to-leading order contributions, heavy quark mass uncertainty, *etc.*

However, some experimental data cannot be described by the original CSM even with this large K-factor. For example, the dominant mechanism in large p_T J/ψ production at $p\overline{p}$ collisions has been found to be parton fragmentation in the framework of CSM by Braaten and Yuan[4]. This fragmentation mechanism has been applied to various studies of prompt charmonium production at the Tevatron[5], and the CDF data on prompt J/ψ production[6] qualitatively fit these theoretical predictions. However, this fragmentation contribution in the CSM was not enough. The measured ψ' production rate at CDF was 30 times larger than the theoretical prediction, which is the so-called ψ' anomaly[7], even after including parton fragmentation contributions. Braaten and Fleming proposed the color-octet mechanism as a possible scenario to solve this problem[8]. They included gluon fragmentation into color-octet $Q\overline{Q}$ pairs in a ${}^3S_1^{(8)}$ state, followed by soft transition of this pair into physical quarkonium. This subprocess is not allowed in the CSM, where only the color-singlet $Q\overline{Q}$ pair with the same spectroscopic state (${}^3S_1^{(1)}$) as that of final quarkonium (J/ψ) can hadronize into a physical quarkonium. This work triggered active studies of the NRQCD color-octet mechanism phenomenology.

Cho and Leibovich considered a large class of color-octet diagrams which can contribute to J/ψ production at hadron colliders[9], below the region where fragmentation is dominant. The same approach has been used[10,11] in inclusive photoproduction of J/ψ. The process, $Z^0 \rightarrow J/\psi + X$, has been studied as another test of color-octet mechanism. In this talk, we briefly summarize results on this process.

2 $Z \rightarrow J/\psi + X$

This study is partly motivated by the recent report by the OPAL collaboration [12] that they have observed a number of events where $Z^0 \rightarrow \Upsilon(nS) + X$ for $n = 1, 2, 3$, which exceeds theoretical expectation by a factor of ~ 10: $\sum_{n=1}^{3} B(Z^0 \rightarrow \Upsilon(nS) + X)_{\rm exp} \simeq (1.2^{+0.9}_{-0.6} \pm 0.2) \times 10^{-4}$, compared to the $b-$quark fragmentation contribution of 1.6×10^{-5} [13]. A similar excess was also observed in prompt J/ψ and ψ' production in Z^0 decays, although the experimental errors are quite large [14] :

$$B(Z^0 \rightarrow J/\psi + X) = (3.0 \pm 2.3) \times 10^{-4}, \qquad (2.6 \times 10^{-5}) \tag{1}$$

where the result for the heavy quark fragmentation[13] is shown in parentheses for comparison.

As a possible solution to this puzzle, the color-octet mechanism was applied. This increases the predicted branching ratio by a factor of ~ 10 or so[15]. The most striking theoretical prediction is the energy distribution of the produced J/ψ[15]. The polar angle distribution of J/ψ's produced via color-octet mechanism is also drastically different from that (via charm-quark fragmentation) in the color-singlet model[16]. These two distributions can provide key tests for the idea of the color-octet gluon fragmentation which was invented to solve the ψ' anomaly at the Tevatron. This idea might be the simultaneous solution to the ψ' anomaly at the Tevatron and the excess of events in the channel $Z^0 \rightarrow \Upsilon + X$ at LEP, if the observed energy and polar angle distributions follow NRQCD predictions[15,16]. Studies of the polarization of the final J/ψ including color-octet contributions have also been carried out[17]. This will be another nice probe to test color-octet mechanism in NRQCD.

Acknowledgments

The author thanks for Professor H. S. Song, Professor Pyungwon Ko and Mr. Seungwon Baek for their enjoyable collaborations on the subjects discussed in this summary. This work is supported by the Korea Science and Engineering Foundation through KOSEF Fellowship and CTP program.

References

1. G.T. Bodwin, E. Braaten and G.P. Lepage, *Phys. Rev.* D **51**, 1125 (1995); Erratum-*ibid.* D **55**, 5853 (1997).
2. E.L. Berger and D. Jones, *Phys. Rev.* D **23**, 1521 (1981).
3. R. Baier and R. Rückl, *Z. Phys.* C **19**, 251 (1983).
4. E. Braaten and T.C. Yuan, *Phys. Rev. Lett.* **71**, 1673 (1993).
5. E. Braaten and M.A. Doncheski, S. Fleming and M.L. Mangano, *Phys. Lett.* B **333**, 548 (1994);
   D.P. Roy and K. Sridhar, *Phys. Lett.* B **339**, 141 (1994);
   M. Cacciari and M. Greco, *Phys. Rev. Lett.* **73**, 1586 (1994).
6. The CDF Collaboration, Fermilab-Conf-94/136-E (1994).
7. The CDF Collaboration, M. Mangano, in *Proceedings of the 27th International Conference on High Energy Physics*, Glasgow, Scotland, 1994, edited by P.J. Bussey and I.G. Knowles (IOP, London, 1995), and references therein.
8. E. Braaten and S. Fleming, *Phys. Rev. Lett.* **74**, 3327 (1995).
9. P. Cho, A.K. Leibovich, *Phys. Rev.* D **53**, 150 (1996); *ibid.* D **53**, 6203 (1996).
10. M. Cacciari and M. Krämer, *Phys. Rev. Lett.* **76**, 4128 (1996);
    J. Amundson, S. Fleming and I. Maksymyk, hep-ph/9601298;
    M. Krämer, *Nucl. Phys.* B **459**, 3 (1996);
    F. Maltoni, M.L. Mangano, A. Petrelli, CERN-TH/97-202, hep-ph/9708349.
11. P. Ko, Jungil Lee and H.S. Song, *Phys. Rev.* D **54**, 4312 (1996).
12. OPAL Collaboration, *Phys. Lett.* B **370**, 185 (1996).
13. E. Braaten, K. Cheung and T.C. Yuan, *Phys. Rev.* D **48**, 4230 (1993).
14. OPAL Collaboration, "Observation of J/ψ and ψ' productions in hadronic Z^0 decays," OPAL Physics Note PN178 (1995).
15. K. Cheung, W.-Y. Keung and T.C. Yuan, *Phys. Rev. Lett.* **76**, 877 (1996);
    P. Cho, *Phys. Lett.* B **368**, 171 (1996).
16. S. Baek, P. Ko, Jungil Lee and H.S. Song, *Phys. Rev.* D **55**, 6839 (1997).
17. S. Baek, P. Ko, Jungil Lee and H.S. Song, *Phys. Lett.* B **389**, 609 (1996).

ASPECTS OF HADRONIC J/ψ PRODUCTION

K. SRIDHAR[a]
Theory Group, Tata Institute of Fundamental Research, Homi Bhabha Road, Bombay 400 005, India.

In this talk, I start with a brief introduction to Non-Relativistic QCD (NRQCD) and its applications to quarkonium physics. This theory has provided a consistent framework for the physics of quarkonia, in particular, the colour-octet Fock components predicted by NRQCD have important implications for the phenomenology of charmonium production in experiments. We discuss the applications of NRQCD to J/ψ production at the Tevatron and tests of the theory in other experiments. Finally, we discuss the production of the $^1P_1$ charmonium state, h_c, at large-p_T at the Tevatron. The observation of h_c is interesting, since its existence has yet to be experimentally confirmed. Moreover, this rate is a ***prediction*** of NRQCD, and the observation of the h_c at the Tevatron can, therefore, be used as a test of NRQCD.

J/ψ production has traditionally been described in terms of the colour–singlet model[1,2], which, in spite of its reasonable success with its predictions at lower energies, fails completely in explaining the data on large p_T J/ψ production at the Tevatron[3]. This has led to a major revision of the theory of quarkonium formation. In the colour–singlet model, the relative velocity v between the charm quarks is neglected. However, for charmonia $v^2 \sim 0.3$, and is not negligible. It is possible to include the effects coming from higher orders in a perturbation series in v in non-relativistic QCD (NRQCD)[4,5].

In this picture, the physical charmonium state can be expanded in terms of its Fock components in a perturbation series in v, and the $c\bar{c}$ states appear in either colour–singlet or colour–octet configurations in this series. The colour–octet $c\bar{c}$ state makes a transition into a physical state by the emission of one or more soft gluons, dominantly via $E1$ or $M1$ transitions with higher multipoles being suppressed by powers of v. The cross section for the production of a meson H then takes on the following factorized form:

$$\sigma(H) = \sum_{n=\{\alpha,S,L,J\}} \frac{F_n}{m^{d_n-4}} \langle \mathcal{O}_\alpha^H(^{2S+1}L_J)\rangle \qquad (1)$$

where F_n's are the short distance coefficients and $\mathcal{O}_n$ are local 4-fermion operators, of naive dimension d_n, describing the long distance physics.

In fact, the importance of the colour–octet components was first noted[6] in the case of $P-$wave charmonium decays, and even in the production case the

[a]sridhar@theory.tifr.res.in

importance of these components was first seen [7] in the production of P−state charmonia at the Tevatron. The surprise was that even for the production of S−states such as the J/ψ or ψ', where the colour–singlet components give the leading contribution in v, the inclusion of sub–leading octet states was seen to be necessary for phenomenological reasons [8]. While the inclusion of the colour–octet components seems to be necessitated by the Tevatron charmonium data, the normalisation of these data cannot be predicted because the the long distance matrix elements are not calculable. The data allow a linear combination of octet matrix elements to be fixed [9,10], and much effort was made recently to understand the implications of these colour–octet channels for J/ψ production in other experiments. We discuss some of these below.

1. The prediction [11,12,13] for prompt J/ψ production at LEP in the colour–singlet model is of the order of 3×10^{-5}, which is almost an order of magnitude below the experimental number for the branching fraction obtained from LEP [14]. Recently, the colour–octet contributions to J/ψ production in this channel have been studied [15,16] and it is found that the inclusion of the colour–octet contributions in the fragmentation functions results in a prediction for the branching ratio of 1.4×10^{-4}, which is compatible with the measured value of the branching fraction from LEP [14].

2. The production of J/ψ in low energy e^+e^- machines can also provide a stringent test of the colour–octet mechanism [17]. In this case, the colour–octet contributions dominate near the upper endpoint of the J/ψ energy spectrum, and the signature for the colour–octet process is a dramatic change in the angular distribution of the J/ψ near the endpoint.

3. One striking prediction of the colour–octet fragmentation process both for $p\bar{p}$ colliders and for J/ψ production at the Z−peak, is that the J/ψ coming from the process $g \to J/\psi X$ is produced in a transversely polarised state [18]. For colour–octet $c\bar{c}$ production, this is predicted to be a 100% transverse polarisation, and heavy quark spin symmetry will then ensure that nonperturbative effects which convert the $c\bar{c}$ to a J/ψ will change this polarisation only very mildly. This spin alignment can, therefore, be used as a test of colour–octet fragmentation.

4. The colour–octet components are found [19] to dominate the production processes in fixed target pp and πp experiments. Using the colour–octet matrix elements extracted from elastic photoproduction data it is possible to get a very good description of the $\sqrt{s}$ dependence and also the x_F and rapidity distributions.

5. The associated production of a $J/\psi+\gamma$ is also a crucial test of the colour–octet components[20] and also of the fragmentation picture[21]. Similar tests can be conceived of with double J/ψ production at the Tevatron[22].

6. J/ψ and ψ' production in pp collisions at centre–of–mass energies of 14 TeV at the LHC also provides a crucial test of colour–octet fragmentation[23].

7. The fragmentation process has also definite predictions for the large p_T inelastic J/ψ photoproduction at HERA[24]. At HERA, the colour–singlet component dominates inelastic J/ψ production but at large z ($z > 0.7$) the colour–octet components start growing rapidly. Such a rapid growth is clearly in conflict with experiment[25]. The applicability of NRQCD factorisation at these large values of z has been questioned[26], and more work needs to be done before the disagreement between theory and the HERA inelastic data can be interpreted as a failure of NRQCD.

One important feature of the NRQCD Lagrangian is that it shows an approximate heavy quark symmetry, which is valid to $O(v^2) \sim 0.3$. An implication of this symmetry is that nonperturbative parameters have a weak dependence on the magnetic quantum number. We show that the production cross section of the ${}^1P_1$ charmonium state, h_c, can be *predicted* in NRQCD, because of the heavy quark symmetry. The Tevatron data on χ_c production fixes[10] the colour–octet matrix element which specifies the transition of a ${}^3S_1$ octet state into a ${}^3P_J$ state. We would expect from the heavy quark spin symmetry of the NRQCD Lagrangian that the matrix element for ${}^1S_0^{[8]} \to h_c$ should be of the same order as that for ${}^3S_1^{[8]} \to {}^3P_1$. The production of the ${}^1P_1$ charmonium state, h_c, is, therefore, a crucial prediction of NRQCD.

The production of the h_c is interesting in its own right : charmonium spectroscopy[27] predicts this state to exist at the centre–of–gravity of the $\chi_c({}^3P_J)$ states. While the E760 collaboration at the Fermilab has reported[28] the first observation of this resonance, its existence needs further confirmation. In the following we report on the calculations for h_c production at the Tevatron presented in Ref.[29].

Being a $P-$state, the leading colour–singlet contribution is already at $O(v^2)$, and at the same order we have the octet production of the ${}^1P_1$ state through an intermediate ${}^1S_0$ state. The octet channel is, therefore, expected to be very important for the production of this state. Once produced, the h_c can be detected by its decay into a $J/\psi + \pi^0$. The subprocesses that we are interested in are the following:

$$g + g \quad \to {}^1P_1^{[1]} + g,$$

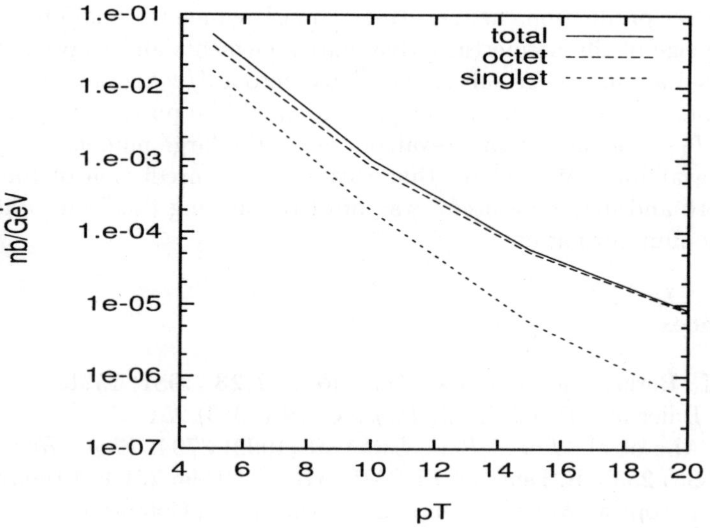

Figure 1: $d\sigma/dp_T$ in nb/GeV for h_c production (after folding in the $\mathrm{Br}(h_c \to J/\psi + \pi)$=0.5) at 1.8 TeV c.m. energy with $-0.6 \leq y \leq 0.6$.

$$\begin{aligned} g+g &\to {}^1S_0^{[8]} + g, \\ q(\bar{q})+g &\to {}^1S_0^{[8]} + q(\bar{q}), \\ q+\bar{q} &\to {}^1S_0^{[8]} + g. \end{aligned} \tag{2}$$

The ${}^1S_0^{[8]} \to h_c$ is mediated by a gluon emission in a $E1$ transition.

The cross section $d\sigma/dp_T$ for h_c production at the Tevatron energy ($\sqrt{s}$ = 1.8 TeV) presented in Ref. [29] is shown in Fig. 1. For 20 pb^{-1} total luminosity, for p_T integrated between 5 and 20 GeV we expect of the order of 650 events in the $J/\psi+\pi$ channel. Of these, the contribution from the colour–singlet channel is a little more than 40, while the octet channel gives more than 600 events. The colour–octet dominance is more pronounced at large p_T. Recent results on J/ψ production from CDF are based on a total luminosity of 110 pb^{-1}. For this sample, more than 3000 events can be expected to come from the decay of the h_c into a J/ψ and a π. With this large event rate, the h_c should certainly be observable if the π^0 coming from its decay can be reconstructed efficiently.

In conclusion, we find that a reasonably large rate for the production of the ${}^1P_1$ is expected at the Tevatron, with a dominant contribution from the colour–octet production channel. Heavy quark symmetry relations allow us to infer the size of the nonperturbative matrix elements and to predict the rate for h_c production and its subsequent decay into a J/ψ and a π. By looking at J/ψ events associated with a soft pion, it should be possible to pin down the elusive ${}^1P_1$ resonance at the Tevatron, given the large number of J/ψ events already available. We believe that this is a firm prediction of the NRQCD framework and may be a useful way of distinguishing this from other models of quarkonium formation.

References

1. E.L. Berger and D. Jones, *Phys. Rev.* **D 23** (1981) 1521.
2. R. Baier and R. Rückl, *Z. Phys.* **C 19** (1983) 251.
3. F. Abe et al., *Phys. Rev. Lett.* **69** (1992) 3704; *Phys. Rev. Lett.* **71** (1993) 2537; R. Demina, FERMILAB-CONF-96-201-E, Presented at the 11th Topical Workshop on Proton-Antiproton Collider Physics (PBARP 96), Padua, Italy, 26th May – 1st June 1996.
4. W.E. Caswell and G.P. Lepage, *Phys. Lett.*, **B 167** (1986) 437.
5. G.T. Bodwin, E. Braaten and G.P. Lepage, *Phys. Rev.* **D 51** (1995) 1125.
6. G.T. Bodwin, E. Braaten and G.P. Lepage, *Phys. Rev.* **D 46** (1992) R1914.
7. E. Braaten, M.A. Doncheski, S. Fleming and M. Mangano, *Phys. Lett.* **B 333** (1994) 548; D.P. Roy and K. Sridhar, *Phys. Lett.* **B 339** (1994) 141; M. Cacciari and M. Greco, *Phys. Rev. Lett.* **73** (1994) 1586.
8. E. Braaten and S. Fleming, *Phys. Rev. Lett.* **74** (1995) 3327.
9. M. Cacciari, M. Greco, M. Mangano and A. Petrelli, *Phys. Lett.* **B 356** (1995) 553.
10. P. Cho and A.K. Leibovich, *Phys. Rev.*, **D 53** (1996) 150; *Phys. Rev.*, **D 53** (1996) 6203.
11. V. Barger, K. Cheung and W.-Y. Keung, *Phys. Rev.* **D 41** (1990) 1541.
12. W.-Y. Keung, *Phys. Rev.* **D 23** (1981) 2072; J.H. Kühn and H. Schneider, *Phys. Rev.* **D 24** (1981) 2996.
13. E. Braaten, K.Cheung and T.C. Yuan, *Phys. Rev.* **D 48** (1993) 4230.
14. O. Adriani et al. (L3 collaboration), *Phys. Lett.* **B 288** (1992) 412; P. Abreu et al. (DELPHI collaboration), *Phys. Lett.* **B 341** (1994) 109; G. Alexander et al. (OPAL collaboration), CERN Preprint CERN-PPE-95-153.
15. K. Cheung, W.-Y. Keung and T.C. Yuan, *Phys. Rev. Lett.* **76** (1996) 877; S. Baek, P. Ko, J. Lee and H.S. Song, *Phys. Lett.* **B 389** (1996)

609.
16. P. Cho, *Phys. Lett.* **B 368** (1996) 171.
17. E. Braaten and Y. Chen, *Phys. Rev. Lett.* **76** (1996) 730.
18. P. Cho and M. Wise, *Phys. Lett.* **B 346** (1995) 129; S. Baek, P. Ko, J. Lee and H.S. Song, *Phys.Rev.* **D 55** (1997) 6839; M. Beneke and I. Rothstein, *Phys. Lett.* **B 372** (1996) 157; M. Beneke and M. Krämer, *Phys. Rev.* **D 55** (1997) 5269.
19. S. Gupta and K. Sridhar, *Phys. Rev.* **D 54** (1996) 5545; M. Beneke and I. Rothstein, *Phys. Rev.* **D 54** (1996) 2005; S. Gupta and K. Sridhar, *Phys. Rev.* **D 55** (1997) 2650.
20. C.S. Kim, J. Lee and H.S. Song, KEK Preprint KEK-TH-474, hep-ph/9610294.
21. D.P. Roy and K. Sridhar, *Phys. Lett.* **B 341** (1995) 413.
22. V. Barger, S. Fleming and R.J.N. Phillips, *Phys. Lett.* **B 371** (1996) 111.
23. K. Sridhar, *Mod. Phys. Lett.* **A 11** (1996) 1555.
24. R.M. Godbole, D.P. Roy and K. Sridhar, *Phys. Lett.* **B 373** (1996) 328.
25. N. Cacciari and M. Krämer, *Phys. Rev. Lett.* **76** (1996) 4128; P. Ko, J. Lee, H.S. Song, *Phys. Rev.* **D 54** (1996) 4312.
26. M. Beneke, I.Z. Rothstein, M.B. Wise, *Phys. Lett.* **B 408** (1997) 373.
27. Y.-P. Kuang, S.F. Tuan and T.-M. Yan, *Phys. Rev.* **D 37** (1988) 1210.
28. T.A. Armstrong et al., *Phys. Rev. Lett.* **69** (1992) 2337.
29. K. Sridhar, *Phys. Rev. Lett.* **77** (1996) 4880.

THE KEK-PS LONG BASELINE NEUTRINO OSCILLATION EXPERIMENT (E362)

MAKOTO SAKUDA
Institute of Particles and Nuclear Studies,
KEK, Tsukuba, 305 Japan
Representing the K2K Collaboration[1]

The KEK-PS E362 experiment will search for neutrino oscillations by measuring the ν_μ flux with a near detector at KEK and with the Super-Kamiokande detector, 250 km away from KEK. A wide-band ν_μ beam with a mean energy of 1.4 GeV will be produced at the KEK proton synchrotron. The experiment can detect neutrino oscillations with $\Delta m^2 > 0.003$ eV^2 and $\sin^2\theta > 0.45$ for $\nu_\mu \to \nu_\tau$ and $\sin^2\theta > 0.1$ for $\nu_\mu \to \nu_e$ at the 3σ confidence level, respectively. It is scheduled to start in early 1999.

1 Introduction

In the standard model the neutrinos are assumed to be massless and there is no mixing between the different neutrino flavors. These two important issues can be tested by searching for neutrino oscillations. The present upper limits from direct mass measurement experiments are 15 eV$/c^2$, 170 keV$/c^2$ (90%C.L.) and 24 MeV$/c^2$ (95%C.L.) for the ν_e, ν_μ and ν_τ [2], respectively. Although drastic improvements in these direct measurements are unlikely, neutrino oscillation experiments can probe a neutrino mass range not accessible by direct experiments.

The probability of neutrino oscillation between two flavor eigenstates, ν_a and ν_b, is given by

$$P(\nu_a \to \nu_b) = \sin^2 2\theta \; \sin^2(\frac{1.27\Delta m^2 L}{E_\nu}) \quad , \tag{1}$$

where θ is the flavor mixing angle, L the distance between the neutrino source and the detector in kilometers, E_ν the neutrino energy in GeV, and $\Delta m^2 = |m_1^2 - m_2^2|$ is the difference in the mass squared of the mass eigenstates in eV^2. The oscillation sensitivity of the detector is roughly given by Δm^2 (eV^2) $\sim$ E_ν (GeV)$/L$ (km), where the probability of Eq. 1 is maximal. The K2K experiment has an average value of $E/L \sim 6 \times 10^{-3}$ eV^2.

The Kamiokande experiment [3] and some underground detectors [4] have reported atmospheric neutrino ratios (ν_μ/ν_e) which are significantly lower than the value predicted by theoretical calculations. Atmospheric neutrinos are produced in cosmic ray interactions in the atmosphere ($p+N \to \pi^{+,-}/K^{+,-}+X \to$

$\nu_\mu/\nu_e + X$). A possible explanation of this deficit was given by Kamiokande in terms of neutrino oscillations, either $\nu_\mu \to \nu_e$ or $\nu_\mu \to \nu_\tau$, with $\Delta m^2 \simeq 10^{-2}$ eV^2 [3]. The 90% allowed region suggested by Kamiokande and the excluded regions by the previous experiments are summarized for the $\nu_\mu \to \nu_\tau$ oscillation [5] and for the $\nu_\mu \to \nu_e$ oscillation [6] in Figs. 1(a) and (b), respectively. Recently, the Super-Kamiokande experiment [7] has confirmed the low ν_μ/ν_e ratio and its zenith angle distribution with better accuracy than Kamiokande. Their preliminary result for the 90%C.L. allowed region for the $\nu_\mu \to \nu_\tau$ oscillation is consistent with the Kamiokande result, but gives a somewhat lower Δm^2, as shown in Fig. 1(a). Further, the CHOOZ experiment [8] has shown no evidence of neutrino oscillations in the disappearance mode $\overline{\nu}_e \to \overline{\nu}_X$ for the parameter region $\Delta m^2 > 0.9 \times 10^{-3}$ eV^2 and $\sin^2\theta > 0.18$, as shown in Fig. 1(b). This excludes the possibility of $\nu_\mu \to \nu_e$ oscillations suggested by Kamiokande. The possibility of $\nu_\mu \to \nu_\tau$ oscillations in the range $\Delta m^2 = 10^{-2} \sim 10^{-3}$ eV^2 still remains, even if these latest results are considered.

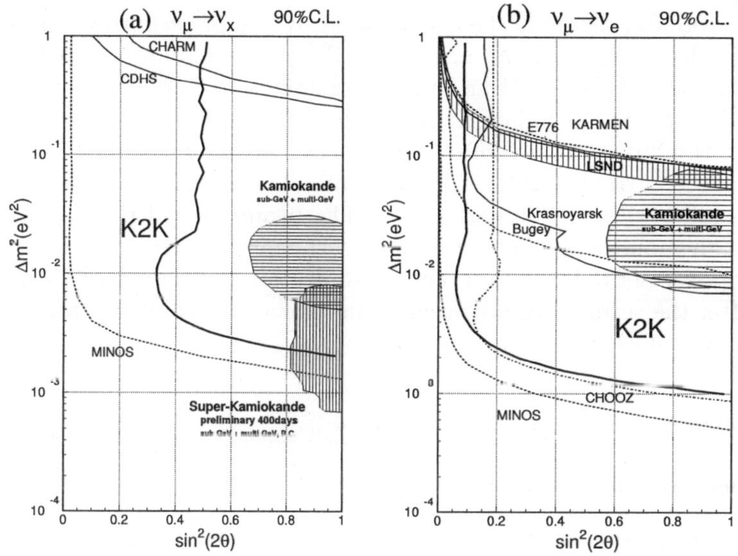

Figure 1: 90% C.L. exclusion contours for (a) $\nu_\mu \to \nu_\tau$ oscillations and (b) $\nu_\mu \to \nu_e$ oscillations, with the allowed regions by Kamiokande. The 90% C.L. allowed region for $\nu_\mu \to \nu_\tau$ oscillations by Super-Kamiokande is preliminary. The contours of K2K are indicated by the solid lines. The MINOS experiment is scheduled to start in 2001.

The KEK-PS E362 experiment [9,1], now called K2K, started construction in March, 1995, in order to detect both $\nu_\mu \to \nu_e$ and $\nu_\mu \to \nu_\tau$ oscillations in the parameter regions suggested by the Kamiokande experiment. It is expected to cover the parameter region $\Delta m^2 > 3 \times 10^{-3}$ eV^2 at 3σ C.L. with 10^{20} protons on target, which is equivalent to 3 years of operation with the expected proton intensity.

2 Neutrino Beam

The neutrino beamline which will be used in the K2K experiment is now being built at KEK for the first time in Japan [10]. A proton beam of 12 GeV will be extracted in a single revolution of 1.1 μsec from the KEK proton synchrotron[11]. The intensity of the beam is expected to be 6 $\times 10^{12}$ protons per burst, where the repetition rate is 1 burst per 2.2 seconds. The proton beam will be transported to an aluminum target embedded in the first magnetic horn[12]. The beam size at the target is about 2 cm in diameter. Only the positively charged particles (π^+ and K^+) produced in the target are focused onto a nearly parallel beam in a toroidal magnetic field generated by a pair of horns and they are guided to a decay region 200m in length. Thus, a wide-band ν_μ beam with a mean energy of 1.4 GeV and an angular divergence of about 3 mrad will be obtained. The ν_μ flux at Super-Kamiokande is enhanced by a factor of 14 compared with the flux without horns. The horns were successfully tested with a nominal pulsed current of 250 kA in fall, 1997. The installation of the magnets in the beamline will start in early 1998. The beamline will be equipped with several beam monitors. They include a secondary emission chamber and a current transformer before the target, an ionization chamber and a Cherenkov counter after the second horn, as well as an ionization chamber after the beam dump. They will assure stable operation of the beamline.

3 Detector

3.1 Design principle

The near neutrino detector at KEK is located 300 m downstream of the target. It is schematically shown in Fig. 2. It comprises a fine-grained detector and 1 kton water Cherenkov detector. The far detector is Super-Kamiokande, which has been operational since April, 1996. The expected event rate for each detector is summarized in Table 1. The main purpose of the near detector is to understand the beam properties, such as the ν_μ flux, the energy spectrum $\Phi(E_{\nu_\mu})$, the $x-y$ profile, and the ν_e contamination in the ν_μ beam. In addition, the 1 kton detector will study neutrino interactions with the neutrino beam

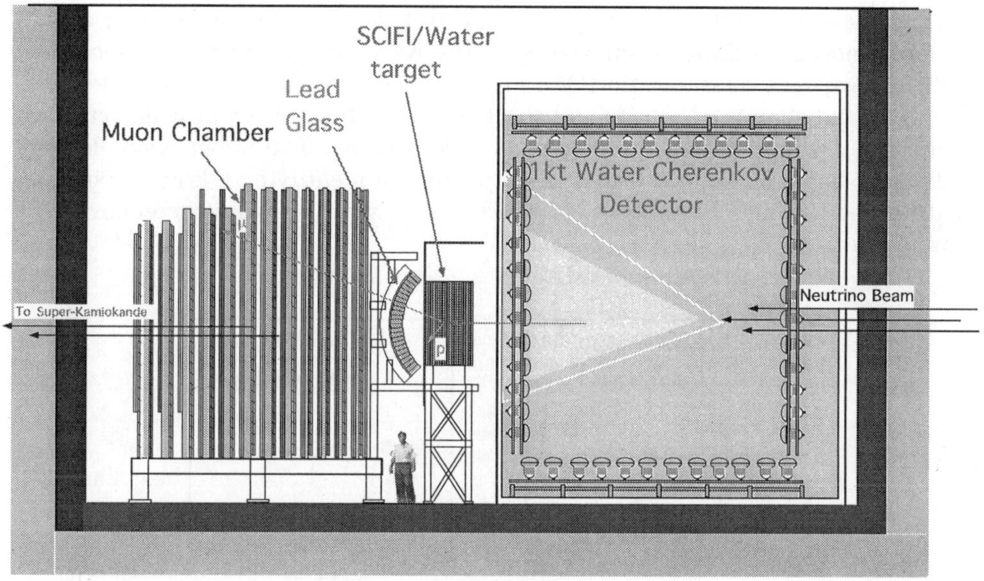

Figure 2: Near neutrino detector. The cylindrical experimental hall is 24m in diameter and 16m deep.

Table 1: Event Rate with near and far detectors at 10^{20} protons on target. The expected flux is also shown. CC, NC and QE denote the charged current, the neutral current and quasi-elastic, respectively. The event rate at the Super-Kamiokande is estimated assuming no oscillations.

	Fine-grained detector	1 kton detector	Super-Kamiokande
Fiducial volume	$\sim$ 4 ton	$\sim$ 21 ton	$\sim$ 22.5 kton
Flux (ν_μ/cm^2)	3×10^{12}	3×10^{12}	2.3×10^6
ν_μ CC events	127K	408K	350
QE events	44K	142K	120
ν_μ NC events	45K	144K	120
Total ν_μ events	172K	552K	470
$\nu_e/\nu_\mu = 1\%$ case			
ν_e CC events	1250	4K	4
QE events	320	1K	1

using the same water Cherenkov-detector technique as the far detector. The main function of Super-Kamiokande in the K2K experiment is to detect the charged-current quasi-elastic (QE) events, $\nu_\mu + n \to \mu^- + p$ and $\nu_e + n \to e^- + p$, and compare the rate and spectrum with those of the near detector in order to see if there is any difference between them. The Super-Kamiokande detector has fully demonstrated the capability of identifying QE events in the atmospheric neutrino analysis, in which events with a single Cherenkov ring were selected.

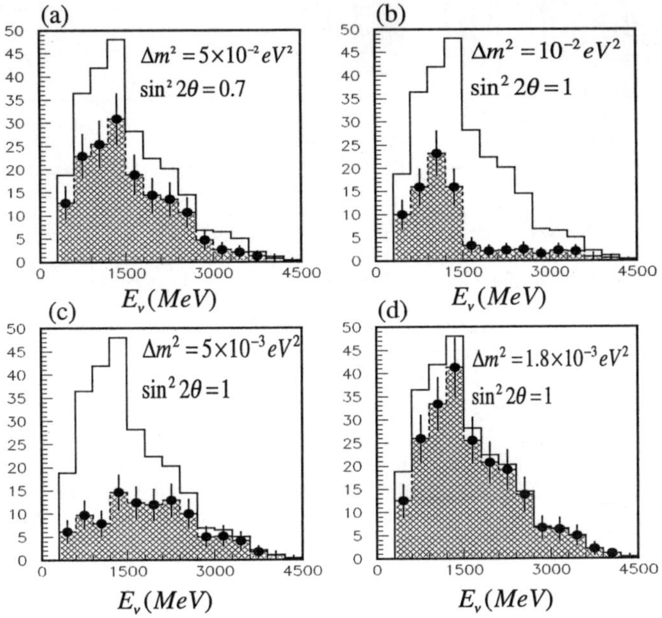

Figure 3: Expected ν_μ energy spectra with Super-Kamiokande at 10^{20} protons on target for various $\nu_\mu \to \nu_\tau$ oscillation parameters (a)-(d) (data points with error bars) and for no oscillations (solid histogram).

Since the beam energy is less than the τ production threshold (3.5 GeV) for $\nu_\tau + n \to \tau^- + p$, we cannot detect τ's, even if $\nu_\mu \to \nu_\tau$ oscillations occur. Therefore, in order to detect $\nu_\mu \to \nu_\tau$ oscillations, we must detect any reduction of the ν_μ beam at Super-Kamiokande as compared to the beam which we expect from the near detector by extrapolation. The expected spectra at the far detector are shown in Fig. 3 for various $\nu_\mu \to \nu_\tau$ oscillation parameters (a)-(d) (points with error bars) and for no oscillations (solid histogram). On the other hand, $\nu_\mu \to \nu_e$ oscillations can easily be detected in the appearance mode. While we expect almost no events with an electron for $E_e > 1$ GeV, we will see more than one hundred events for the maximal mixing case.

The fine-grained detector is designed so that it can measure the charged-current QE events $\nu_\mu + n \to \mu^- + p$ and $\nu_e + n \to e^- + p$. The neutrino flux $\Phi(E_{\nu_\mu})$ can be obtained as

$$\Phi(E_{\nu_\mu}) = \frac{N(E_{\nu_\mu})}{\sigma_{QE}(E_{\nu_\mu}) N_{target}\, \epsilon}\ , \qquad (2)$$

where $N(E_{\nu_\mu})$ is the number of observed QE events between E_{ν_μ} and $E_{\nu_\mu} + \Delta E$ (after background subtraction), $\sigma_{QE}(E_{\nu_\mu})$ is the QE cross-section, N_{target} is the number of target neutrons and ϵ is the detection efficiency. The cross section $\sigma_{QE}(E_\nu)$ is calculated using a dipole form factor from a vector mass (M_V) of 0.84 GeV/c^2 and axial vector mass (M_A) of 1.09±0.03±0.02 GeV/c^2[13]. The uncertainty in M_A corresponds to a ±3% error in σ_{QE} for $Q^2 > 0.2$ GeV$^2/c^2$.

The neutrino energy (E_{ν_μ}) is calculated from the measured muon energy (E_μ) and the muon scattering angle (θ_μ) with respect to the neutrino beam direction:

$$E_{\nu_\mu} = \frac{m_N E_\mu - m_\mu^2/2}{m_N - E_\mu + p_\mu \cos\theta_\mu}\ , \qquad (3)$$

where m_N (m_μ) is the mass of the neutron (muon). The reconstruction of the neutrino energy is essential in the analysis of neutrino oscillation. The energy determination at $E_\nu < 1$ GeV is limited by the Fermi motion of the neutron in oxygen, $\Delta E \sim 100$ MeV, which is ignored in Eq. 3.

Our goal is to measure the ν_μ flux with an accuracy better than 10%. As can be seen from Eq. 2, the efficiency of the detector and the fiducial volume must be known to 5% accuracy or better. Most importantly, the separation of the QE events from other, inelastic, events must be good enough to determine $N(E_{\nu_\mu})$ unambiguously. Equations similar to Eqs. 2 and 3 hold for the ν_e flux $\Phi(E_{\nu_e})$ and the E_{ν_e} determination, where measurements of E_e and θ_e are required. A small ν_e contamination (1%) in the ν_μ beam originates from μ and K decays.

3.2 Fine-grained Detector

Considering the above requirements, we have designed the fine-grained detector as follows. It consists of the water target, a scintillating fiber (sci-fi) detector, scintillating counters surrounding the sci-fi detector, lead glass counters, and a muon range detector.

The sci-fi tracking detector [14] consists of twenty 2.4 m × 2.4 m modules. The structure of each module is a 1.6 cm thick honeycomb panel. A double-layered sci-fi ribbon (= doublet) is glued in the horizontal direction on one side and in the vertical direction on the other side of the panel. Each module is placed after a 6 cm thick water target of 2.4 m × 2.4 m area, contained in 1.8 mm thick aluminum tubes with a cross section of 6 cm × 16 cm (see Fig. 4). The scintillating fibers are 0.7 mm in diameter. We obtain a detection efficiency of $> 99\%$ and a position resolution of $\sim 280\ \mu$m for a sci-fi doublet using cosmic ray muons. The scattering angle of the charged tracks can be measured by the sci-fi detector with an accuracy better than 1 degree. The present performance of the sci-fi detector, although still preliminary, is good enough for reconstructing the charged tracks, and thus for separating quasi-elastic events from inelastic events and for determining the fiducial volume of neutrino interactions.

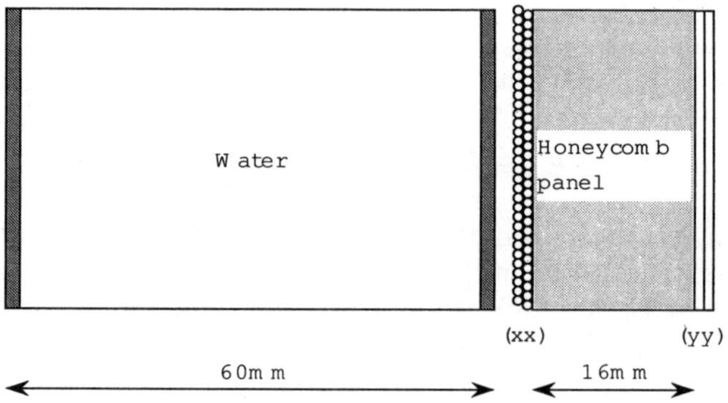

Figure 4: Set of a water target and a sci-fi module. Only a part of the cross section is shown. The sci-fi detector comprises 20 sets.

The muon range detector consists of twelve iron plates and drift chambers. The energy of the muon (E_μ) is measured by the muon range detector with an accuracy of $\sim 8\%$ for $E_\mu = 0.5 \sim 3$ GeV. Electrons are identified by lead glass counters. The energy resolution of the lead glass counter is $\Delta E/E = 10\%/\sqrt{E}$.

According to a detector simulation, we can determine the beam flux $\Phi(E_{\nu_\mu})$ with $\sim 8\%$ accuracy and the neutrino energy with better accuracy than 10% in the region $E_{\nu_\mu} = 0.5 \sim 3.0$ GeV. We can also measure the ν_e/ν_μ ratio with $\pm 0.3\%$ statistical errors at $\nu_e/\nu_\mu = 1\%$.

The data acquisition system at the near detector is simple since the event rate is low. Beam-associated data will be taken once for every trigger gate, synchronized with the beam extraction. Both the near and far detectors record the GPS time clock so that the far detector can identify those events at the offline stage which occur during the beam gate.

Even if systematic effects of the near and far detectors are well understood, there is another source of systematic errors in the $\nu_\mu \to \nu_\tau$ oscillation search: the uncertainty in the extrapolation of the beam flux from the near detector to the far detector. In order to reduce the systematics, the beam properties must be well measured and understood with the near detector and beam monitors. In estimating the contours in Fig. 1, we assumed an overall uncertainty of 10% in addition to the systematic errors of the detectors.

4 Summary

The beamline and the near detector of the K2K experiment will be completed in December, 1998. It will be the first accelerator experiment to search for $\nu_\mu \to \nu_\tau$ and $\nu_\mu \to \nu_e$ oscillations in the parameter region of $\Delta m^2 > 10^{-3}$ eV^2 in 1999.

References

1. KEK-PS E362, now called K2K (Kek To Kamioka), is a collaboration of Japan, Korea and the United States. The following institutions are members of the K2K experiment: ICRR, KEK, Kobe, Niigata, Okayama, Tohoku, Tokai, Chonnam, Dongshin, Korea Univ., Seoul Nat.Univ., Boston, Los Alamos, SUNY, Hawaii, UC Irvine, Univ. of Washington.
2. Review of Particle Physics, *Phys. Rev.* D **54**, 1 (1996).
3. K.S.Hirata *et al*, *Phys. Lett.*B **205**, 416 (1988); K.S.Hirata *et al*, *Phys. Lett.*B **2280**, 146 (1992); Y.Fukuda *et al*, *Phys. Lett.*B **335**, 237 (1994).
4. D.Casper *et al* (IMB), *Phys. Rev. Lett.* **66**, 2561 (1991); R.Becker-Szendy *et al* (IMB), *Phys. Rev.* D **46**, 3720 (1992); W.W.M Allison et al.(Soudan2), Phys.Lett.B391,491(1997).
5. F.Dydak *et al* (CDHS), *Phys. Lett.*B **134**, 281 (1984); F. Bergsma *et al* (CHARM), *Z. Phys.* C **40**, 171 (1988); MINOS collab.,"P-875:A long baseline neutrino oscillation experiment at Fermilab", NuMI-L-63, Feb. 1995.
6. G. Zacek *et al* (Gosgen), *Phys. Rev.* D **34**, 2621 (1986); B. Armbruster *et al* (KARMEN), Nucl.Phys.B(Proc.Suppl.)38,235(1995); G.S.Vidyakin *et al* (Krasnoyarsk), JETP Lett. 59,390(1994); B.Achkar *et al* (Bugey), *Nucl. Phys.* B **434**, 503 (1995); L.Borodovsky *et al* (E776), *Phys. Rev.*

Lett. **68**, 274 (1992); C.Athanassopoulos *et al* (LSND),UCRHEP-E197, Sep. 1997.

7. M.Nakahata, this proceedings.
8. M. Apollonio *et al* (CHOOZ), HEP-EX/9711002, November 1997.
9. K. Nishikawa et al., E362 KEK-PS proposal, March, 1995;K.Nishikawa, Nucl.Phys.B(Proc.Suppl.)59(1997)289.
10. H. Noumi *et al*, *Nucl.Instrum. Methods* A **398**, 399 (1997).
11. H. Sato, "Performance of the KEK 12 GeV-PS and Upgrade", in Proceedings of the 16th RCNP Osaka Int. Symp. on Multi-GeV High Performance Accelerators and Related Technology", Osaka, March 1997.
12. Y. Yamanoi et al., "Large Horn Magnets at the KEK Neutrino Beam Line", KEK Preprint 97-225, November 1997.
13. L.A. Ahrens *et al* (E734),*Phys. Lett.*B **202**, 284 (1988).
14. A. Suzuki, in Proceedings of SCIFI97 Conference on Scintillating and Fiber Detectors, at Notre Dame, November 2-6, 1997.

CP- AND T-VIOLATION EFFECTS IN LONG BASELINE NEUTRINO OSCILLATION EXPERIMENTS

MASAFUMI KOIKE

Institute for Cosmic Ray Research, University of Tokyo, Midori-cho, Tanashi, Tokyo 188, Japan

JOE SATO

Department of Physics, University of Tokyo, Hongo, Bunkyo-ku, Tokyo 133, Japan

We examine how large $CP-$ and $T-$violation effects are allowed in long baseline neutrino experiments with three generations of neutrinos, considering both the solar neutrino deficit and the atmospheric neutrino anomaly. We consider two cases: Adopting only the mass parameter suggested by the atmospheric neutrino anomaly, which leads to an energy-independent reduction of the solar neutrino flux, a large CP-violation effect may be obtained. Adopting different mass parameters for the solar and atmospheric neutrino deficit, sizable $CP-$violation effects can be seen. In the latter case, however, $CP-$violation cannot be observed directly due to matter effect. Two methods are shown to separate a pure CP violating effect from matter effects in this case: One is to observe envelopes of the curves of oscillation probabilities as functions of the neutrino energy; the merit of this method is that only a single detector is needed to determine the presence of CP violation. The other is to compare experiments with at least two different baseline lengths; this has the merit that it needs only a narrow energy range of oscillation data.

1 Introduction

CP and T violation is a fundamental and important problem of particle physics and cosmology. CP violation has been observed only in the hadron sector, and it is very hard for us to understand where the CP violation originates. We can expect invaluable clues to the origin of CP violation and to physics beyond the Standard Model if we observe CP violation in the lepton sector through neutrino oscillation experiments.

The neutrino oscillation search is a powerful experiment which can examine the masses and/or mixing angles of neutrinos. Several underground experiments, in fact, have shown a lack of solar neutrinos[1,2,3,4], and an anomaly in the atmospheric neutrino fluxes,[5,6,7 a] strongly indicating neutrino oscillations.[10,11,12] The solar neutrino deficit implies $10^{-5} \sim 10^{-4}$ eV^2 as a difference of the masses squared (δm^2), while the atmospheric neutrino anomaly suggests δm^2 around $10^{-3} \sim 10^{-2}$eV^2.[10,11,12] It is, however, possible to explain both experiments by only one δm^2 scale around $\mathcal{O}(10^{-3} \sim 10^{-2}$ eV^2).[22]

[a]Some experiments have not observed the atmospheric neutrino anomaly.[8,9]

These results encourage us to perform long baseline neutrino oscillation experiments, which will be operated in the near future.[13,14] It is now desirable to examine the possibility to observe not only neutrino oscillations but also CP or T violation by long baseline experiments.[15,16,17,18,19,20,21]

In this paper we review our papers [16,17,18,19] to show how large CP and T violation effects we may see in long baseline neutrino experiments with three generations of neutrinos, considering both the solar neutrino deficit and the atmospheric neutrino anomaly. In Sec. 2 we briefly review CP and T violation in neutrino oscillations. We then consider how large T and CP violating effects can be. The case where both δm^2's are $\mathcal{O}(10^{-2}\ \text{eV}^2)$ is considered in Sec. 3. Sec. 4 treats the "disparate" case, where the two δm^2's are $\mathcal{O}(10^{-2}\ \text{eV}^2)$ and $\mathcal{O}(10^{-4}\ \text{eV}^2)$. In Sec. 5 we summarize our work and give discussions.

2 Formulation of CP and T Violation in Neutrino Oscillation

Let us briefly review CP and T violation in neutrino oscillation [24,25,26] to clarify our notation.

We assume three generations of neutrinos which have mass eigenvalues $m_i (i = 1, 2, 3)$ and mixing matrix U relating the flavor eigenstates $\nu_\alpha (\alpha = e, \mu, \tau)$ and the mass eigenstates in the vacuum $\nu'_i (i = 1, 2, 3)$ as $\nu_\alpha = U_{\alpha i} \nu'_i$. We parameterize U [27,28,29] using Gell-Mann matrices λ_i's as

$$U = \mathrm{e}^{\mathrm{i}\psi\lambda_7} \cdot \mathrm{diag}(1, 1, \mathrm{e}^{\mathrm{i}\delta}) \cdot \mathrm{e}^{\mathrm{i}\phi\lambda_5} \cdot \mathrm{e}^{\mathrm{i}\omega\lambda_2}. \tag{1}$$

The evolution equation in matter for the flavor eigenstate vector $\nu \equiv (\nu_\mathrm{e}, \nu_\mu, \nu_\tau)^\mathrm{T}$ is given by

$$\mathrm{i}\frac{\mathrm{d}\nu}{\mathrm{d}x} = \frac{1}{2E}\tilde{U}\mathrm{diag}(\tilde{m}_1^2, \tilde{m}_2^2, \tilde{m}_3^2)\tilde{U}^\dagger \nu \equiv H\nu \tag{2}$$

(tilde indicates quantities in matter in the following), with [30,31]

$$\tilde{U}\mathrm{diag}(\tilde{m}_1^2, \tilde{m}_2^2, \tilde{m}_3^2)\tilde{U}^\dagger = U\mathrm{diag}(0, \delta m_{21}^2, \delta m_{31}^2)U^\dagger + \mathrm{diag}(a, 0, 0). \tag{3}$$

Here $\delta m_{ij}^2 = m_i^2 - m_j^2$ and

$$a \equiv 2\sqrt{2}G_\mathrm{F} n_\mathrm{e} E = 7.56 \times 10^{-5}\mathrm{eV}^2 \frac{\rho}{\mathrm{g\,cm^{-3}}} \frac{E}{\mathrm{GeV}}, \tag{4}$$

where n_e is the electron density and ρ is the matter density. The solution of Eq.(2) is then

$$\nu(x) = \mathrm{T}\exp\left(-\mathrm{i}\int_0^x \mathrm{d}s H(s)\right)\nu(0) \equiv S(x)\nu(0), \tag{5}$$

giving the oscillation probability as

$$P(\nu_\alpha \to \nu_\beta; E, L) \;=\; |S_{\beta\alpha}(L)|^2 \quad (\alpha, \beta = \mathrm{e}, \mu, \tau). \tag{6}$$

The oscillation probability for the antineutrinos $P(\bar{\nu}_\alpha \to \bar{\nu}_\beta)$ is obtained by replacing $a \to -a$ and $U \to U^*$(i.e. $\delta \to -\delta$) in Eq.(6).

We assume in the following that the matter density is independent of space and time for simplicity.[b] In this case we obtain T-violation effect as[32]

$$P(\nu_\alpha \to \nu_\beta; E, L) - P(\nu_\beta \to \nu_\alpha; E, L) = 4Jf, \tag{7}$$

where

$$J \;\equiv\; -\mathrm{Im}\, \tilde{U}_{\beta 1}\tilde{U}^*_{\beta 2}\tilde{U}^*_{\alpha 1}\tilde{U}_{\alpha 2} = \pm s_{\tilde{\delta}} c^2_{\tilde{\phi}} s_{\tilde{\phi}} c_{\tilde{\psi}} s_{\tilde{\psi}} c_{\tilde{\omega}} s_{\tilde{\omega}}, \tag{8}$$

$$f \;\equiv\; \sin\Delta_{21} + \sin\Delta_{32} + \sin\Delta_{13} = -4\sin\frac{\Delta_{21}}{2}\sin\frac{\Delta_{32}}{2}\sin\frac{\Delta_{13}}{2}, \tag{9}$$

with $\Delta_{ij} \equiv (\delta\tilde{m}^2_{ij}L)/(2E)$ and $c_{\tilde{\psi}} = \cos\tilde{\psi}, s_{\tilde{\phi}} = \sin\tilde{\phi}$, etc. The $+$ $(-)$ sign in Eq.(8) is taken when α, β are in cyclic (anti–cyclic) order, i.e. $+$ for $(\alpha, \beta) = (e, \mu), (\mu, \tau)$ or (τ, e).

In the vacuum the CPT theorem relates CP violation and T violation as

$$P(\nu_\alpha \to \nu_\beta) - P(\bar{\nu}_\alpha \to \bar{\nu}_\beta) = P(\nu_\alpha \to \nu_\beta) - P(\nu_\beta \to \nu_\alpha). \tag{10}$$

3 Case of Comparable Mass Differences

Acker and Pakvasa[22] stated the possibility of explaining both solar and atmospheric neutrino experiments by one δm^2 scale of $\mathcal{O}(10^{-3} \sim 10^{-2}\ \mathrm{eV}^2)$. We first examine this "comparable mass differences" case. We use a parameter set $(\delta m^2_{21}, \delta m^2_{31}) = (3.8, 1.4) \times 10^{-2}\mathrm{eV}^2$, $(\omega, \phi, \psi) = (19°, 43°, 41°)$ with arbitrary δ, derived by Yasuda[12] through an analysis of the atmospheric neutrino anomaly. We need not distinguish T violation and CP violation in this mass region, since the matter effect is negligibly small, and equation (10) holds.

Using Eqs.(7) and (10), this parameter set gives the CP-violation effect

$$\begin{aligned} P(\nu_\alpha \to \nu_\beta) - P(\bar{\nu}_\alpha \to \bar{\nu}_\beta) \;&=\; 0.22(\sin 3.8y + \sin 2.4y - \sin 1.4y)\sin\delta \\ &=\; 0.22 f(y)\sin\delta, \end{aligned} \tag{11}$$

where $y = 2.5(L/100\ \mathrm{km})/(E/\mathrm{GeV})$. Figure 1 shows the oscillatory part $f(y)$.

[b] In case the matter density varies spatially, one can use an averaged value $\langle a \rangle$ in place of a.[17,19]

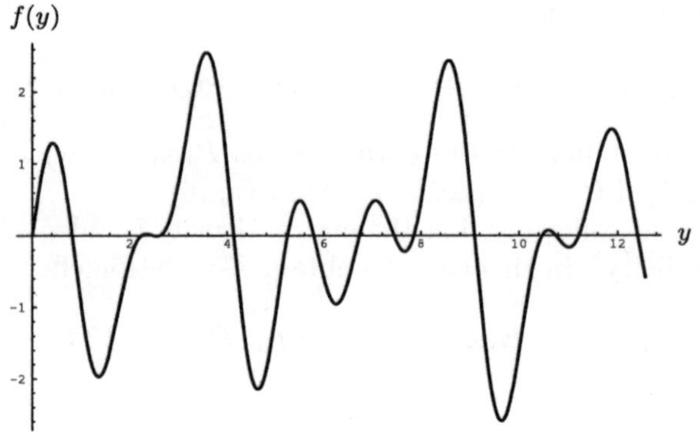

Figure 1: Graph of $f(y)$. There are high peaks (positive or negative) at $y = 0.42, 1.4, 3.6, 4.6\cdots$. Values of $f(y)$ at peaks averaged over an energy spread of $10\% \sim 20\%$ are $\langle f(0.42)\rangle = 1.2 \sim 1.3, \langle f(1.4)\rangle = -1.9 \sim -1.8, \langle f(3.6)\rangle = 2.2 \sim 1.4, \langle f(4.6)\rangle = -1.5 \sim -0.40\cdots$.

It has many peaks showing the possibility to observe a large $CP-$violating effect. For example, we may see a very large difference between the transition probabilities,

$$P(\nu_\alpha \to \nu_\beta) - P(\bar{\nu}_\alpha \to \bar{\nu}_\beta) \sim 0.4 \sin\delta, \tag{12}$$

for $L = 250$ km (the K2K experiment) and $E \sim 4.5$ GeV corresponding to $y \sim 1.4$. Hence it will be possible to detect CP violation if we have large $\sin\delta$.

In general the atmospheric neutrino anomaly indicates a large mixing angle. Thus we may see a large $CP-$violating effect when we have comparable mass differences. In this respect we note that the long baseline experiments are very desirable.

4 Case of Disparate Mass Differences

Next we consider a "disparate mass difference" case where two δm^2 scales are $\mathcal{O}(10^{-5} \sim 10^{-4}\ \mathrm{eV}^2)$ and $\mathcal{O}(10^{-3} \sim 10^{-2}\ \mathrm{eV}^2)$. [c] We see in this case that we cannot ignore the matter effect a given by Eq.(4), since it has the same order of magnitude as the smaller δm^2.

Assuming $a, \delta m^2_{21} \ll \delta m^2_{31}$, we can calculate the oscillation probabilities

[c] Some other authors [15,20,21] have also discussed the possibility to observe CP violation in long baseline neutrino oscillation experiments, but they adcpted large δm^2's of $\mathcal{O}(1\ \mathrm{eV}^2)$ and $\mathcal{O}(10^{-2}\ \mathrm{eV}^2)$, suggested by LSND experiments [35] and atmospheric neutrino observations.[10,11,12]

up to the first order in a and δm^2_{21}.[18] We obtain, for example,[d]

$$\Delta P \equiv P(\nu_\mu \to \nu_{\rm e}; L) - P(\bar{\nu}_\mu \to \bar{\nu}_{\rm e}; L) = \Delta P_1 + \Delta P_2 + \Delta P_3 \qquad (13)$$

with

$$\Delta P_1 = 16\frac{a}{\delta m^2_{31}} \sin^2 \frac{\delta m^2_{31} L}{4E} c^2_\phi s^2_\phi s^2_\psi (1 - 2s^2_\phi), \qquad (14)$$

$$\Delta P_2 = -4\frac{aL}{2E} \sin \frac{\delta m^2_{31} L}{2E} c^2_\phi s^2_\phi s^2_\psi (1 - 2s^2_\phi), \qquad (15)$$

and

$$\Delta P_3 = -8\frac{\delta m^2_{21} L}{2E} \sin^2 \frac{\delta m^2_{31} L}{4E} s_\delta c^2_\phi s_\phi c_\psi s_\psi c_\omega s_\omega. \qquad (16)$$

We now investigate how we can divide ΔP into a pure $CP-$violating part (ΔP_3) and a matter effect part ($\Delta P_1 + \Delta P_2$). We introduce two methods for this purpose in the following.

One method is to observe the pattern of the envelope of ΔP, and to separate ΔP_3 from it. Considering the energy dependence of $a(\propto E)$, We see that $\Delta P_1/L$, $\Delta P_2/L$ and ΔP_3 depend on a variable L/E alone. Each of them has its characteristic envelope as seen in Fig. 2. It is thus possible to separate these three functions and determine the $CP-$violating effect ΔP_3. Such characteristic envelope behavior enables us to determine whether a $CP-$violating effect is present or not even if ΔP is smeared due to the energy spread of the neutrino beam.[e] This point is readily seen in Fig.3. This method has the merit that the pure CP violating effect can be determined with a single detector.

The other method is to separate the pure $CP-$violating effect by comparing experiments with two different L's. Recalling that ΔP is a function of L/E apart from the matter effect factor $a(\propto E)$, we see that the difference $P(\nu_\mu \to \nu_e; E_1, L_1) - P(\nu_\mu \to \nu_e; E_2, L_2)$ with $L_1/E_1 = L_2/E_2$ is due only to terms proportional to "a". We obtain ΔP_3 by subtracting these terms ($\Delta P_1 + \Delta P_2$) from ΔP as[f]

$$\Delta P_3(E_1 L_1) = \left[\Delta P(E_1 L_1) - \frac{L_1}{L_2 - L_1} \{\Delta P(E_2 L_2) - \Delta P(E_1 L_1)\}\right]_{L/E=\text{const}} \qquad (17)$$

[d] ΔP's for other processes can be written down similarly. It is also straightforward to apply the following arguments to other processes like $\nu_\mu \to \nu_\tau$. We present the cases of $\nu_\mu \to \nu_{\rm e}$ and $\bar{\nu}_\mu \to \bar{\nu}_{\rm e}$ as examples.

[e] This is also true for observations of T-violation.

[f] We can use $2\{P(E_2 L_2) - P(E_1 L_1)\}$ in place of $\Delta P(E_2 L_2) - \Delta P(E_1 L_1)$ in Eq.(17). This form does not require $P(\bar{\nu}_\mu \to \bar{\nu}_e; L_2)$.

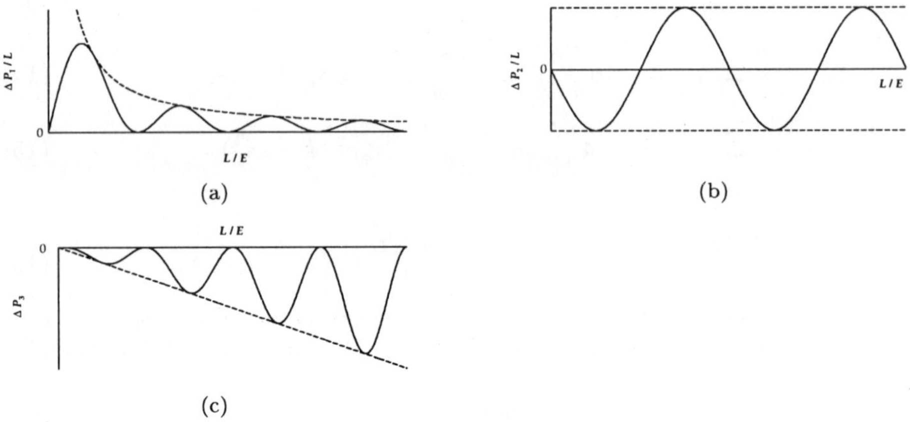

Figure 2: The oscillation behaviors of $\Delta P_1/L$(a), $\Delta P_2/L$(b) and ΔP_3(c) for $s_\delta c_\phi^2 s_\phi c_\psi s_\psi c_\omega s_\omega > 0$. Note the difference of the envelopes.

Figure 3: The oscillation probabilities for $\delta = \pi/2$ (a) and $\delta = 0$ (b). $P(\nu_\mu \rightarrow \nu_e), P(\bar{\nu}_\mu rightarrow \bar{\nu}_e)$ and $\Delta P(\nu_\mu \rightarrow \nu_e)$ are given by a broken line, a dotted line and a solid line, respectively. Here $\rho = 2.34\,\mathrm{g/cm^3}$ and $L = 250$ km (the distance between KEK and Super–Kamiokande) are taken. Other parameters are fixed at the following values which are consistent with the solar and atmospheric neutrino experiments: [11] $\delta m_{21}^2 = 10^{-4}\,\mathrm{eV}^2, \delta m_{31}^2 = 10^{-2}\,\mathrm{eV}^2, s_\psi = 1/\sqrt{2}, s_\phi = \sqrt{0.1}$ and $s_\omega = 1/2$.

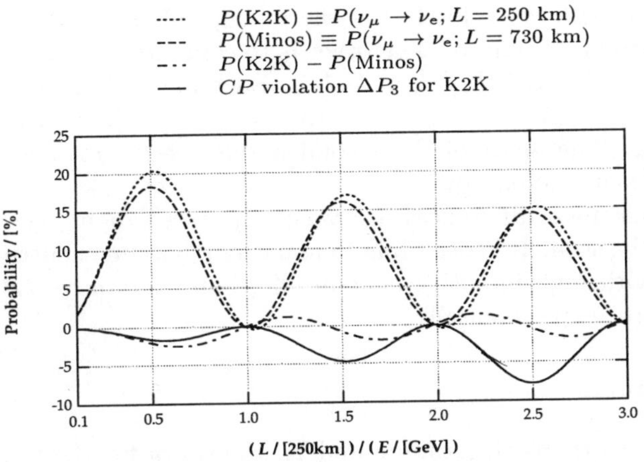

Figure 4: The oscillation probability $P(\nu_\mu \to \nu_e)$ for the K2K experiment with $L = 250$ km (broken line), and for the Minos experiment with $L = 730$ km (dotted line). Masses and mixing angles are the same as in Fig. 3(a). Their difference, which consists only of matter effect, is shown by the dot-dashed line. The pure CP−violating effect in the K2K experiment determined by Eq.(17) is shown by the solid line.

(Here $\Delta P(E_1 L_1) \equiv \Delta P(E_1, L_1)$, etc.). This method has the merit that it there is no need to observe the envelope nor many oscillation bumps in the low energy range. In Fig. 4 we compare $P(\nu_\mu \to \nu_e)$ for $L = 250$ km (K2K) with that for $L = 730$ km (Minos) in a case with the same neutrino masses and mixing angles as those in Fig. 3(a). We also show the pure CP−violating effect obtained using Eq.(17). Note that a pure CP−violating effect of 5% or more can be seen by lowering the neutrino energy.

5 Summary and Discussions

We have investigated CP and T violation in neutrino oscillation and analyzed how large the violation can be, taking into account the solar neutrino deficit and the atmospheric neutrino anomaly.

We first examined the case of comparable mass differences, where $\delta m^2_{21}, \delta m^2_{31}$ and δm^2_{32} are all in the range 10^{-3}eV^2 to 10^{-2}eV^2. We found a possibility of observing large CP−violating effects in long baseline experiments.

We then examined the case where δm^2_{21} is much smaller than δm^2_{31}. We have shown two methods to distinguish pure CP violation from matter effect.

One method is to observe closely the energy dependence of ΔP including the envelope of oscillation bumps. The other is to compare results of two different distances L_1 and L_2 with $L_1/E_1 = L_2/E_2$. Each method has both merits and demerits. The first one has the merit that we need experiments with only a single detector. The merit of the second is that we do not need to survey a wide range of neutrino energy.

Even if the mass differences are very disparate, about 5% or more CP violation may be seen. It is desirable to make long baseline neutrino oscillation experiments with high intensity neutrino flux, and to study CP violation in the lepton sector experimentally.

Acknowledgments

We would like to thank J. Arafune and K. Hagiwara for their valuable comments and encouragement. We are very grateful to M. Sakuda, A. Sakai and M. Komazawa.

1. GALLEX Collaboration, P. Anselmann *et al.*, *Phys. Lett.* B **35**, 237 (1995).
2. SAGE Collaboration, J. N. Abdurashitov *et al.*, *Phys. Lett.* B **328**, 234 (1994).
3. Kamiokande Collaboration, Y. Suzuki, *Nucl. Phys.* B(Proc. Suppl.) **38**, 54 (1995).
4. Homestake Collaboration, B. T. Cleveland *et al.*, *Nucl. Phys.* B(Proc. Suppl.) **38**, 47 (1995).
5. Kamiokande Collaboration, K.S. Hirata *et al.*, *Phys. Lett.* B **205**, 416 (1988); *ibid.* B **280**, 146 (1992); Y. Fukuda *et al.*, *Phys. Lett.* B **335**, 237 (1994); Y. Totsuka, invited talk at the 18th International Symposium on Lepton-Photon Interaction, July 28 – August 1, 1997 Hamburg.
6. IMB Collaboration, D. Casper *et al.*, *Phys. Rev. Lett.* **66**, 2561 (1991); R. Becker-Szendy *et al.*, *Phys. Rev.* D **46**, 3720 (1992).
7. SOUDAN2 Collaboration, T. Kafka, *Nucl. Phys.* B(Proc. Suppl.) **35**, 427 (1994); M. C. Goodman, *ibid.* **38** (1995) 337; W. W. M. Allison *et al.*, *Phys. Lett.* B **391**, 491 (1997)
8. NUSEX Collaboration, M. Aglietta *et al.*, Europhys. Lett. **8**, 611 (1989); *ibid.* **15**, 559 (1991).
9. Fréjus Collaboration, K. Daum *et al.*, *Z. Phys.* C **66**, 417 (1995).
10. G. L. Fogli, E. Lisi, D. Montanino and G. Scioscia, *Phys. Rev.* D **55**, 4385 (1997).
11. G. L. Fogli, E. Lisi and D. Montanino, *Phys. Rev.* D **54**, 2048 (1996); G. L. Fogli, E. Lisi and D. Montanino, *Phys. Rev.* D **49**, 3626 (1994).

12. O. Yasuda, preprint TMUP-HEL-9603, hep-ph/9602342.
13. K. Nishikawa, INS-Rep-924 (1992).
14. S. Parke, Fermilab-Conf-93/056-T (1993), hep-ph/9304271.
15. M. Tanimoto, *Phys. Rev.* D **55**, 322 (1997); M. Tanimoto, *Prog. Theor. Phys.* **97**, 901 (1997).
16. J. Arafune and J. Sato, *Phys. Rev.* D **55**, 1653 (1997).
17. J. Sato, hep-ph/9701306.
18. J. Arafune, M. Koike and J. Sato, *Phys. Rev.* D **56**, 3093 (1997)
19. M. Koike and J. Sato, in *Proceedings of the 5'th KEK Meeting on CP Violation and its Origin*, edited by K. Hagiwara, KEK Proceedings 97-12 (hep-ph/9707203).
20. H. Minakata and H. Nunokawa, Preprint UWThPh-1997-11, DFTT 26/97, hep-ph/9705300; H. Minakata and H. Nunokawa, Preprint TMUP-HEL-9705, FTUV/97-31,IFIC/97-30, hep-ph/9706281.
21. S. M. Bilenky, C. Giunti and W. Grimus, to appear in the proceedings of XXXIInd Rencontres de Moriond (Preprint UWThPh-1997-11, DFTT 26/97, hep-ph/9705300).
22. A. Acker and S. Pakvasa, *Phys. Lett.* B **397**, 209 (1997).
23. P. I. Krastev, *Nuovo Cimento* A **103**, 361 (1990).
24. For a review, M. Fukugita and T. Yanagida, in *Physics and Astrophysics of Neutrinos*, edited by M. Fukugita and A. Suzuki (Springer-Verlag, Tokyo, 1994).
25. S. M. Bilenky and S. T. Petcov, *Rev. Mod. Phys.* **59**, 671 (1987).
26. S. Pakvasa, in *High Energy Physics – 1980*, Proceedings of the 20th International Conference on High Energy Physics, Madison, Wisconsin, edited by L. Durand and L. Pondrom, AIP Conf. Proc. No. 68 (AIP, New York, 1981), Vol. 2, pp. 1164.
27. L. -L. Chau and W. -Y. Keung, *Phys. Rev. Lett.* **59**, 671 (1987).
28. T. K. Kuo and J. Pantaleone, *Phys. Lett.* B **198**, 406 (1987).
29. S. Toshev, *Phys. Lett.* B **226**, 335 (1989).
30. L. Wolfenstein, *Phys. Rev.* D **17**, 2369 (1978).
31. S. P. Mikheev and A. Yu. Smirnov, *Sov. J. Nucl. Phys.* **42**, 913 (1985).
32. V. Barger, K. Whisnant and R. J. N. Phillips, *Phys. Rev. Lett.* **45**, 2084 (1980).
33. N. Cabibbo, *Phys. Lett.* B **72**, 333 (1978).
34. K. Nishikawa, Private communication.
35. LSND Collaboration, C. Athanassopoulos *et al.*, *Phys. Rev.* C **54**, 2685 (1996); *Phys. Rev. Lett.* **77**, 3082 (1996); *ibid.* **75**, 2650 (1995); hep-ph/9706006.

RECENT RESULTS FROM THE LSND EXPERIMENT AND EVIDENCE FOR NEUTRINO OSCILLATION

H.J.KIM [a]

Dept. of Physics, Seoul National Univ, Seoul, Korea and Dept. of Physics, Louisiana State Univ., Baton Rouge, La 70803 , USA

ν_eC and ν_μC scattering have been studied at LAMPF with the Liquid Scintillator Neutrino Detector. The measured $\nu_e\ ^{12}C \rightarrow e^-\ ^{12}N_{gs}$ and $\nu_e\ ^{12}C \rightarrow e^-X$ cross sections agree well with previous measurements and theoretical predictions. The $\nu_\mu\ ^{12}C \rightarrow \mu^-\ ^{12}N_{g.s.}$ cross section is also consistent with theory. However, the measured $\nu_\mu\ ^{12}C \rightarrow \mu^-X$ inclusive cross section is about half the FGM and CRPA model predictions. A search for $\bar{\nu}_\mu \rightarrow \bar{\nu}_e$ oscillations has been performed at LAMPF using $\bar{\nu}_\mu$ from μ^+ decays at rest from the 1993 to 1995 data sample. An excess is observed which, if attributed to $\bar{\nu}_\mu \rightarrow \bar{\nu}_e$ oscillations, corresponds to an oscillation probability (averaged over the experimental acceptance) of $(0.31 \pm 0.12 \pm 0.05)$ %. The preliminary data of 1996 and 1997 show an event excess similar to what was observed in 1993-1995.

Low energy neutrino-nucleus scattering is interesting because it can give improved insight into nuclear structure. Also inelastic neutrino scattering cross sections are ingredients of various astrophysical calculations. Four different neutrino-nucleus scattering processes have been studied and cross sections were obtained. Furthermore, a measurement of $^{12}C(\nu_e, e^-)^{12}N_{g.s.}$ can provide important confirmation of our understanding of the neutrino beam and the detector performance.

Neutrino mass is a central issue for both particle physics and cosmology. The existence of neutrino oscillations would imply that neutrinos have mass and that there is mixing among the different flavors of neutrinos. To search for neutrino oscillations, an experiment using the Liquid Scintillator Neutrino Detector (LSND) [1] was carried out by looking for $\bar{\nu}_e$ events in excess of those expected from conventional processes.

At LAMPF, $\bar{\nu}_\mu$ are produced predominantly from the decay chain that starts with π^+ produced by 800 MeV protons incident on a water target followed by $\pi^+ \rightarrow \mu^+\nu_\mu$, $\mu^+ \rightarrow e^+\nu_e\bar{\nu}_\mu$. The $\bar{\nu}_e$ produced from the π^- decay chain are suppressed by approximately 7.5×10^{-4} due to the lower π^- production cross section, as well as π^- absorption and μ^- capture in the beam stop. A detailed Monte Carlo simulation, gives a value of 7.8×10^{-4} for the flux ratio of $\bar{\nu}_e$ to $\bar{\nu}_\mu$.

Presented data are from 1993-1995, except for preliminary decay-at-rest

[a] Representing the LSND collaboration [1].

results from 1996 and 1997.

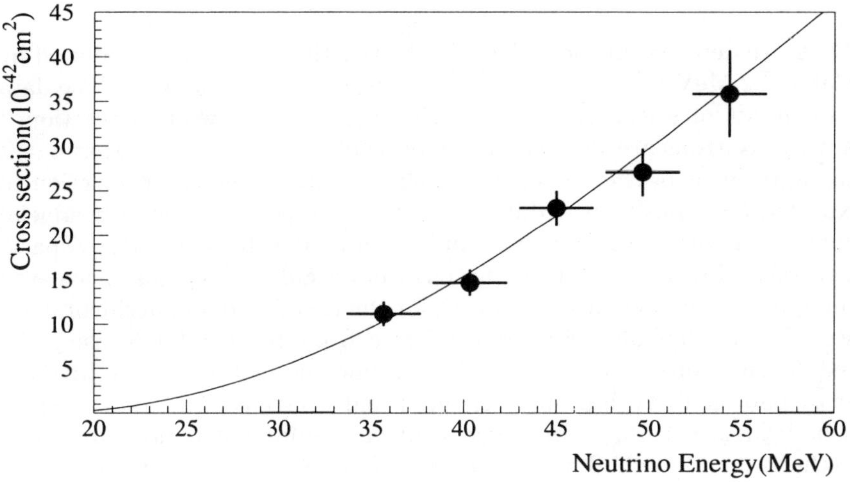

Figure 1: The measured and expected (solid line) cross section for the process $^{12}C(\nu_e, e^-)^{12}N_{g.s.}$.

The process $^{12}C(\nu_e, e^-)^{12}N_{g.s.}$ has been measured with a clean sample of 500 events for which both the e^- and the e^+ from the beta decay of the $^{12}N_{g.s.}$ are detected. For this process cross section calculations using empirical form factors are expected to be very reliable. The flux averaged cross section is measured to be $(9.1 \pm 0.4 \pm 0.9) \times 10^{-42} cm^2$ in good agreement with other experiments [2] and theoretical expectations [3]. Fig. 1 shows that the measured cross section agrees with the expected energy dependence. The angular and energy distributions of the electron also agree well with theoretical expectations. The excellent agreement found with expectations and with previous experiments for the reaction $^{12}C(\nu_e, e^-)^{12}N_{g.s.}$ provides important confirmation of our understanding of the neutrino beam and the detector performance [5,6]. The $^{12}C(\nu_e, e^-)^{12}N^*$ cross section was measured to be $(6.0 \pm 0.6 \pm 0.8) \times 10^{-42} cm^2$ which agrees well with the KARMEN experiment [8] and a Continuum Random Phase Approximation (CRPA) calculation [9].

The reaction $\nu_\mu +^{12} C \rightarrow^{12} N_{g.s.} + \mu^-$ is identified by the triple coincidence of a $\mu-$, $e+$ from a decay of $^{12}N_{g.s.}$ and e^- from μ^- decay. The observed flux-averaged cross section, $(6.6 \pm 1.0 \pm 1.0) \times 10^{-41} cm^2$ [5,7], agrees well with reliable theoretical expectations [10]. A measurement was also obtained for the inclusive cross section to all accessible ^{12}N states, $^{12}\mathrm{C}(\nu_\mu, \mu^-)\mathrm{X}$ by the double coincidence of a $\mu-$, $e-$. This flux-averaged cross section is $(11.2 \pm 0.3 \pm 1.8) \times 10^{-40}$ cm^2, which is approximately half of that given by a recent CRPA calculation [9].

The $\bar{\nu}_e$ are detected in the LSND via the reaction $\bar{\nu}_e\, p \to e^+\, n$ followed by a correlated 2.2 MeV γ from the neutron capture reaction, $np \to d\gamma$. The data analysis consists of selecting events consistent with this two-fold reaction. [4] Events with positrons are selected on the basis of an ID parameter that quantifies the agreement of the space, time, and Čerenkov cone reconstruction to that expected for a positron. Additional cuts are imposed to define a fiducial volume and to reduce various cosmic and beam-related backgrounds. Separation of correlated neutron-capture photons from accidental signals is achieved using an approximate likelihood ratio, R, for the correlated and accidental hypotheses. These likelihoods are formed for each photon candidate using the expected distributions in positron-to-photon time and distance along with the number of photon PMT hits for correlated and accidental photon. A clean sample of $^{12}\mathrm{C}(\nu_\mathrm{e}, \mathrm{e}^-)^{12}\mathrm{N}_{\mathrm{g.s.}}$ for which neutron production is not possible is used to check the R method and indeed found no correlated component [6].

The backgrounds to $\bar{\nu}_e p \to e^+ n$ followed by n capture fall into three general classes: beam-off events (cosmic ray induced), beam-related events with correlated neutrons, and beam-related events with an accidental γ. The cosmic ray background to beam-on events is 0.07 times the number of beam-off events which pass the same criteria. The major sources of beam-induced backgrounds are from μ^- DAR, discussed above, and from π^- DIF in the beam stop. The latter results in a background from $\bar{\nu}_\mu\, p$ interactions where the final μ^+ is missed, and its Michel decay positron is mistaken for a primary $\bar{\nu}_e\, p$ event. These $\bar{\nu}_\mu$ backgrounds are estimated using the detector Monte Carlo simulation [1]. The backgrounds with accidental γ overlap are greatly reduced by selection on the R parameter. Details of all backgrounds considered are presented in Ref. 4.

In the largely background-free region of $36 < E_e < 60$ MeV, we find 22 signal events coincident with a γ of $R > 30$. The e^+ (beam-off subtracted) energy distribution for these events is shown in Fig. 2. The measured beam-unrelated contamination of this sample is 2.5 ± 0.4 events. The calculated beam-related background is 2.1 ± 0.4 events. The probability that this signal is due to a statistical fluctuation of these backgrounds is 4.1×10^{-8}.

To determine the oscillation probability we fit the overall R distribution in the full energy range $20 < E_e < 60$ MeV. A fit to the full energy range gives an oscillation probability of $(0.31 \pm 0.12 \pm 0.05)$ %.

For simplicity we present the results in the two-generation formalism, in

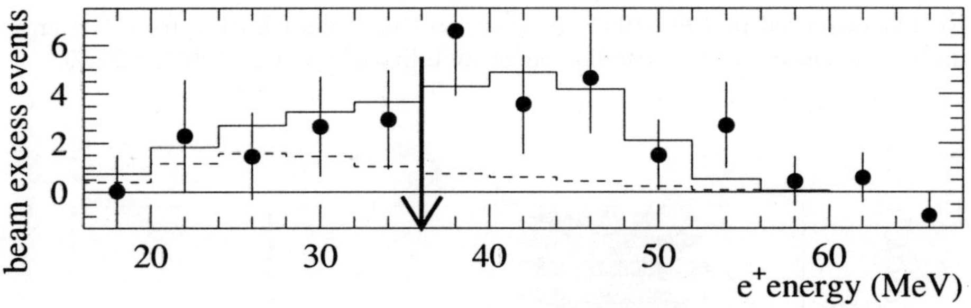

Figure 2: The energy distribution of the beam excess data along with the estimated ν background (dashed) and the expected distribution for ν background plus a large-Δm^2 signal (solid). The arrow indicates the e^+ energy cut.

which the mixing probability is written as

$$P = \sin^2 2\theta \sin^2(1.27\Delta m^2 L/E_\nu), \tag{1}$$

where θ is the mixing angle, Δm^2 is the difference of the squares of the two mass eigenstates in eV^2, L is the distance from neutrino production in meters, and E_ν is the neutrino energy in MeV. An overall likelihood fit has been made to determine favored regions in the Δm^2 versus $\sin^2 2\theta$ parameter space for two-neutrino mixing. The fit was made to distributions in the observed event energy, the neutron likelihood ratio R, the reconstructed direction of the electron relative to the neutrino beam direction, and the distance of the primary event from the beam stop neutrino source. The beam-related and cosmic ray backgrounds were added to the expected neutrino oscillation signal, and a likelihood was calculated for a range of Δm^2 versus $\sin^2 2\theta$ values. Fig. 3 shows regions which are within 2.3 and 4.5 log-likelihood units of the maximum (referred to as the 90% or 99% likelihood regions, respectively) and 90% C.L. limits from KARMEN at ISIS[8] (dashed curve), E776 at BNL[11] (dotted curve), and the Bugey reactor experiment[12] (dot-dashed curve).

These results are clear evidence of anomalous production of $\bar{\nu}_e$ and may be interpreted as evidence for $\bar{\nu}_\mu \to \bar{\nu}_e$ oscillations[4] within the favored range of Fig. 3.

We have also performed a very preliminary global analysis of $\bar{\nu}_\mu \to \bar{\nu}_e$ oscillation results taking into account all 1993-1997 data. Using the same analysis as in [Ref. 4], the 1996 and 1997 data show an event excess similar to that observed in 1993-1995. A χ^2 fit to the R distribution from the entire data set gives a $\bar{\nu}_\mu \to \bar{\nu}_e$ oscillation probability of $(0.31 \pm 0.09 \pm 0.05)\%$.

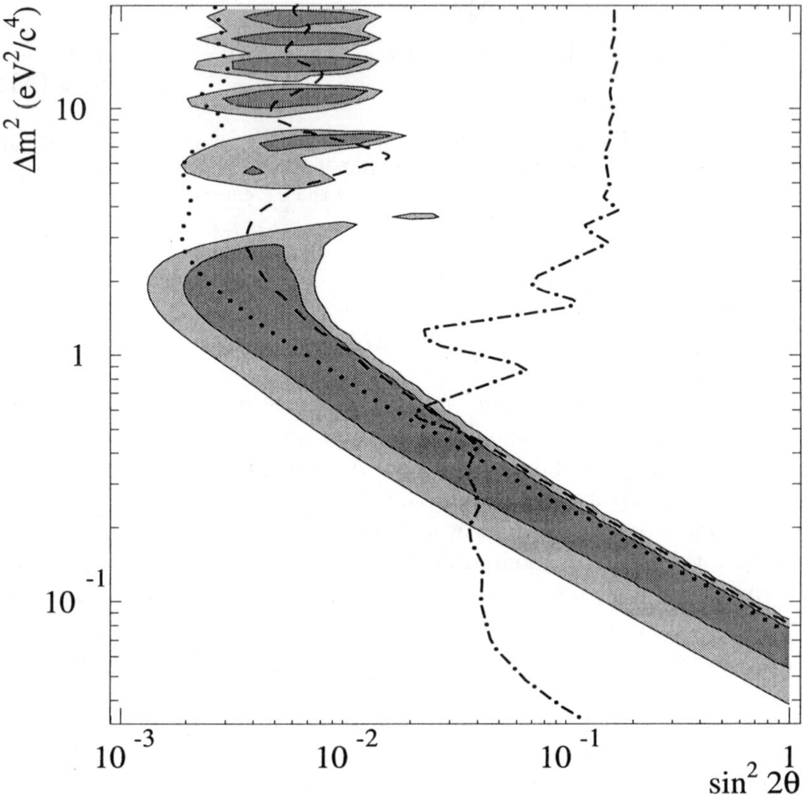

Figure 3: Plot of the LSND Δm^2 *vs* $\sin^2 2\theta$ favored regions. The shaded regions are the 90% or 99% likelihood regions. Also shown are 90% C.L. limits from KARMEN (dashed curve), E776(dotted curve), and the Bugey experiment (dot-dashed curve).

References

1. The LSND Collaboration: C. Athanassopoulos, L. B. Auerbach, R. L. Burman, I. Cohen, D. O. Caldwell, B. D. Dieterle, J. B. Donahue, A. M. Eisner, A. Fazely, F. J. Federspiel, G. T. Garvey, M. Gray, R. M. Gunasingha, R. Imlay, K. Johnston, H. J. Kim, W. C. Louis, R. Majkic, J. Margulies, K. McIlhany, W. Metcalf, G. B. Mills, R. A. Reeder, V. Sandberg, D. Smith, I. Stancu, W. Strossman, R. Tayloe, G. J. VanDalen, W. Vernon, N. Wadia, J. Waltz, Y-X. Wang, D. H. White, D. Works, Y. Xiao, S. Yellin, LA-UR-96-1327, NIM **A388**, 149 (1997).
2. D. A. Krakauer *et. al.* Phys. Rev. **C45**, 2450 (1992), G. Drexlin, KARMEN Collaboration, data presented at PANIC96, to be published in the Proceedings.
3. M. Fukugita et al., *Phys. Lett.* B **212**, 139 (1988), T. W. Donnelly and R. D. Peccei, Phys. Rep. **50**, 1 (1979); S. L. Mintz and M. Pourkaviani, J. Phys. **G16**, 569(1990),
4. C. A. Athanassopoulos *et. al.*, Phys. Rev. **C54**, 1359 (1996).
5. H.J. Kim *et. al*, Proceedings of the 14th International Conference on Particles and Nuclei (PANIC 96), 583 (1996).
6. C. Athanassopoulos *et al.*, Phys. Rev. **C55**, 2078 (1997).
7. C. Athanassopoulos *et al.*, Phys. Rev. **C56**, 2806 (1997).
8. B. Zeitnitz *et. al.*, Prog. Part. Nucl. Phys. **32**, 351 (1994).
9. E. Kolbe, K. Langanke and S. Krewald, Phys. Rev. **C49**, 1122 (1994); E. Kolbe *et al.*, Phys. Rev. **C52**, 3437 (1995).
10. J. Engel, E. Kolbe, K. Langanke and P. Vogel, Phys. Rev. **C54**, 2740 (1996).
11. L. Borodovsky *et. al.*, Phys. Rev. Lett. **68**, 274 (1992).
12. B. Achkar *et. al.*, Nucl. Phys. **B434**, 503 (1995).

PULSAR VELOCITY WITH THREE-NEUTRINO OSCILLATIONS IN NON-ADIABATIC PROCESSES

J. SONG
Center for Theoretical Physics, Seoul National University, Seoul 151-742, Korea

C.W. KIM, J.D. KIM
Department of Physics and Astronomy, The Johns Hopkins University, Baltimore, Maryland 21218, USA

The Kusenko-Segrè mechanism to explain the observed proper motion of pulsars is examined in three-generation mixing of neutrinos and in a non-adiabatic case. We have shown that a general treatment with three neutrinos does not modify the result of the two generation treatment if the standard neutrino mass hierarchy is assumed. It is also shown that possible non-adiabatic processes decrease the pulsar velocity, whereas an excess of ν_e flux over other flavor neutrino fluxes increases it.

In a recent paper, Kusenko and Segrè [1] (KS) have proposed a new explanation for the peculiar velocities of pulsars by considering magnetically distorted neutrino oscillations. Many mechanisms[2] for this phenomenon have been suggested by taking into account some asymmetry during the supernova collapse and explosion. However, most of them have difficulties in explaining the observed average pulsar velocity 450 ± 90 km/s[3]. The KS mechanism begins with the fact that neutrinos emitted during the cooling of a protoneutron star have roughly 100 times the momentum of the proper motion of the pulsar so that a neutrino momentum asymmetry of $\Delta k/k \sim 1$ % can explain the observed motion of pulsars. Therefore, it is worthwhile to study the KS mechanism in a more general case for supernovae and neutrinos.

First, let us re-examine the KS mechanism in which only $\nu_e \leftrightarrow \nu_\tau$ oscillations are considered. The difference in the mean free paths of ν_e and ν_τ renders the ν_e-sphere, S_e, to be located at a larger radius than the ν_τ-sphere, S_τ. If resonant conversion between ν_e and ν_τ occurs outside S_τ but inside S_e, the ν_e converted from ν_τ is absorbed by the background medium whereas the ν_τ converted from ν_e escapes freely. In effect, the ν_τ's propagate out not from S_τ, but from the resonance surface which is distorted in the presence of a magnetic field. The corresponding resonance condition is [4]

$$\Delta_0^2 \cos 2\theta = 2\sqrt{2} G_F N_e(r) k(r) + \sqrt{2} e G_F \left(\frac{3N_e(r)}{\pi^4} \right)^{\frac{1}{3}} \vec{k}(r) \cdot \vec{B} \tag{1}$$

where $\Delta_0^2 \equiv m(\nu_3)^2 - m(\nu_1)^2$, $N_e = (n_{e^-} - n_{e^+})$, θ is the neutrino mixing angle in vacuum, G_F the Fermi constant, $\vec{B}$ the magnetic field of pulsars, $k \simeq |\vec{k}|$ the energy of relativistic neutrinos, ϕ the angle between $\vec{B}$ and $\vec{k}$, and r the radial coordinate. Around typical neutrino spheres with $\rho \approx 10^{11\sim 12}$ g/cm^3 and the electron fraction per baryon $Y_e \approx 0.1$, the order of magnitude of the second term of the right hand side in Eq.(1) is much smaller than the first term, so that a perturbation calculation can be applied. To the zeroth order, the resonance condition is satisfied at r_0 as $\Delta_0^2 \cos 2\theta = 2\sqrt{2} G_F N_e(r_0) k(r_0)$. The presence of a $\vec{B}$ field modifies the resonance position into $r(\phi) = r_0 + \delta_r \cos\phi$. We observe that the neutrino energy k is also dependent on r, which was neglected in the KS mechanism. In leading order, we obtain

$$\left(\left| \frac{\mathrm{d} \ln N_e}{dr} \right|_{r_0} + \left| \frac{\mathrm{d} \ln k}{dr} \right|_{r_0} \right) \delta_r = \frac{e}{2} \left(\frac{3}{\pi^4 N_e^2} \right)^{\frac{1}{3}} B. \qquad (2)$$

Assuming blackbody radiation for neutrinos so that $|\mathrm{d}\ln k/\mathrm{d}r| = 4|\mathrm{d}\ln T/\mathrm{d}r|$, defining the scale heights as $h_T \equiv |\mathrm{d}\ln T/\mathrm{d}r|_{r_0}^{-1}$ and $h_{N_e} \equiv |\mathrm{d}\ln N_e/\mathrm{d}r|_{r_0}^{-1}$ and using $N_e \approx \mu_e^3/3\pi^2$ where μ_e is the chemical potential of the degenerate relativistic background electrons, the total neutrino momentum asymmetry is:

$$\begin{aligned} \frac{\Delta k}{k} &= \frac{1}{6} \frac{\int_0^\pi \mathcal{F}_\nu(r_0 + \delta_r \cos\phi) \cos\phi \sin\phi \mathrm{d}\phi}{\int_0^\pi \mathcal{F}_\nu(r_0 + \delta_r \cos\phi) \sin\phi \mathrm{d}\phi} \qquad (3) \\ &\approx 0.01 \left(\frac{1}{1 + (h_T/4h_{N_e})} \right) \left(\frac{B}{3 \times 10^{15}\mathrm{Gauss}} \right) \left(\frac{\mu_e}{11\mathrm{MeV}} \right)^{-2}, \end{aligned}$$

where all six species of neutrinos are approximated to equally contribute to the total energy of a supernova. Note that we have not used the relation $\mathrm{d}N_e/\mathrm{d}T \approx (\partial N_e/\partial T)_{\mu_e}$ in Ref.[1] which is, strictly speaking, not valid inside protoneutron stars as pointed out in Ref.[5]. Numerical supernova calculations[6] in a model of fixed baryon mass $1.4 M_\odot$ shows $h_{N_e}/h_T \sim 0.3$, implying that the incorporation of the r-dependence of the neutrino energy requires higher B values in order to explain the observed birth velocity of pulsars.

Now, let us discuss the problem in a three-neutrino generation scheme with a mass hierarchy that can accommodate the solar neutrino data[7]. For simplicity, we assume two natural conditions: (1) $m_1 \ll m_2 \ll m_3$; (2) neutrino mixing angles in vacuum are small. It is well known then that three neutrino oscillations in matter separate into two parts, each being similar to two neutrino oscillations in matter. We call R_h the resonance surface which occurs at higher density, and R_l the one at lower density. The conditions for each resonance are[8]

$$\begin{aligned} A_{R_h} &\cong \left[m_3^2 - \frac{1}{2} \left\{ m_1^2(1 + \cos\theta_{12}) + m_2^2(1 - \cos\theta_{12}) \right\} \right] \cos 2\theta_{13}, \qquad (4) \\ A_{R_l} &\cong (m_2^2 - m_1^2) \cos 2\theta_{12}, \end{aligned}$$

where A is the right hand side of Eq.(1).

A neutrino sphere may be considered as an imaginary surface such that inside neutrinos are trapped due to the interactions with medium, but outside they escape freely. Since the temperature ($T \approx 3$ MeV) around the neutrino spheres is too low to thermally create μ or τ, the ν_μ-sphere is expected to coincide with the ν_τ-sphere, S_τ. As can be seen in Eq.(4), the locations of resonance surfaces depend on the neutrino mass and mixing angles. According to a comprehensive analysis in a hierarchical three-generation scheme, the MSW solution to the solar neutrino problem suggests $m_2^2 - m_1^2 \approx 10^{-5}\mathrm{eV}^2$ [9] and the atmospheric neutrino anomaly leads to $m_3^2 \gtrsim 10^{-3}\mathrm{eV}^2$ [9]. Therefore, the density around neutrino spheres is too high for the lower resonance to occur, implying that the lower resonance surface R_l lies outside of S_e. Within the current experimental limit of m_3, R_h can lie between S_e and S_τ [10].

In the following we assume that R_l for $\nu_e \leftrightarrow \nu_\mu$ is located outside S_e, but R_h for $\nu_e \leftrightarrow \nu_\tau$ lies between S_e and S_τ. At R_h, ν_τ will be converted into ν_e and absorbed, whereas ν_e will turn into ν_τ and propagate out. The ν_μ does not undergo any resonant conversion at R_h. Even if neutrino flavors are resonantly converted again at R_l, this conversion does not affect the pulsar velocities because the momentum transfer of neutrinos is independent of neutrino flavors. This is in contrast to the case at R_h where ν_τ propagates out but ν_e cannot. Once neutrinos escape from S_e, therefore, flavor-changing of the neutrinos does not affect the kick velocity of pulsars.

Next, we discuss the effects of possible non-adiabatic neutrino oscillations on the birth velocity of pulsars. If the oscillation process is non-adiabatic, a level-crossing takes place with the level-crossing (Landau-Zener) probability $P_{LZ} = \exp(-\frac{\pi}{4}Q)$, where $Q = \Delta_0^2 S_{2\theta}^2 h_{N_e}/kC_{2\theta}$ is the adiabaticity parameter in the two generation case. For $\Delta_0^2 \sim 10^4$ eV^2, $h_{N_e} \sim 6$ km, and $k \sim 10$ MeV, the non-adiabatic effect becomes substantial when $\theta \gtrsim 10^{-4}$.

Let us consider a general non-adiabatic case with a possibility that the initial fluxes of ν_e's and ν_τ's are different. Since ν_μ and ν_τ are produced via neutral currents but ν_e via both charged and neutral currents, the ν_e flux is larger than other neutrino fluxes. We introduce the parameter α_{ν_e} such that neutrino fluxes are given by $\mathcal{F}_{\nu_e} = (1+\alpha_{\nu_e})\mathcal{F}$ and $\mathcal{F}$ is the common flux factor and $\alpha_{\nu_e} \approx 5 \sim 10\%$[11]. The neutrino momentum asymmetry is then given by

$$\left(\frac{\Delta k}{k}\right)_{\mathrm{NA}} \approx \frac{\mathcal{P}_{\mathrm{NA}}(\nu_e \to \nu_\tau)(1+\alpha_{\nu_e})}{(6+\alpha_{\nu_e})} \frac{\int \mathcal{F}(r_0 + \delta_r \cos\phi)\cos\phi \,\mathrm{d}\cos\phi}{\int \mathcal{F}(r_0 + \delta_r \cos\phi)\mathrm{d}\cos\phi}. \tag{5}$$

In the small vacuum mixing case ($\theta_f \sim 0, \theta_i \sim \pi/2$) and small α_{ν_e}, we have

$$\left(\frac{\Delta k}{k}\right)_{\mathrm{NA}} \approx \left(\frac{\Delta k}{k}\right)(1 - P_{LZ})\left(1 + \frac{5}{6}\alpha_{\nu_e}\right), \tag{6}$$

where $(\Delta k/k)$ is given by Eq.(3). This shows that the non-adiabaticity has a tendency to decrease the pulsar velocity and extremely non-adiabatic processes ($P_{LZ} = 1$) cause no kick velocity due to the absence of flavor conversion. The pulsar velocity increases with α_{ν_e}. This is also expected because the more ν_e's are produced inside the core, the more they turn into ν_τ's, leading to a larger birth velocity of the pulsar. Thus, the effect of non-adiabaticity is opposite to that of the excess flux of ν_e.

In summary, we have studied the Kusenko-Segrè mechanism in more general conditions. First, the r-dependent neutrino energy has been considered, which becomes relatively important in the case $h_{N_e}/h_T > 1$. The KS mechanism is also studied in a three generation scheme and in a non-adiabatic case. While the three generation treatment leaves the KS result intact, the non-adiabaticity decreases the kick velocity but the excess flux of ν_e increases it.

Acknowledgments

JS acknowledges support from the Korean Science and Engineering Foundation (KOSEF) through the Center for Theoretical Physics (CTP).

References

1. A. Kusenko and G. Segrè, *Phys. Rev. Lett.* **77**, 4872 (1996).
2. J.R. Gott, J.E. Gunn and J.P. Ostriker, *Astrophys.J.* **160**, L91 (1970); E.R. Harrison and E. Tademaru, *Astrophys.J.* **201**, 447 (1975); H.Th. Janka and E. Müller, *Astron.& Astrophys.* **290**, 496 (1994); A. Vilenkin, *Astrophys.J.* **451**, 700 (1995).
3. A.G. Lyne and D.R. Lorimer, *Nature* **369**, 127 (1994).
4. J.C. D'olivo and J.F. Nieves, *Phys. Rev.* D **40**, 3679 (1989); S. Esposito and G. Capone, *Z. Phys.* C **70**, 70 (1996).
5. Y.-Z. Qian, astro-ph/9705055.
6. A. Burrows and J.M. Lattimer, *Astrophys.J.* **307**, 178 (1986).
7. S.M. Bilenky *et al.* *Phys. Lett.* B **356**, 273 (1995).
8. T.K. Kuo and J. Pantaleone, *Phys. Rev. Lett.* **59**, 1805 (1986).
9. G.L. Fogli, E. Lisi and D. Montanio, *Phys. Rev.* D **49**, 3626 (1994) and references therein.
10. Particle Data Group, *Phys. Rev.* D **54**, (1996).
11. R. Mayle *et al.* *Astrophys.J.* **318**, 288 (1987).

NEUTRINO MASS MATRIX, MIXING AND OSCILLATION

KYUNGSIK KANG
Department of Physics, Brown University, Providence, RI 02912, USA

SIN KYU KANG
School of Physics, Korea Institute for Advanced Study, Seoul, Korea

We investigate the phenomenological consequences of lepton mass matrices that originate from a family permutation symmetry and its suitable breaking. Adopting the recently proposed new mass matrix for the charged leptons and a Majorana neutrino mass matrix considered before, we find that the resulting lepton mixing matrix is consistent with the current data on various types of neutrino oscillation experiments, except the Chlorine data and the LSND result, and that three neutrinos are almost degenerate in mass. Three neutrinos can account for about 50% of the hot dark matter(HDM) when the neutrino mass matrix is constrained by the small angle MSW solution for the solar neutrino problem, atmospheric neutrino data of the Super-Kamiokande group, and the limits on the electron-type neutrino mass due to the non-observation of neutrino-less double beta decays.

(Presented by Kyungsik Kang)

1 Introduction

The flavor problem, concerning the hierarchical pattern of fermion masses as well as flavor mixing, remains one of the most fundamental problems in particle physics. Various attempts towards an understanding of flavor mixing, with a quark mass matrix ansatz satisfying the calculability condition, are presented in another session [1]. Though the simple Fritzsch-type mass matrix had attracted a great deal of attention, it has been ruled out[2], because it predicts the top quark mass to be below 100 GeV.

Nevertheless, the Fritzsch-type mass matrix is very attractive due to its simplicity. Thus, the next move is to generalize and modify it while maintaining the calculability property. Recently we have proposed a generalized hermitean mass matrix ansatz and made a systematic phenomenological study [3]:

$$M_H = \begin{pmatrix} 0 & A & 0 \\ A & D & B \\ 0 & B & C \end{pmatrix}, \tag{1}$$

which is consistent with experiment, including a heavy top and maximal CP violation, for the range $|w \equiv B/D| = 0.97 - 1.87$. Setting $D = 0$ recovers the original Fritzsch ansatz. As shown in Refs. 1 and 3, this form can be achieved by successive breaking of maximal permutation symmetry corresponding to a "democratic mass matrix". It generalizes various specific forms of mass matrices proposed by others, where D and B are related to each other in a particular way, but mostly outside the range of w that we found. We will assume the same form of mass matrix for the charged lepton sector.

On the other hand, all neutrino masses are zero and lepton numbers are exactly conserved in the context of the standard model (SM). Strictly speaking, it has not been established yet that the neutrino masses are non-zero and hierarchical. However, the current experimental anomalies of solar [4,5,6,7] and atmospheric [8,9,10] neutrinos lead us to speculate that neutrinos may be massive and have mixing, as they can be interpreted as signals for neutrino oscillations. If we assume that only two neutrino flavors participate in the oscillations, the parameter space consists of one mixing angle and one mass-squared difference. On the other hand, oscillations among three neutrino flavors may provide a simultaneous solution for the solar and atmospheric neutrino problems. In this case, we have two degrees of freedom in the choice of mass-squared differences. Several authors have suggested [11,12] that almost degenerate neutrinos are needed to accommodate the solar and atmospheric neutrino observations as well as the cosmological constraint that arises when we regard neutrinos as candidates for hot dark matter (HDM) within the three flavor framework.

We will show that an almost degenerate scenario among three flavors indeed follows from the standpoint of mass matrix ansatz. In order to do so, we will construct neutrino mass matrices that are constrained by the solar and atmospheric neutrino deficits and the non-observation of neutrinoless double beta decay. We note that both charged and neutral lepton mass matrices can originate from the permutation symmetry and its suitable breaking.

2 Mass matrix for charged leptons

The new class of mass matrix of Eq.(1) is obtained from the "democratic mass matrix" with the maximal $S(3)_L \times S(3)_R$ symmetry, which is broken successively down to $S(2)_L \times S(2)_R$ and $S(1)_L \times S(1)_R$, followed by a unitary transformation with $U = (u_1^T, u_2^T, u_3^T)$, where $u_1 = (\frac{1}{\sqrt{2}}, \frac{1}{\sqrt{6}}, \frac{1}{\sqrt{3}}), u_2 = (-\frac{1}{\sqrt{2}}, \frac{1}{\sqrt{6}}, \frac{1}{\sqrt{3}}), u_3 = (0, -\frac{2}{\sqrt{6}}, \frac{1}{\sqrt{3}})$. As discussed in Refs. 1 and 3, the (2,2) element is related to the (2,3) element in Eq. (1) by $w \equiv B/D = (k+1)/\sqrt{2}(2k-1)$ in the hierarchical mass eigenstate.

We assume a quark–charged lepton symmetry, so that the mass matrix for the charged lepton sector has exactly the same form as the quark mass matrix

(1). The mass matrix M_H for $K = U_L^l U_R^{l\dagger} = diag[-1,1,1]$ can then be written as [1,3]

$$M_H = \begin{pmatrix} 0 & \sqrt{\frac{m_1 m_2 m_3}{m_3-\epsilon}} & 0 \\ \sqrt{\frac{m_1 m_2 m_3}{m_3-\epsilon}} & m_2 - m_1 + \epsilon & w(m_2 - m_1 + \epsilon) \\ 0 & w(m_2 - m_1 + \epsilon) & m_3 - \epsilon \end{pmatrix}; \tag{2}$$

the small parameter ϵ is related to w, $w \simeq \pm\frac{\sqrt{\epsilon m_3}}{m_2}\left(1 + \frac{m_1}{m_2} - \frac{m_2}{2m_3}\right)$, whose range is to be determined from experiment.

Then, the diagonalizing matrix U_L^l can be written as

$$U_L^l = U_{23}(\theta_{23}) \cdot U_{12}(\theta_{12}) \tag{3}$$

where

$$U_{12} = \begin{pmatrix} \cos\theta_{12} & \sin\theta_{12} & 0 \\ -\sin\theta_{12} & \cos\theta_{12} & 0 \\ 0 & 0 & 1 \end{pmatrix}, \qquad U_{23} = \begin{pmatrix} 1 & 0 & 0 \\ 0 & \cos\theta_{23} & \sin\theta_{23} \\ 0 & -\sin\theta_{23} & \cos\theta_{23} \end{pmatrix} \tag{4}$$

Since the (1,1), (1,3) and (3,1) elements of M_H are zero, we may put $U_{13}(\theta_{13}) = 1$ without loss of generality. The mixing angles θ_{12} and θ_{23} can be written to a very good approximation as $\tan\theta_{12} = \sqrt{\frac{m_1}{m_2}}$ and

$$\tan\theta_{23} = \frac{1}{2w}\left[\left(1 + \frac{m_1 - m_2}{m_3}\right) - \sqrt{\left(1 + \frac{m_1 - m_2}{m_3}\right)^2 + 4w^2\left(\frac{m_1 - m_2}{m_3}\right)}\right] \tag{5}$$

For negative D, the real symmetric matrix M_H can be diagonalized as $U_L^l M_H U_L^{l\dagger} = diag[m_1, -m_2, m_3]$, thus reversing the signs of both m_1 and m_2 in the above equations.

In both cases above it turns out that the experimentally allowed range of w for the quark sector is $0.97 \leq |w| \leq 1.87$ in the leading approximation [3]. Thus, the same range of w is assumed also for the charged lepton sector. However, physical observables such as survival and transition probabilities for ν_α's turn out to be insensitive to the precise value of w in the allowed range, as discussed in the following.

3 Mass matrix for neutrinos

It is evident that the mass matrix of the charged lepton sector is not appropriate for the neutrino sector. The neutrino oscillation experiments do not seem to support such hierarchical pattern for neutrino masses unlike the quark and charged lepton masses, but rather nearly degenerate neutrinos within the three flavor framework [11,12]. In this section, we will show how such an almost degenerate neutrino scenario follows from the neutrino mass matrix, which is clearly different from the approach used by others [12]. We propose for the neutrino sector the permutation symmetry among three family indices rather than the flavor democracy. It can be represented by the following matrix in the *symmetry basis* of $S(3)$[11],

$$\widetilde{M}_\nu^{(0)} = c_\nu \begin{pmatrix} 1 & r & r \\ r & 1 & r \\ r & r & 1 \end{pmatrix} . \tag{6}$$

After the unitary transformation with $V = (v_1^T, v_2^T, v_3^T)$ where $v_1 = (\frac{1}{\sqrt{3}}, \frac{1}{\sqrt{6}}, \frac{1}{\sqrt{2}})$, $v_2 = (\frac{1}{\sqrt{3}}, \frac{1}{\sqrt{6}}, -\frac{1}{\sqrt{2}})$ and $v_3 = (\frac{1}{\sqrt{3}}, -\frac{2}{\sqrt{6}}, 0)$, the matrix (6) becomes

$$M_\nu^{(0)} = c_\nu \begin{pmatrix} 1+2r & 0 & 0 \\ 0 & 1-r & 0 \\ 0 & 0 & 1-r \end{pmatrix} . \tag{7}$$

Note that all three neutrinos are degenerate either for $r = 0$ or $r = -2$, while two neutrinos are always degenerate for any r [13], so that only one Δm_ν^2 is available.

In order to confront the above mass matrix ansatz with experimental and cosmological observations, we use three observations which may be accounted for by assuming massive neutrinos: (1) solar neutrino data from four different experiments, HOMESTAKE [4], GALLEX [5], SAGE [6], and Kamiokande II-III [7], (2) atmospheric neutrino data measured by three experiments, Super-Kamiokande [8], Soudan2 [9], and IMB [10], (3) the likely need of neutrinos as a candidate of dark matter [14]. As is well known, the solar neutrino deficit can be explained through the MSW mechanism if $\Delta m^2_{solar} \simeq 6 \times 10^{-6}$ eV^2 and $\sin^2 2\theta_{solar} \simeq 7 \times 10^{-3}$ (small angle case), or $\Delta m^2_{solar} \simeq 9 \times 10^{-6}$ eV^2 and $\sin^2 2\theta_{solar} \simeq 0.6$ (large angle case), or through the just-so vacuum oscillations if $\Delta m^2_{solar} \simeq 10^{-10}$ eV^2 and $\sin^2 2\theta_{solar} \simeq 0.9$ [15]. The atmospheric neutrino data of Super-Kamiokande are consistent with $\Delta m^2_{32} \approx (2-8) \times 10^{-3} eV^2$ and $\sin^2 \theta_{32} = 0.85 - 1$, at 90%CL. If light neutrinos account for the HDM of the universe, one has to require [14] $\sum_{i=1,2,3} |m_{\nu_i}| \sim 6$ eV. Thus we see that all three neutrinos may be almost degenerate in their masses, with $m_{\nu_i} \sim$ a few eV, rather than $m_{\nu_1} \ll m_{\nu_2} \ll m_{\nu_3}$, as assumed sometimes in three neutrino mixing scenarios. As discussed above, in order to account for the solar and atmospheric neutrino oscillations simultaneously, two separate $\Delta m^2_{ij} = m^2_{\nu_i} - m^2_{\nu_j}$

scales are required. The two Δm_{ij}^2 scales can be generated from our neutrino mass matrix (7) by lifting the degeneracy between the second and third neutrinos, i.e., by breaking $S(3)$ to $S(2)$. This can be achieved by allowing the (3,2) and (2,3) elements of the matrix (7) to be nonzero. The resulting matrix for neutrinos in the flavor basis then becomes

$$M_\nu = c_\nu \begin{pmatrix} 1+2r & 0 & 0 \\ 0 & 1-r & \epsilon_\nu \\ 0 & \epsilon_\nu & 1-r \end{pmatrix}, \tag{8}$$

which can be diagonalized by $U_\nu = (u_1^T, u_2^T, u_3^T)$, where $u_1 = (1,0,0), u_2 = (0, \frac{1}{\sqrt{2}}, \frac{1}{\sqrt{2}}), u_3 = (0, \frac{1}{\sqrt{2}}, -\frac{1}{\sqrt{2}})$, with the resulting eigenvalues $m_{\nu_i}/c_\nu = (1+2r), (1-r\mp\epsilon_\nu)$. Note that U_ν is independent of neutrino mass eigenvalues, or equivalently, of c_ν, ϵ_ν and r. This is because the mass matrix M_ν given by (8) has a residual $S(2)$ symmetry acting upon the first and the second family indices. Alternatively, one may take nonzero (2,1) and (1,2) elements of the matrix (7), but such case turns out to be inconsistent with the atmospheric data from Super-Kamiokande.

One can solve for c_ν, r and ϵ_ν by requiring three conditions, $\Delta m_{solar}^2 = 10^{-5}$ eV^2, $\Delta m_{atmos}^2 = 10^{-2}$ eV^2 and the neutrino mass bound due to non-observation of the neutrino-less double beta decay[16], $\langle m_\nu \rangle = |\sum_{i=1}^3 \eta_i V_{\nu_1 i}^2 m_i| \leq 1 \sim 2$ eV, where $\eta_i = \pm 1$ depending on the CP property of ν_i. Note that we solve the solar neutrino problem by adopting the MSW mechanism. Then we get $r \simeq -1.9925, c_\nu \simeq 0.3345$ and $\epsilon_\nu \simeq 0.0075$, leading to $|m_1| \simeq 0.998301$ eV, $|m_2| \simeq 0.998305$ eV and $|m_3| \simeq 1.0033$ eV. We note that the breaking parameter ϵ_ν turns out to be much smaller than the parameter r.

4 Neutrino mixing matrix and predictions

Combining U_L^l given by Eq.(3) with U_ν that diagonalizes the neutrino mass matrix (8), we get the flavor mixing matrix in the lepton sector,

$$V_\nu \equiv U_L^{l^\dagger}\ U_\nu = \frac{1}{\sqrt{2}} \begin{pmatrix} \sqrt{2}c_{12} & s_{12}(-c_{23}+s_{23} & s_{12}(c_{23}+s_{23}) \\ \sqrt{2}s_{12} & c_{12}(c_{23}-s_{23}) & c_{12}(-c_{23}-s_{23}) \\ 0 & s_{23}+c_{23} & -s_{23}+c_{23} \end{pmatrix}, \tag{9}$$

where we have abbreviated $\cos\theta_{ij}$ and $\sin\theta_{ij}$ as c_{ij} and s_{ij} respectively. For the whole range of $0.97 \leq |w| \leq 1.87$, the neutrino mixing matrix is given by

$$|V_\nu| = \begin{pmatrix} 0.9952 & -0.0453 & 0.0521 \\ 0.0692 & 0.6521 & -0.7494 \\ 0.0000 & 0.7531 & 0.6551 \end{pmatrix} \text{ to } \begin{pmatrix} 0.9952 & -0.0440 & 0.0506 \\ 0.0692 & 0.6326 & -0.7271 \\ 0.0000 & 0.7307 & 0.6356 \end{pmatrix}. \tag{10}$$

Note that our lepton mixing matrix predicts zero for the $(V_\nu)_{31}$ element, i.e., the tau-ν_e coupling is forbidden. Because of the smallness of the (1,2) and (1,3) elements in (10), $\langle m_\nu \rangle \simeq m_1 \simeq 1$ eV. From this mixing matrix, we obtain $\sin^2 2\theta_{12} \simeq 0.0054 - 0.0061$ which is consistent with the small angle MSW solution and $\sin^2 2\theta_{23} \simeq 0.85 - 0.96$ consistent with the Super-Kamiokande result.

The predicted probability of $\nu_e - \nu_\mu$ oscillation is given by $P(\nu_e - \nu_\mu) \sim 4|V_{e3}|^2|V_{\mu 3}|^2 sin^2\left(\frac{1.27\Delta m_{32}^2 L}{E}\right) \sim 0.006\ sin^2\left(\frac{1.27\Delta m_{32}^2 L}{E}\right)$. In addition, we see that the three neutrinos can account for about 50% of the HDM, because we get $\sum |m_{\nu_i}| \sim 3$ eV whereas the HDM mass limit is 6 eV.

5 Conclusion

In conclusion, we investigated in this paper phenomenological consequences of lepton mass matrix ansätze with the minimal number of parameters, three each in the charged lepton and Majorana neutrino mass matrices M_H and M_ν, with permutation symmetries among three generations $S(3)_L \times S(3)_R$ and $S(3)$ that are broken down to $S(2)_L \times S(2)_R$ and $S(1)_L \times S(1)_R$ and $S(2)$ and $S(1)$, respectively. We find the mass matrix ansätze Eqs.(1) and (8) lead to a lepton mixing matrix which can accommodate the solar and atmospheric neutrino observations as well as the constraint from non-observation of neutrino-less beta decays. Three light Majorana neutrinos can provide about 50% of the HDM, with $\sum |m_{\nu_i}| \sim 3$ eV.

Note added : After the presentation of this talk, the results of the long baseline reactor experiment CHOOZ appeared [24], which are consistent with our results.

References

1. S.K. Kang's talk in the proceeding.
2. K. Kang and S. Hadjitheodoridis, Phys. Lett. B**193**, 504 (1987); K. Kang, J. Flanz, and E. Paschos, Z. Phys. C **55**, 75 (1992). See also H. Harari and and Y. Nir, Phys. Lett. B**195**, 586 (1987).
3. K. Kang and S.K. Kang, Phys. Rev. D**56**, 1511 (1997), hep-ph/9704253.
4. R. Davis et al., Proc. 21st Int. Cosmic Ray Conf. Univ. of Adelaide, ed. R.J. Protheroe, Vol. 12, 143 (1990).
5. P. Anselmann *et al.*, Phys. Lett. B**285**, 376 (1992) *ibid.* **285**, 390 (1992); *ibid.* **314**, 445 (1993); *ibid.* **327**, 377 (1994).

6. A.I. Abazov *et al.*, Phys. Rev. Lett. **67**, 3332 (1991).
7. Y. Totsuka's talk at the 1997 Lepton-Photon Symp. July 28-August 1, 1997, Hamburg, Germany.
8. K.S. Hirata *et al.*, Phys. Lett. B**205**, 416 (1988); *ibid.* **280**, 146 (1988); Y. Fukuda *et al.*, *ibid.* **335**, 237 (1994).
9. P.J. Litchfield, Proc. Int. Europhys. Conference on High Energy Physics, Marseilles, 557 (1993). (Editions Frontieres, ed., J. Carr and M. Perrottet.)
10. R. Becker-Szendy *et al.*, Phys. Rev. D**46**, 3720 (1992); D. Casper *et al.*, Phys. Rev. Lett. **66**, 2561 (1992).
11. K. Kang, S.K. Kang, J.E. Kim and P. Ko, Mod. Phys. Lett. A **12**, 1175 (1997), hep-ph/9611369; hep-ph/9706535.
12. D. Caldwell and R.N. Mohapatra, Phys. Rev. D**48**, 3259 (1993); *ibid.***50**, 3477 (1994); S.T. Petcov and A.Y. Smirnov, Phys. Lett. B**322**, 109 (1994); A. Joshipura, Z. Phys. C**64**, 31 (1994); Phys. Rev. D**51**, 1321 (1995); A. Ioannissyan and J.W.F. Valle, Phys. Lett. B**332**, 93 (1994); D.G. Lee and R.N. Mohapatra, *ibid.* **329**, 463 (1994); C.Y. Cardall and G.M. Fuller, Nucl. Phys. Proc. Supp. **51B**, 259 (1996); A. Acker and S. Pakvasa, Phys. Lett. **B 397**, 209 (1997).
13. K. Kang, J. E. Kim and P. Ko, Z. Phys. C**72**, 671 (1996).
14. G. Gelmini and E. Roulet, Rep. Prog. Phys. **58**, 1207 (1995).
15. Particle Data Group, R.M.Barnett *et al.*, Phys Rev. D**54**, 1 (1996).
16. H.V. Klapdor-Kleingrothaus *et al.*, Prog. Part. Nucl. Phys. **32**, 261 (1994), and references therein. The 130Te experiment with more uncertain nuclear matrix element calculations could favor $\langle m_\nu \rangle \sim 4$eV. See E. Garcia, in TAUP93 Workshop, Gran Sasso, Italy, 1993.
17. B. Bodman *et al.*, Phys. Lett. B**280**, 198 (1992).
18. Y. Declais, LAPP-EXP-94-08; Y. Declais, et al., Phys. Lett. B**338**, 383 (1994); B.A. Achkar et al., Nucl. Phys. B**354**, 503 (1995).
19. J.-L. Vuilleumier *et al.*, Phys. Lett. B**114**, 298 (1982); K. Gabathuler *et al.*, *ibid.* **138**, 449 (1984); V. Zacek *et al.*, *ibid.***164**, 193 (1985); G. Zacek *et al.*, Phys. Rev. D**34**, 2621 (1986).
20. G.S. Vidyakin *et al.*, JETP Lett. **55**, 206 (1992); *ibid.* **59**, 391 (1994).
21. P. Berge *et al.*, Z. Phys. C**35**, 443 (1987).
22. F. Bergsma *et al.*, Phys. Lett. B**142**, 103 (1984).
23. Fox example, G. Gelmini and E. Roulet, CERN-TH-7541/94 (1994).
24. M. Apollonio et al., hep-ex/9711002.

GRAVITATIONAL WAVES FROM A PARTICLE IN CIRCULAR ORBIT AROUND A ROTATING BLACK HOLE

HIDEYUKI TAGOSHI
National Astronomical Observatory, Mitaka, Tokyo 181

SHUHEI MANO and EIICHI TAKASUGI
Department of Physics, Osaka University, Toyonaka 560

When a particle moves around a Kerr black hole, it radiates gravitational waves. Some of those waves are absorbed by the black hole. We calculate such absorption of gravitational waves induced by a particle in circular orbit on a equatorial plane around a Kerr black hole. We obtain formulas for the luminosity to $O((v/c)^8)$ compared to the lowest order. Using the energy flux formulas for the black hole absorption, we investigate its effect on the orbital evolution of coalescing compact binaries.

1 Introduction

Among the possible sources of gravitational waves, coalescing compact binaries are the most promising candidates for detection by ground based laser interferometric detectors such as LIGO, VIRGO, GEO600, and TAMA. When a signal of gravitational waves is detected, we will try to extract parameters for binaries, such as masses and spins etc., from inspiral wave forms using the matched filtering technique, from which parameters for binaries are determined by cross–correlating the noisy signal from detectors with theoretical templates. If the signal and the templates lose phase with each other by one cycle over $\sim 10^3 - 10^4$ cycles as the waves sweep through the interferometer band, their cross correlation will be significantly reduced. This means that, in order to extract the information optimally, we need to make theoretical templates which are accurate to better than one cycle during an entire sweep through the interferometer's band[1].

The standard method to calculate inspiraling wave forms from coalescing binaries is the post–Newtonian expansion of the Einstein equations, in which the orbital velocity v of the binaries are assumed to be small compared to the speed of light. In the case $\mu \ll M$ with μ being the mass of the orbiting particle and M the black hole mass, gravitational waves can be calculated up to a very high order of post–Newtonian corrections by using a relatively simple algorithm. The gravitational radiation at infinity was calculated using this method to $O(v^8)$ by Tagoshi, Shibata, Tanaka, and Sasaki[2] in a Kerr black hole case.

Here we report on our recent work on the effect of black hole absorption[3]. In this paper, we discussed that, when a black hole is rotating, black hole absorption appears at $O(v^5)$ and the effect of black hole absorption on the orbital evolution of coalescing compact binaries becomes important. To evaluate this effect, we used the analytic solutions of the homogeneous Teukolsky equation formulated by Mano, Suzuki, and Takasugi[4].

2 The Teukolsky formalism

In the Teukolsky formalism[5], the gravitational perturbations of the Kerr black hole are described by a function $\psi_4 = -C_{\alpha\beta\gamma\delta}n^\alpha \bar{m}^\beta n^\gamma \bar{m}^\delta$. $C_{\alpha\beta\gamma\delta}$ is the Weyl tensor, $n^\alpha = ((r^2+a^2), -\Delta, 0, a)/(2\Sigma)$, $m^\alpha = (ia\sin\theta, 0, 1, i/\sin\theta)/(\sqrt{2}(r + ia\cos\theta))$. We use the Boyer–Lindquist coordinate (t, r, θ, ϕ) in this paper and $\Sigma = r^2 + a^2\cos^2\theta$, $\Delta = r^2 - 2Mr + a^2$, and a is the spin parameter of a Kerr black hole. We decompose ψ_4 into Fourier–harmonic components according to

$$(r - ia\cos\theta)^4\psi_4 = \sum_{\ell m}\int d\omega e^{-i\omega t + im\varphi}\ {}_{-2}S_{\ell m}(\theta)R_{\ell m\omega}(r), \tag{1}$$

where the radial function $R_{\ell m\omega}$ satisfies the Teukolsky equation with $s = -2$ as

$$\Delta^2\frac{d}{dr}\left(\frac{1}{\Delta}\frac{dR_{\ell m\omega}}{dr}\right) - V(r)R_{\ell m\omega} = T_{\ell m\omega}, \tag{2}$$

The potential $V(r)$ is given by

$$V(r) = -\frac{K^2 + 4i(r-M)K}{\Delta} + 8i\omega r + \lambda, \tag{3}$$

where $K = (r^2 + a^2)\omega - ma$ and λ is the eigenvalue of ${}_{-2}S^{a\omega}_{\ell m}$. The source term $T_{\ell m\omega}$ is given in [2].

From the solutions of the homogeneous Teukolsky equation ($T = 0$) which satisfy suitable boundary conditions, a solution of the Teukolsky equation in Eq.(2) which has purely outgoing property at infinity and has purely ingoing property at the horizon can be found by using the Green function method:

$$R_{\ell m\omega} = \frac{1}{W}\left\{R^{\rm up}\int_{r_+}^{r} dr' R^{\rm in}T_{\ell m\omega}\Delta^{-2} + R^{\rm in}\int_{r}^{\infty} dr' R^{\rm up}T_{\ell m\omega}\Delta^{-2}\right\}, \tag{4}$$

where the Wronskian W is given by

$$W = 2i\omega D_{\ell m\omega}A^{\rm in}_{\ell m\omega}. \tag{5}$$

Then, ψ_4 has an asymptotic property at the horizon as

$$R_{\ell m\omega}(r \to r_+) \to \tilde{Z}^{\rm H}_{\ell m\omega}\Delta^2 e^{-ikr^*}, \tag{6}$$

where $R^{\rm in}$ and $R^{\rm up}$ are solutions of the homogeneous equation which behave as an incoming form at the outer horizon and an outgoing form at infinity, respectively, and

$$\tilde{Z}^{\rm H}_{\ell m\omega} = \frac{C_{\ell m\omega}}{2i\omega D_{\ell m\omega}A^{\rm in}_{\ell m\omega}}\int_{r_+}^{\infty} dr' R^{\rm up} T_{\ell m\omega}\Delta^{-2}. \tag{7}$$

The energy flux formula of waves going into a Kerr black hole is given by[6]

$$\left(\frac{dE_{\rm hole}}{dtd\Omega}\right) = \sum_{\ell m}\int d\omega \frac{{}_2S^2_{\ell m}}{2\pi}\frac{128\omega k(k^2+4\epsilon^2)(k^2+16\epsilon^2)(2Mr_+)^5}{|C|^2}|\tilde{Z}^{\rm H}_{\ell m\omega}|^2, \tag{8}$$

where the constant C is given in Ref.[6].

As for the analytical method to obtain the solutions of the homogeneous Teukolsky equation and the explicit forms of these solutions, we recommend to see the papers in Ref.[4].

3 The ingoing wave luminosity

By using the source term $T_{\ell m\omega}$ calculated in Ref.[2] and the solutions by Mano, Suzuki and Takasugi, we obtain

$$\begin{aligned}\left(\frac{dE}{dt}\right)_{\rm H} = & \left(\frac{dE}{dt}\right)_N v^5\left[-\frac{1}{4}q - \frac{3}{4}q^3 + \left(-\frac{33}{16}q^3 - q\right)v^2\right. \\ & + \left(\frac{55}{36}q^2 - \frac{79}{6}q^4 + \frac{1}{2} + 6q^3B_2 + \frac{13}{2}\kappa q^2 + 3q^4\kappa + \frac{1}{2}\kappa + 2qB_2\right)v^3 \\ & \left. + \left(-\frac{4651}{336}q^3 - \frac{17}{56}q^5 - \frac{43}{7}q\right)v^4\right],\end{aligned} \tag{9}$$

where $(dE/dt)_N$ is the Newtonian quadrupole luminosity at infinity:

$$\left(\frac{dE}{dt}\right)_N = \frac{32\mu^2M^3}{5r_0^5} = \frac{32}{5}\left(\frac{\mu}{M}\right)^2 v^{10}.$$

Here we only show the formula up to $O((v/c)^4)$, although we evaluated the luminosity to $O((v/c)^8)$ compared to the lowest order[7].

q	$(1.4M_\odot, 10M_\odot)$	$(1.4M_\odot, 40M_\odot)$	$(1.4M_\odot, 70M_\odot)$
0	$< 10^{-2}$	0.02	0.1
0.9	0.7	1.7	2.4
−0.9	0.3	0.45	0.39

Table 1: The difference of the total number of cycles N caused by the black hole absorption effect for typical neutron star (NS) – black hole (BH) binaries with mass $(M_{\rm NS}, M_{\rm BH})$.

We see that the absorption effect starts at $O(v^5)$ beyond the quadrupole formula in the case $q \neq 0$. If we set $q = 0$, the above formula reduces to the case of the Schwarzschild black hole,

$$\left(\frac{dE}{dt}\right)_{\rm H} = \left(\frac{dE}{dt}\right)_N \left(v^8 + O(v^{10})\right), \tag{10}$$

derived by Poisson and Sasaki[8]. We also notice that Eq.(9) can be negative when $q > 0$. This is an example of the super–radiance phenomenon.

4 The orbital evolution of coalescing compact binaries

Using the above results, we estimate the effect of the black hole absorption on the orbital evolution of coalescing compact binaries. The total number of cycles N of gravitational waves from an inspiraling binary is given by

$$N = \int_{v_f}^{v_i} dv \frac{\Omega}{\pi} \frac{dE/dv}{dE/dt}, \tag{11}$$

where $v_i = (M/r_i)^{1/2}$, $v_f = (M/r_f)^{1/2}$, and r_i and r_f are the initial and final orbital separation of the binary. The energy loss rate is given by $dE/dt = (dE/dt)_{\rm H} + (dE/dt)_\infty$ where $(dE/dt)_\infty$ is the luminosity at infinity.

We adopt $r_f = 6M$ and r_i is the one at which the frequency of the wave is 10 Hz. We evaluate ΔN, which is the difference of the total number of cycles N predicted by the formulas for dE/dt with and without the black hole absorption effect. The value of ΔN is calculated explicitly as

$$\Delta N = \left| \int_{v_f}^{v_i} dv \frac{\Omega}{\pi} \frac{dE/dv}{(dE/dt)_{\rm H} + (dE/dt)_\infty} - \int_{v_f}^{v_i} dv \frac{\Omega}{\pi} \frac{dE/dv}{(dE/dt)_\infty} \right|. \tag{12}$$

The results for typical NS–BH systems are given in Table 1.

We find that the effect of black hole absorption is very small when $q = 0$. We also find that the black hole absorption is more important for $q > 0$ than for

$q < 0$. ΔN becomes smaller for $q < 0$ since the total number of cycles is much smaller than for $q > 0$. Further, in the case $q > 0$, the black hole absorption is more important when the mass of the black hole is large. The reason is as follows: when the black hole mass becomes larger, the initial orbital radius (for fixed initial wave frequency) becomes smaller and v_i becomes larger. When $q > 0$, this effect is dominant. Hence, although the total cycle N become small when the black hole mass is large, the effect of the black hole absorption becomes larger.

5 Summary

We have calculated the gravitational wave energy which is absorbed by a rotating black hole, and obtained luminosity formulas to $O(v^1 3)$ beyond the Newtonian quadrupole luminosity at infinity. We find that black hole absorption occurs at $O(v^5)$ beyond the quadrupole formula, which is $O(v^3)$ lower than in the non–rotating case. Using this result, we calculated the effect on the number of cycles of waves emitted by binaries in the laser interferometer's band. We found that the effect of the black hole absorption is more important in the case when $q > 0$ (this is the case when a particle moves toward the same direction as the black hole rotation). We also found that the black hole absorption is important in the case when the mass of black hole is large.

Gal'tsov gave a formula for the black hole absorption from the radiation reaction force on the particle. Gal'tsov's formula[8] is expressed as

$$\frac{dE}{dt} = \left(\frac{dE}{dt}\right)_N \frac{v^5}{2} \left\{ v^3 \left(1 + \sqrt{1-q^2}\right) - \frac{q}{2} \right\} \left(1 + 3q^2\right) . \tag{13}$$

This formula agrees with our result only at the lowest order.

References

1. C. Cutler et al., Phys. Rev. Lett. **70**, 2984 (1993).
2. H. Tagoshi, H. Shibata, T. Tanaka and M. Sasaki, Phys. Rev. **D 54**, 1439 (1996)
3. H. Tagoshi, S. Mano and E. Takasugi, Prog. Theor. Phys. **98**, 829 (1997)
4. S. Mano, H. Suzuki and E. Takasugi, Prog. Theor. Phys. **95**, 1079 (1996); S. Mano and E. Takasugi, Prog. Theor. Phys. **97**, 213 (1997).
5. S.A.Teukolsky, Astorophys. J. **185**, 635 (1973).
6. S.A.Teukolsky and W.H. Press, Astorophys. J. **193**, 443 (1974).
7. E.Poisson and M.Sasaki, Phys. Rev. **D 51**, 5753 (1995).
8. D.V. Gal'tsov, J. Phys.**15**, 3737 (1982).

STRING PHENOMENOLOGY

C. MUÑOZ

Departamento de Física Teórica C–XI
Universidad Autónoma de Madrid, Cantoblanco, 28049 Madrid, Spain
e–mail: cmunoz@delta.ft.uam.es

String theory is trying to answer the question *why* in the Standard Model. This attempt is very briefly reviewed in this talk. After discussing the phenomenological aspects of string theory, the study of soft terms is emphasized as a possible connection between theory and experiment. Special attention is paid to the issues of flavor changing neutral currents, loop effects, and M–theory versus weakly coupled string theory.

1 Introduction

Dyson, in an entertaining article written in 1953 [1], drew the following analogy between quantum field theory and 19th–century chemistry: "The latter described the properties of the chemical elements and their interactions: How the elements behave; it did not try to explain why a particular set of elements, each with its particular properties, exists. To answer the question *why*, completely new sciences were needed: atomic and nuclear physics. (...) Quantum field theory treats elementary particles just as 19th–century chemists treated the elements. The theory is in its nature descriptive and not explanatory. It starts from the existence of a specified list of elementary particles, with specified masses, spins, charges and specified interactions with one another. All these data are put into the theory at the beginning. The purpose of the theory is simply to deduce from this information what will happen if particle A is fired at particle B with a given velocity. We are not yet sure whether the theory will be able to fulfill even this modest purpose completely. Many technical difficulties have still to be overcome."

Now, in 1998, we do know that the theory fulfilling the modest purpose mentioned by Dyson is the *Standard Model.* What 19th–century chemistry did with chemical elements, the 20th–century Standard Model does with elementary particles. It describes how they behave, but does not try to explain why a particular set of elementary particles, each with its particular properties, exists. To answer the question *why*, it seems that new sciences are not needed, as atomic and nuclear physics in the case of chemistry, but just new theories. This is precisely the purpose of the *string theory*.

2 String Phenomenology

In string theory, elementary particles are not point–like objects but extended, string–like objects. It is surprising that this apparently small change may allow us to answer fundamental questions that in the context of a quantum field theory of point–like particles cannot even be posed: why do we live in four dimensions? Why is the gauge group $SU(3) \times SU(2) \times U(1)$? Why are there three families of particles? Why is the pattern of quark and lepton masses so strange? Why is the fine structure constant given by $\alpha = 1/137$?, etc. For example, with respect to the last two questions, let us remark that in string theory there are no free parameters, all coupling constants are in fact not constants but *fields*. In particular, the measured experimental values of the gauge coupling constants and Yukawa couplings correspond to the vacuum expectation values (VEVs) of those fields. A gauge singlet field S, called dilaton, determines the gauge couplings, whereas gauge singlet fields T_i, called moduli, determine the Yukawa couplings. These fields are generically present in four–dimensional models: the dilaton arises from the gravitational sector of the theory and the moduli parametrize the size and shape of the compactified variety (let us recall that string theory can only be quantized in ten dimensions).

Although the most phenomenologically promising string theory, the heterotic string, has a gauge group $E_8 \times E_8$ in ten dimensions, there exist different methods to build standard–like models (i.e. gauge group $SU(3) \times SU(2) \times U(1)$ with three families) in four dimensions [2]: orbifolds, Calabi–Yau spaces, fermionic strings, covariant lattices, Gepner models, etc. In spite of these achievements it is fair to say that the Standard Model has not been obtained yet. In this sense the problem of string theory is that it is too ambitious: the correct model must reproduce not only the gauge group and families of the standard model, but also the correct values of the gauge coupling constants, the correct mass hierarchy for quark and leptons, a realistic Kobayashi–Maskawa matrix, etc., i.e. the nineteen parameters fixed by experiment in the Standard Model. In order to reach this goal, thousands of models (vacua) can be built due to the fact that there are many consistent ways to compactify the extra dimensions. Many models have a number of families different from three, no appropriate gauge groups, no appropriate matter, etc., but others have the gauge group of the Standard Model or grand unified gauge groups, and three families of particles. Then, using the experimental results available (e.g. fermion masses or mixing angles), it is in principle possible to discard most of the models in order to obtain the Standard Model.

Unfortunately, the model space is so huge that this type of analysis is very cumbersome in practice. Another possibility is to select the correct model (vacuum) using some (for the moment unknown) *dynamical* mechanism. In this sense duality symmetries may help to solve the problem. The five consistent string theories, type I with gauge symmetry $SO(32)$, heterotics with gauge symmetries $E_8 \times E_8$ or $SO(32)$, type IIA and type IIB, are in fact connected by dualities and maybe the thousands of vacua of these theories are also connected. Thus due to these connections some mechanism could select one point in the huge model space. Besides, it has recently been proposed that the five string theories can be derived from a single theory, called M−-theory. This might be the underlying fundamental theory of elementary particles. For example, the strong coupling limit of the ten–dimensional $E_8 \times E_8$ heterotic string may be obtained by compactifying the eleven–dimensional M−-theory on an orbifold. The resulting theory, unlike the weakly coupled heterotic string, provides an attractive framework for the unification of couplings [3]. Although the quantum version of $M-$theory is yet to be built, some properties are already known. It is related to membranes, and its low–energy limit is eleven–dimensional supergravity (SUGRA).

3 "Soft" Phenomenology

We saw in the previous section that finding a connection between theory and experiment (i.e. the Standard Model) in strings is possible in principle but difficult in practice. Here we will discuss a possible way out of this problem. Since the soft SUSY–breaking parameters can be computed in principle in the context of (super)string models, a comparison between these predictions about soft terms and the experimentally observed SUSY spectrum would allow us to test string (or $M-$) theory. This is the path, followed in the last years by several groups, that I will try to review very briefly in this section. Let me mention at this point that I will concentrate in subsect. 3.1 on weakly coupled heterotic string models leaving the discussion of strongly coupled strings (from $M-$theory) for subsect. 3.2.

3.1 From Superstrings

As is well known the soft parameters can be computed in generic hidden sector SUGRA models [4]. They depend on three functions, which determine the full $N = 1$ SUGRA Lagrangian: the Kähler potential K, the superpotential W, and the gauge kinetic function f. However, one can think of many possible SUGRA models (with different K, W and f) leading to *different* results for the soft terms. This arbitrariness can be ameliorated in SUGRA models derived from superstring theory, where K, f, and the hidden sector are more constrained. These models have a natural hidden sector built–in: the dilaton field S and the moduli fields T_i. The associated effective SUGRA Kähler potentials, at the superstring tree level, are of the type:

$$K = -\ln(S+\overline{S}) + \hat{K}(T_i,\overline{T}_i) + \tilde{K}_{\alpha\overline{\beta}}C^\alpha\bar{C}^{\overline{\beta}} + \frac{1}{4}\tilde{K}_{\alpha\overline{\beta}\gamma\overline{\delta}}C^\alpha\bar{C}^{\overline{\beta}}C^\gamma\bar{C}^{\overline{\delta}}$$
$$+ \left[\frac{1}{2}Z_{\alpha\beta}C^\alpha C^\beta + \frac{1}{2}Z_{\alpha\beta\overline{\gamma}}C^\alpha C^\beta\bar{C}^{\overline{\gamma}} + \frac{1}{6}Z_{\alpha\beta\gamma\overline{\delta}}C^\alpha C^\beta C^\gamma\bar{C}^{\overline{\delta}} + h.c.\right] \quad (1)$$

where the coefficient functions depend on T_i and $\overline{T}_i$. The first piece is the usual term corresponding to the complex dilaton S that is present for any compactification. The second piece is the Kähler potential of the moduli fields, which in general depends on the compactification scheme and can be a complicated function. The same comment applies to the coefficient functions $\tilde{K}_{\alpha\overline{\beta}}(T_i,\overline{T}_i)$, $Z_{\alpha\overline{\beta}}(T_i,\overline{T}_i)$, etc. Finally, for any four–dimensional superstring the tree–level gauge kinetic function is independent of the moduli sector and is simply given by $f_a = k_a S$. Then, without specifying the SUSY–breaking mechanism (i.e. W), just assuming that the auxiliary fields of S and T_i, F^S and F^i respectively, are the seed of supersymmetry breaking, interesting predictions for this simple class of models can be obtained [4]. This is the approach that we will discuss here.

Let us focus first on the very interesting limit where the dilaton S is the only source of SUSY breaking. At superstring tree level the dilaton couples in a universal manner to all particles; therefore this limit is compactification *independent*. In particular, since $\tilde{K}_{\alpha\overline{\beta}}(T_i,\overline{T}_i)$ in (1) is independent on S, the soft scalar masses turn out to be universal $m_\alpha = m_{3/2}$. Because of the simplicity of this scenario the predictions about the low–energy spectrum are quite precise [4]. For example, for the first two generations, squarks are almost degenerate with gluinos and much heavier than sleptons.

In general the moduli fields, T_i, may also contribute to SUSY breaking, and in that case their effects on soft parameters must also be included. To illustrate the main features of mixed dilaton/moduli–dominated SUSY breaking, let us concentrate first on the simple situation of diagonal moduli and matter metrics. This is the case for most $(0,2)$ symmetric Abelian orbifolds which have $\tilde{K}_{\alpha\overline{\beta}} = \delta_{\alpha\beta}\Pi_i(T_i+\overline{T}_i)^{n^i_\alpha}$ where n^i_α are the modular weights of the matter fields C^α. Then e.g. the scalar masses are given by: $m^2_\alpha = m^2_{3/2} + \sum_i \frac{n^i_\alpha}{(T_i+\overline{T}_i)^2}|F^i|^2$.

Universality is a desirable property for phenomenological reasons, particularly to avoid flavor changing neutral currents (FCNC). Notice however that the above scalar masses show in general a *lack of universality* due to the modular weight dependence[5]. So, even with diagonal matter metrics, FCNC effects may appear. This potential problem seems to be endemic in superstring models[6]. E.g. in Calabi–Yau compactifications, the metrics are generically off–diagonal, and therefore the mass eigenvalues are typically non–degenerate[7].

In fact, this situation may even be aggravated once one includes loop effects. On the one hand, superstring loop effects may spoil the universality in the dilaton–dominated SUSY–breaking limit due to an S–dependent contribution in $\tilde{K}_{\alpha\overline{\beta}}$ (1) [8]. On the other hand, the soft parameters are usually computed at the tree level of SUGRA interactions. However, as has been shown explicitly [9], there can be a significant modification of this procedure due to quadratically divergent SUGRA one–loop effects. In particular, due to the contributions of the pieces depending on $Z_{\alpha\beta\overline{\gamma}}$ and $\tilde{K}_{\alpha\overline{\beta}\gamma\overline{\delta}}$ in K (1), the scalar masses will have a generic matrix structure with non–degenerate eigenvalues[a]. Let us remark that results from loop effects by SUGRA interactions are valid in general, and therefore can be applied to any superstring theory, and in particular to $M-$theory.

Nevertheless, we recall that the low–energy running of the scalar masses has to be taken into account. In particular, in the squark case, for gluino masses heavier than (or of the same order as) the scalar masses at the boundary scale, there are large flavour–independent gluino loop contributions which are the dominant source of scalar masses. This situation is very common e.g. in orbifold models. This effect can therefore help in fulfilling the FCNC constraints[10].

3.2 From M–Theory

Given the theoretical and phenomenological virtues of $M-$theory mentioned in sect. 2, it is crucial to examine how the pattern of soft terms changes when one moves from the weakly coupled heterotic string limit to the $M-$theory limit. For example, higher order corrections to the Kähler potential in the $M-$theory limit imply a large $S-$dependent contribution to $\tilde{K}_{\alpha\overline{\beta}}$ (1)[11]. Under the same assumption as in the previous subsection, i.e. that SUSY is spontaneously broken by the auxiliary components of the bulk S and T_i superfields, it has been shown explicitly[12] that there can be a sizable difference between the heterotic string limit and the $M-$theory limit even in the overall pattern of soft terms.

[a] It is worth noticing here that in the case of the μ term these quantum corrections do not only modify the already known contributions from $\tilde{\mu}_{\alpha\beta}$ and $Z_{\alpha\beta}$ in the tree–level matching condition[4], but also provide *new* sources of the μ term, naturally of order the weak scale[9]. These sources depend upon the coefficients $Z_{\alpha\beta\overline{\gamma}}$ and $Z_{\alpha\beta\gamma\overline{\delta}}$ in K, and also on possible coefficients $\tilde{f}_{a\alpha\beta}$ in f_a.

With respect to the issue of FCNC, due to the above mentioned $S-$dependent contribution, there can be a large violation of scalar mass universality even in the dilaton–dominated scenario. Notice that a similar effect mentioned in the previous subsection is expected to be much smaller, since it is due to string loops.

4 Final comments

Let us hope that the comment about chemists that Dyson wrote in the article mentioned in the Introduction – "Looking backward, it is now clear that 19th–century chemists were right to concentrate on the *how* and to ignore the *why*. They did not have the tools to begin to discuss intelligently the reasons for the individualities of the *elements*. They had to spend a hundred years building up a good quantitative descriptive theory before they could go further. And the result of their labors, the classical *science of chemistry*, was not destroyed or superseded by the later insight that *atomic physics* gave." – will also be written by somebody in the next century about 20th–century physicists substituting elements by *elementary particles*, science of chemistry by *Standard Model* and atomic physics by *string theory*.

References

1. Reproduced in the book: "Particles and Fields", Scientific American (1980) 17.
2. For a recent review, see: F. Quevedo, hep-ph/9707434, and references therein.
3. For a review, see: K.R. Dienes, Phys. Rep. **287** (1997) 447, and references therein.
4. For a recent review, see: A. Brignole, L.E. Ibáñez and C. Muñoz, hep-ph/9707209, and references therein.
5. L.E. Ibáñez and D. Lüst, Nucl. Phys. **B382** (1992) 305.
6. For a recent review, see: C. Muñoz, hep-ph/9710388, and references therein.
7. H.B. Kim and C. Muñoz, Z. Phys. **C75** (1997) 367.
8. J. Louis and Y. Nir, Nucl. Phys. **B447** (1995) 18.
9. K. Choi, J.S. Lee and C. Muñoz, hep-ph/9709250.
10. A. Brignole, L.E. Ibáñez and C. Muñoz, Nucl. Phys. **B422** (1994) 125; P. Brax and M. Chemtob, *Phys. Rev.* **D51** (1995) 6550
11. A. Lukas, B.A. Ovrut and D. Waldram, hep-th/9710208.
12. K. Choi, H.B. Kim and C. Muñoz, hep-th/9711158.

MODEL BUILDING USING SUSY QCD

NAOYUKI HABA
Faculty of Engineering, Mie University, Tsu Mie 514, Japan

In $N = 1$ Supersymmetric gauge theories, we can calculate non-perturbative effects in the superpotential. It is interesting to apply this effect to phenomenology. We show two models here. One is a low energy supersymmetry (SUSY) breaking scenario without messenger sector, and the second is a composite model generating Yukawa interaction dynamically.

1 Introduction

Recently, we learned to calculate non-perturbative effects in $N = 1$ SUSY gauge theories [1]. Many people are trying to apply these effects to phenomenology. We will show two models among them. One is a low energy SUSY breaking scenario without messenger sector, and the second is a composite model generating Yukawa couplings dynamically.

We consider the gauge group $SU(N_c)$ with N_f (anti-)fundamental matter representations $Q, (\overline{Q})$. The non-perturbative effects, which determine the vacuum structures, strongly depend on the number N_f. We will show model building for $N_f = N_c - 1$ and $N_f = N_c + 1$. Each case induces a non-perturbative superpotential [a]

$$W_{dyn} = \frac{\Lambda^{b_0}}{\det M} \qquad (N_f = N_c - 1), \tag{1}$$

$$W_{dyn} = \frac{1}{\Lambda^{b_0}}(BM\overline{B} - \det M) \qquad (N_f = N_c + 1). \tag{2}$$

2 A model of gauge mediation

2.1 *Why do we need a theory of SUSY breaking?*

Before showing a model, let us consider briefly why we a need SUSY breaking theory. SUSY guarantees the smallness of the Higgs mass through the chiral symmetry of fermions. This is the main motivation of introducing SUSY from the viewpoint of phenomenology. The standard model (SM) can be extended to the supersymmetric standard model (SSM), and the minimal extension is so-called the minimal supersymmetric standard model (MSSM). If SUSY is

[a] Here $M \equiv Q\overline{Q}$, $B \equiv \epsilon Q \cdots Q, \overline{B} \equiv \epsilon\overline{Q} \cdots \overline{Q}$. Since the superfields are bosons, the gauge invariant operators B, $\overline{B}$ exist only in the region $N_c \leq N_f$.

exact, the number of free parameters 19(1) in the SM [b, c] does not increase in the MSSM. However, since SUSY is broken in our real world, the number of free parameters increases greatly. The total number of parameters becomes 125(44) in the MSSM [d]. Most come from SUSY breaking parameters and especially from their flavor indices. Furthermore, if these SUSY breaking parameters are arbitrary, MSSM induces too large flavor changing neutral current (FCNC) in $K^0 - \overline{K^0}$ and $\mu \to e\gamma$, and too large a electric dipole moments (EDM) of the neutron and the electron. Thus, if there exists SUSY in our world, there must exist an underlying theory of SUSY breaking, which should naturally lead to highly degenerated squark/slepton masses $\delta\tilde{m}^2/\tilde{m}^2 < O(10^{-3})$ for the supersymmetric version of GIM cancellation to work, and small CP phases (less than $O(10^{-2})$) in SUSY breaking parameters [e].

2.2 How is SUSY breaking communicated to our world?

The underlying theory of SUSY breaking might be a model of spontaneous SUSY breaking. However, the tree level supertrace formula $\mathrm{Str}M^2 = 0$ shows that squarks/sleptons which are lighter than quarks/leptons appear when SUSY is broken spontaneously. There are three ways to circumvent this formula. Correspondingly there are three ways to mediate SUSY breaking to our world, that is,
(a): Gravity mediated scenario (Hidden sector scenario) [2],
(b): Anomalous $U(1)$ gauge scenario [3],
(c): Gauge mediated scenario (Visible sector scenario) [4].
The 125(44) parameters in the MSSM decrease in each scenario. They are (a): 22(3), (b): 34(3) [f], and (c): 22(3) [g]. Each model has both merits and demerits. Here we concentrate on the scenario (c), which has the merits of calculability and naturally gives small FCNC since gauge interactions are flavor blind.

---

[b] The number in ( ) shows the number of CP phases.

[c] Gauge couplings 3; quark/lepton masses 9; KM matrix 4(1); Higgs mass 1; Higgs self-coupling 1; θ 1.

[d] Gauge couplings 3; quark/lepton masses 9; KM matrix 4(1); θ 1; μ 2(1); B 2(1); $m^2_{H_u}$ 1; $m^2_{H_d}$ 1; ξ 1; gaugino mass M_λ 2(1) $\times$ 3; squark/slepton mass $\tilde{m}^2$ 9(3) $\times$ 5; A 18(9) $\times$ 3; minus 4 phases (field redefinitions), which is easily obtained by counting the reduction of the global symmetry $U(1)_B U(1)^3_L U(1)_{PQ} U(1)_R \to U(1)_B U(1)_L$ by introducing μ and SUSY breaking parameters.

[e] A few TeV superparticle mass can also avoid large EDM.

[f] Here we count anomalous $U(1)$ charges as parameters.

[g] Here we count μ 2(1); B 2(1).

2.3 A SUSY breaking model without messenger sector

Dynamical SUSY breaking can explain the large hierarchy between the Planck scale and the weak scale. Eq.(1) shows that in the case of $N_f = N_c - 1$ the vacuum can be lifted dynamically. SUSY is broken in general when there are no flat directions and the continuous global symmetry is spontaneously broken [h]. This is the case for the 3-2 model [6] [i], whose superpotential is

$$W_{3-2} = \lambda X_1 + \frac{2\Lambda^7}{X_3}. \tag{3}$$

The second term is induced from non-perturbative effects as in Eq.(1). The X_i ($i = 1 \sim 3$) are moduli fields [j]. Since we analyze the potential in the Higgs picture for the calculability, where the gauge coupling is weak enough to use the canonical Kähler potential, λ should be small, of $O(10^{-7})$. Such a small parameter can be avoided by building a modified 3-2 model [k, l] as

$$W'_{3-2} = \lambda' \frac{X_1 X_3}{M_{\text{Planck}}} + \frac{2\Lambda^7}{X_3}. \tag{4}$$

Now let us introduce the vector-like fields $q, \bar{q}$ and $l, \bar{l}$, which are the components of $\mathbf{5}$, $\bar{\mathbf{5}}$ of $SU(5)$. By a suitable R charge assignment of vector-like fields, we can build a simple model of gauge mediation (c) without messenger sector [7], whose superpotential is

$$W = W'_{3-2} + h_1 \frac{X_3}{M^3_{\text{Planck}}} q\bar{q} + h_2 \frac{X_3}{M^3_{\text{Planck}}} l\bar{l}\,. \tag{5}$$

Flavor blind squark/slepton and gaugino masses are generated by loop diagrams involving gauge interactions. This is one of the simplest models of gauge mediation (c), which naturally avoids large FCNC since $\delta\tilde{m}^2/\tilde{m}^2 \simeq \tilde{m}^2_{\text{gravity}}/\tilde{m}^2_{\text{loop}} \simeq O(10^{-4})$. The demerits of this model are that μ and B are arbitrary parameters and that the color conserving minimum might not be the true vacuum [m].

[h] This is because the Goldstone boson cannot have a supersymmetric partner[5].

[i] The 3-2 model is the SM without the right-handed electron.

[j] Three moduli fields remain in the Higgs phase: 14 (degrees of freedom of fields) − 8 ($SU(3)$ gauge boson) − 3 ($SU(2)$ gauge boson).

[k] It might not be so unnatural to introduce the scale M_{Planck} in the superpotential, since supergravity always exists. The massless Goldstone fermion can not be absorbed unless we consider supergravity theory.

[l] W'_{3-2} is obtained by changing the R charge assignment of $X_{1,2}$. In this case $\lambda' \simeq O(1)$.

[m] We can easily show that there is no SUSY vacuum in this model.

3 A composite model

Next we would like to consider an application of the case $N_f = N_c + 1$. This is an example of confinement where only M, B, and $\overline{B}$ are massless fields below the strong dynamics scale. Here we consider a sufficiently high energy scale to neglect SUSY breaking effects. People build composite models by using Eq.(2) [8,9]. Here we show a model [10, n] which generates Yukawa couplings dynamically. Our motivation is very simple: when we build a preon model whose "meson" M is a Higgs field and the "(anti-)baryon" B ($\overline{B}$) is a quark/lepton, the first term of Eq.(2) becomes a Yukawa coupling.

We introduce the gauge group $SU(7)_H$ and $N_f = 8$ preons P_i, $\overline{P_i}$, which also have SM gauge quantum numbers as shown in Table 1. When the $SU(7)_H$ gauge group becomes strongly interacting, the confinement states M, B and $\overline{B}$ appear as in Table 2 and Table 3.

	$SU(7)_H$	$SU(3)_C$	$SU(2)_L$	$U(1)_Y$	$U(1)_B$	$U(1)_L$
P_{Q}	$\square$	$\square^*$	$\square$	$-1/3$	$-1/21$	$2/7$
P_ℓ	$\square$	1	$\square$	1	$2/7$	$-5/7$
$\bar{P}_{\mathrm{U}}$	$\square^*$	$\square$	1	$4/3$	$1/21$	$-2/7$
$\bar{P}_{\mathrm{D}}$	$\square^*$	$\square$	1	$-2/3$	$1/21$	$-2/7$
$\bar{P}_{\mathrm{N}}$	$\square^*$	1	1	0	$-2/7$	$5/7$
$\bar{P}_{\mathrm{E}}$	$\square^*$	1	1	-2	$-2/7$	$5/7$

Table 1: Preons

	$SU(7)_H$	$SU(3)_C$	$SU(2)_L$	$U(1)_Y$	$U(1)_B$	$U(1)_L$	baryon ($\times 1/\Lambda^6$)
Q	1	$\square$	$\square$	$1/3$	$1/3$	0	$P_{\mathrm{Q}}^5 P_\ell^2$
l	1	1	$\square$	-1	0	1	$P_{\mathrm{Q}}^6 P_\ell$
$\bar{U}$	1	$\square^*$	1	$-4/3$	$-1/3$	0	$\bar{P}_{\mathrm{U}}^2 \bar{P}_{\mathrm{D}}^3 \bar{P}_{\mathrm{E}} \bar{P}_{\mathrm{N}}$
$\bar{D}$	1	$\square^*$	1	$2/3$	$-1/3$	0	$\bar{P}_{\mathrm{U}}^3 \bar{P}_{\mathrm{D}}^2 \bar{P}_{\mathrm{E}} \bar{P}_{\mathrm{N}}$
$\bar{N}$	1	1	1	0	0	1	$\bar{P}_{\mathrm{U}}^3 \bar{P}_{\mathrm{D}}^3 \bar{P}_{\mathrm{E}}$
$\bar{E}$	1	1	1	2	0	1	$\bar{P}_{\mathrm{U}}^3 \bar{P}_{\mathrm{D}}^3 \bar{P}_{\mathrm{N}}$

Table 2: "Baryons" and "anti-baryons"

Eq.(2) becomes

$$\begin{aligned} W_{\mathrm{dyn}} &= y_u \bar{U}(\bar{\Phi}_{\mathrm{u}} + \bar{\phi}_{\mathrm{u}})Q + y_d \bar{D}(\bar{\Phi}_{\mathrm{d}} + \bar{\phi}_{\mathrm{d}})Q + y_n \bar{N}(\bar{\Phi}_{\mathrm{u}} + \bar{\phi}_{\mathrm{u}})\ell + y_e \bar{E}(\bar{\Phi}_{\mathrm{d}} + \bar{\phi}_{\mathrm{d}})\ell \\ &+ \alpha_u \bar{U} G_{\mathrm{u}} Q + \alpha_d \bar{D} G_{\mathrm{d}} Q + \beta_u \bar{U} \bar{Y} \ell + \beta_d \bar{D} \bar{X} \ell + \beta_n \bar{N} X Q + \beta_e \bar{E} Y Q - [\mathrm{Det}]. \quad (6) \end{aligned}$$

[n] This model is a modified version of Ref [9].

	$SU(7)_H$	$SU(3)_C$	$SU(2)_L$	$U(1)_Y$	$U(1)_B$	$U(1)_L$	meson ($\times 1/\Lambda$)
$\bar{\Phi}_{\rm u}$	1	1	$\Box$	1	0	0	$P_{\rm Q}\bar{P}_{\rm U}$
$\bar{\Phi}_{\rm d}$	1	1	$\Box$	-1	0	0	$P_{\rm Q}\bar{P}_{\rm D}$
$G_{\rm u}$	1	**Ad**	$\Box$	1	0	0	$P_{\rm Q}\bar{P}_{\rm U}$
$G_{\rm d}$	1	**Ad**	$\Box$	-1	0	0	$P_{\rm Q}\bar{P}_{\rm D}$
X	1	$\Box^*$	$\Box$	$-1/3$	$-1/3$	1	$P_{\rm Q}\bar{P}_{\rm N}$
$\bar{X}$	1	$\Box$	$\Box$	$1/3$	$1/3$	-1	$P_\ell\bar{P}_{\rm D}$
Y	1	$\Box^*$	$\Box$	$-7/3$	$-1/3$	1	$P_{\rm Q}\bar{P}_{\rm E}$
$\bar{Y}$	1	$\Box$	$\Box$	$7/3$	$1/3$	-1	$P_\ell\bar{P}_{\rm U}$
$\bar{\phi}_{\rm u}$	1	1	$\Box$	1	0	0	$P_\ell\bar{P}_{\rm N}$
$\bar{\phi}_{\rm d}$	1	1	$\Box$	-1	0	0	$P_\ell\bar{P}_{\rm E}$

Table 3:. "Mesons"

It shows that the Yukawa couplings are really generated dynamically. This model predicts the existence of four Higgs doublets. Unfortunately, the color octet Higgs G and lepto-quark X, Y are also generated as "meson" fields, and $W_{\rm dyn}$ contains their Yukawa couplings. Since they are all massless, we must introduce

$$W_{tree} = g_G \frac{P_{\rm Q}P_{\rm Q}\bar{P}_{\rm U}\bar{P}_{\rm D}}{M} + g_\mu \frac{P_\ell P_\ell \bar{P}_{\rm E}\bar{P}_{\rm N}}{M} + g_x \frac{P_{\rm Q}P_\ell\bar{P}_{\rm D}\bar{P}_{\rm N}}{M} + g_y \frac{P_{\rm Q}P_\ell\bar{P}_{\rm U}\bar{P}_{\rm E}}{M} \quad (7)$$

by hand in order to give them masses. Eq.(7) is the most general gauge invariant quartic superpotential, which also generates μ terms ($\mu \simeq \Lambda^2/M$).

This model[o] can be extended to a 3 generation model. Possible extensions are (1): $SU(7)_H \to SU(7)_H^3$, (2): $SU(7)_H \to SU(23)_H$[p], (3): The first and second generations are elementary and the third generation and Higgs are composite[q].

4 Summary

We have shown two models of phenomenological application of non-perturbative effects in SUSY gauge theories. One is a low energy SUSY breaking scenario without messenger sector, and the second is a composite model gener-

[o] The anomalies in this model are the same as in the SM, $SU(2)_L^2 U(1)_{B+L}, U(1)_Y^2 U(1)_{B+L}$.

[p] This corresponds to the extension of the global symmetry $U(1)_B U(1)_L \to SU(3)_F U(1)_B U(1)_L$.

[q] The idea (3) comes from the fact that preons (Table 1) are just like "mirror fields". Mirror fields are conjugated states of ordinary fields, which appear in string theory.

ating Yukawa interaction dynamically. It is interesting to try to apply non-perturbative effects in SUSY QCD to other phenomenologies.

Acknowledgments

It is my pleasure to thank the organizers of PPPP'97 (APCTP) for their kind hospitality and for giving me the opportunity to present this work. I would like to thank T. Matsuoka, N. Okamura and N. Maru for collaboration on these studies. I would also like to thank K. Hagiwara for continuous encouragements.

References

1. *E.g.*, K. Intriligator and N. Seiberg, Nucl.Phys. **B431**, 551 (1994).
2. *E.g.*, H. P. Nilles, Phys. Rep. **110**, 1 (1984).
3. *E.g.*, G. Dvali and A. Pomarol, Phys. Rev. Lett. **77**, 3728 (1996).
4. *E.g.*, M. Dine and A. E. Nelson, Phys. Rev. **D48**, 1277 (1993).
   M. Dine, A. E. Nelson, and Y. Shirman, Phys. Rev. **D51**, 1362 (1995).
   M. Dine, A. E. Nelson, Y. Nir, and Y. Shirman, Phys. Rev. **D53**, 2658 (1996).
5. A. E. Nelson and N. Seiberg, Nucl. Phys. **B416**, 46 (1994).
6. I. Affleck, M. Dine, and N. Seiberg, Nucl. Phys. **B256**, 557 (1985).
7. N. Haba, N. Maru, and T. Matsuoka, Phys. Rev. **D56**, 4207 (1997).
8. M. J. Strassler, Phys. Lett. **376B**, 119 (1996).
   A. E. Nelson and M. J. Strassler hep-ph/9607362.
9. M. A. Luty and R. N. Mohapatra, Phys. Lett. **B396**, 161 (1997).
   M. A. Luty hcp-ph/9611387.
10. N. Haba and N. Okamura, hep-ph/9709239.

SUPERSYMMETRY BREAKDOWN AND SOFT TERMS IN M−THEORY

MASAHIRO YAMAGUCHI
Department of Physics, Tohoku University
Sendai 980-8578, Japan

Spontaneous breakdown of supersymmetry in heterotic M−theory is discussed in the context of gaugino condensate. Using a macroscopic approach to derive the effective four–dimensional theory, we evaluate the gravitino mass and the soft masses. It turns out that the situation is very similar to the case of the weakly coupled heterotic string, with one important difference: in the M−theory case the gaugino can acquire a mass comparable to the gravitino mass, which avoids problems associated with a small gaugino mass in the weak coupling case.

Since its discovery, the $E_8 \times E_8$ heterotic string theory has been an attractive framework for a unified theory including gravity. The theory is consistent only in ten dimensional space–time; thus an extra six–dimensional space has to be compactified on a tiny manifold. In the conventional weak coupling heterotic string theory, the compactification scale is around the string scale of $\sim 10^{18}$ GeV. This high energy scale for compactification immediately raises the question how to explain the gauge coupling constant unification at $\sim 10^{16}$ GeV, suggested by the LEP experiment.

Compactification at the GUT scale, on the other hand, leads to a strongly coupled string theory. This can be understood if one notices that above the compactification scale the string coupling constant grows as the sixth power of the energy scale, and therefore it becomes much larger than unity if the compactification scale differs from the string scale by more than one order of magnitude.

Hořava and Witten [1] proposed that this strongly coupled $E_8 \times E_8$ (ten-dimensional) heterotic string theory can be well described by M−theory compactified on $R^{10} \times S^1/Z_2$. M−theory is basically eleven-dimensional supergravity. In the present case, it must be on a Z_2 orbifold in order to obtain $N = 1$ supersymmetry in ten dimensions. At each boundary of S^1/Z_2, an E_8 super Yang-Mills theory has to be attached to achieve anomaly cancellation. This M−theory description opens up the possibility to investigate the strong coupling limit of heterotic string theory [2].

When compactified to four dimensions, the following picture naturally emerges [2,3]: the GUT-scale $M_{GUT} = 3 \times 10^{16}$ GeV is identified with $1/R$ where $V = R^6$ is the volume of the compact six–dimensional manifold. The gauge coupling constant α_{GUT} is $1/25$ and the correct value of the four–dimensional reduced Planck mass $M_P = 2.4 \times 10^{18}$ GeV is obtained if one chooses the length of the eleventh–dimensional interval to be $R_{11} \sim 6R$. The fundamental scale of $M-$theory $M_{11} = \kappa^{-2/9}$ has to be two times larger than M_{GUT} (κ is the eleven-dimensional Newton constant).

The above picture naturally explains unification of the running gauge couplings in the minimal supersymmetric standard model. The $M-$theory description of the heterotic string is thus phenomenologically very attractive, motivating its phenomenological study (Ref. 4 and references therein). Here we shall consider supersymmetry breaking by gaugino condensate and discuss the structure of soft masses (*i.e.*, supersymmetry–breaking gaugino masses and scalar masses). We take a macroscopic approach: we investigate an effective four–dimensional supergravity as the low-energy limit of $M-$theory. We refer the reader to Ref. 5 as a complementary approach to ours.

In the following, we shall briefly describe our results[4]. In heterotic $M-$theory, one has an eleven–dimensional bulk and two ten–dimensional boundaries. In the eleven–dimensional bulk, there are elfbein fields (or metric fields g_{MN}), rank three anti–symmetric tensor fields C_{LMN}, and gravitino fields ψ^M. Attached on each boundary are E_8 gauge fields A_M and gaugino fields χ. Hořava and Witten obtained the effective action of $M-$theory in expansions of $\kappa^{2/3}$. In our analysis, we start with this effective action and consider compactification to four dimensions. We take a simple compactification procedure following the method of reduction and truncation in Ref. 6. For the metric we consider

$$g_{MN} = \begin{pmatrix} e^{-\gamma}e^{-2\sigma}g_{\mu\nu} & & \\ & e^{\sigma}\delta_{mn} & \\ & & e^{2\gamma}e^{-2\sigma} \end{pmatrix}, \tag{1}$$

where M, N=1,..., 11, μ, ν=1, ..., 4 and m, n=5, ..., 10. The resulting four–dimensional effective theory is described by an $N = 1$ supergravity, which is specified by a Kähler potential, a superpotential and a gauge kinetic function. At the leading order in $\kappa^{2/3}$, one finds the Kähler potential to be

$$K = -\log(S + \bar{S}) - 3\log(T + \bar{T} - 2C_i\bar{C}_i) \tag{2}$$

with

$$S = \frac{2}{(4\pi)^{2/3}}(e^{3\sigma} + i24\sqrt{2}D), \tag{3}$$

$$T = \frac{\pi^2}{(4\pi)^{4/3}}(e^{\gamma} + i6\sqrt{2}C_{11}) \tag{4}$$

where the fields D and C_{11} are defined from the C_{LMN} fields, and C_i is a matter field which is originally a ten–dimensional gauge field at the E_6 boundary. One can also find the gauge kinetic function to be $f = S$.

An important correction emerges at the next to leading order. The point is that, at this order, the Bianchi identity for the field strength G_{KLMN} made out of the C_{LMN} fields must be modified, leading to non–vanishing G fields. This complicates conditions for unbroken supersymmetry since the G appear in the supersymmetry transformation of the gravitino fields. Witten [2] showed that unbroken supersymmetry requires that the background metric g_{MN} depends on the coordinate of the eleventh dimensional interval x^{11}. To define the S and T fields in four dimensions, we average over the x^{11} coordinate. For the standard embedding in which one of the E_8 is broken to E_6, we find that the gauge kinetic function is

$$f_6 = S + \alpha T, \;\; f_8 = S - \alpha T, \tag{5}$$

at the E_6 and E_8 side, respectively. Here α is a dimensionless constant of order unity. Using our values for R, R_{11} *etc.*, we find that the correction αT is numerically comparable to the leading order contribution S. As the gauge kinetic function does not receive corrections beyond one loop, we expect that our result for the gauge kinetic functions will make sense. We can also obtain a similar correction to the Kähler potential.

We are now ready to discuss supersymmetry breaking. We assume here that a gaugino condensate forms at the E_8 boundary, $\langle\bar{\chi}\chi\rangle = \Lambda^3$ with Λ being the energy scale at which the gaugino condenses. The supersymmetry transformation laws for the eleven-dimensional gravitinos are

$$\begin{aligned}\delta\psi_A &= D_A\eta + \frac{\sqrt{2}}{288}G_{IJKL}\left(\Gamma_A^{IJKL} - 8\delta_A^I\Gamma^{JKL}\right)\eta \\ &\quad -\frac{1}{576\pi}\left(\frac{\kappa}{4\pi}\right)^{2/3}\delta(x^{11})\,(\bar{\chi}^a\Gamma_{BCD}\chi^a)\,(\Gamma_A^{BCD} - 6\delta_A^B\Gamma^{CD})\,\eta + \ldots \qquad (6)\\ \delta\psi_{11} &= D_{11}\eta + \frac{\sqrt{2}}{288}G_{IJKL}\left(\Gamma_{11}^{IJKL} - 8\delta_{11}^I\Gamma^{JKL}\right)\eta \\ &\quad +\frac{1}{576\pi}\left(\frac{\kappa}{4\pi}\right)^{2/3}\delta(x^{11})\,(\bar{\chi}^a\Gamma_{ABC}\chi^a)\,\Gamma^{ABC}\eta + \ldots . \qquad (7)\end{aligned}$$

From the above, we can expect that the gaugino condensate leads to supersymmetry breaking. Furthermore, the Hořava–Witten action has a perfect–square structure which relates G_{ABC11} with the gaugino condensate[7]. A careful computation shows that $\delta\psi_A$ vanishes in the leading order approximation where effects of the modified Bianchi identity are not considered. On the other hand, the supersymmetry variation of ψ_{11} does not vanish at the E_8 boundary and thus it is the goldstino, accounting for supersymmetry breaking.

In order to reduce to four dimensions, one averages over x^{11}. With the help of supersymmetry transformation laws of the elfbein fields, we can show that in leading order,

$$\langle F_T \rangle \sim \frac{\int dx^{11} \delta\psi_{11}}{\int dx^{11}} \neq 0, \quad \langle F_S \rangle = 0. \tag{8}$$

Note that this result is very similar to the case of weakly coupled string theory. The gravitino mass can be evaluated as:

$$m_{3/2} = \frac{\langle F_T \rangle}{\langle T + \bar{T} \rangle} \sim \frac{\Lambda^3}{m_{Pl}^2} \tag{9}$$

where m_{Pl} is the four–dimensional Planck scale 1.2×10^{19} GeV. This expression for the gravitino mass is a standard one, which can be expected in general supergravity models and is indeed realized in the weakly coupled string case. Thus to obtain a gravitino mass in, say, the TeV range, one has to adjust $\Lambda \sim 10^{13}$ GeV, as usual. Note that at the next to leading order, eqs. (8) will receive corrections, but the expression for the gravitino mass presented above will not change drastically.

We now turn to supersymmetry–breaking gaugino masses and scalar masses. First consider the gaugino mass. It is given by

$$m_{1/2} = \left(\frac{\partial f_6}{\partial S} F_S + \frac{\partial f_6}{\partial T} F_T \right) / 2\mathrm{Re} f_6. \tag{10}$$

In leading order $f_6 = S$ and $F_S = 0$, therefore the gaugino mass vanishes. However, in next–to–leading order the gauge kinetic function is corrected to eq. (5). Furthermore, F_S becomes non–zero, $\sim \alpha T/S \, m_{3/2}$. Thus the gaugino acquires a mass at this order:

$$m_{1/2} \sim \frac{\alpha T}{S} m_{3/2}. \tag{11}$$

As was discussed, $\alpha T/S$ is numerically rather close to unity. Thus the gaugino mass is generally comparable to the gravitino mass in strongly coupled string theory. This contrasts with the weak coupling case. In the latter case, though the gaugino also gets a non–vanishing mass at one–loop order, it is numerically much smaller than the gravitino mass.

Similarly we can compute the scalar mass by using the general formula:

$$m_0^2 = m_{3/2}^2 - F^i \bar{F}^{\bar{j}} \frac{Z_{i\bar{j}} - Z_i Z^{-1} Z_{\bar{j}}}{Z} \tag{12}$$

where i, $j = S$, T and Z is the S, T dependent coefficient of the kinetic term of the matter field C. An inspection of eq. (2) shows that the scalar mass vanishes in leading order. Again the next–to–leading order contribution gives a scalar mass of the order of the gravitino mass.

Summarizing the above, we conclude that, in the strong coupling ($M-$theory) regime, supersymmetry breaking via a gaugino condensate leads to the following mass pattern:

$$m_0 \sim m_{1/2} \sim m_{3/2}. \tag{13}$$

This diminishes the long–standing difficulty associated with the gaugino condensation scenario in the weakly coupled case, where the gaugino mass is much smaller than the other supersymmetry–breaking masses, leading to phenomenological problems, such as a fine–tuning problem with very heavy scalar masses, and too large a relic abundance of neutralinos. Superparticle phenomenology based on heterotic $M-$theory is more promising and thus its further study should be encouraged.

Acknowledgments

The author acknowledges Hans–Peter Nilles and Marek Olechowski for enjoyable collaborations. The results presented here are based on works with them. This work was partially supported by the Grant-in-Aid for Scientific Research from the Ministry of Education, Science and Culture of Japan No. 09640333.

References

1. P. Hořava and E. Witten, *Nucl. Phys.* B **460**, 506 (1996); *Nucl. Phys.* B **475**, 94 (1996).
2. E. Witten, *Nucl. Phys.* B **471**, 135 (1996).
3. T. Banks and M. Dine, *Nucl. Phys.* B **479**, 173 (1996).
4. H.P. Nilles, M. Olechowski and M. Yamaguchi, *Phys. Lett.* B **415**, 24 (1997); hep-th/9801030.
5. E.A. Mirabelli and M.E. Peskin, hep-th/9712214.
6. E. Witten, *Phys. Lett.* B **155**, 151 (1985).
7. P. Hořava, *Phys. Rev.* D **54**, 7561 (1996).

M–THEORY AND DYNAMICS OF SUPERSYMMETRIC GAUGE THEORIES

SOONKEON NAM

Department of Physics and Research Institute of Basic Sciences
Kyung Hee University, Seoul, 130-701, Korea

The study of brane configurations has played a rich role in nonperturbative dynamics of SUSY gauge theories. We consider brane configurations in M-theory describing $N = 1$ supersymmetric gauge theories and using the parametric representation of the brane configurations; we can calculate the superpotentials for various cases including multiple gauge groups or fermions. For $SU(n)$ $N = 1$ SQCD with N_f fermion case ($N_f < N_c$), we find that the superpotential from M-theory and the gauge theory agree precisely.

1 Introduction

Among many outstanding problems in particle physics, search for the explanation of the quark confinement, mechanism for SUSY breaking, and a candidate for the unified theory of everything have been discussed widely. Recently there has been much progress in these subjects through the uses of brane configurations parametrizing gauge theories.

2 Generalities of $N = 1$ SUSY Gauge Theories

Let us start by explaining the meaning of effective Lagrangian description of strongly coupled quantum field theory, with the example of $SU(3)$ gauge theory with $N_f = 3$ quarks. Low energy physics is best described by mesons and baryons. So if we can write down an effective Lagrangian in terms of these composite fields, which gives the most general form of their interactions consistent with symmetries, then we are in business. This is of course very difficult to realize in general. With SUSY the situation improved a lot and sometimes, holomorphy and global symmetries are enough to let us determine the renormalized effective superpotential W_{eff} completely.

The SUSY effective theory is described by the following Lagrangian:

$$\mathcal{L} = \int d^4\theta K(\Phi^\dagger, e^V \Phi) + \int d^2\theta \tau(\Phi) WW + h.c. + \int d^2\theta W(\Phi) + h.c. \quad (1)$$

Here τ and W are holomorphic functions of the chiral superfields.

Now we want to impose $R-$symmetry on the superpotential, which is a rotational symmetry in the grassmannian direction.

$$\theta \to e^{-i\alpha}\theta, \quad W \to e^{2i\alpha}W. \tag{2}$$

Here the gauge invariant composite superfields are the mesons

$$M_{ij} = Q_i\overline{Q}_j \ \ (i,j = 1,\cdots N_f). \tag{3}$$

The quantity that is invariant under the global $SU(N_f) \times SU(N_f)$ symmetry is $\det M$. So the form of the superpotential as a holomorphic function, with $R-$charge $= 2$ and mass dimension 3 which respects the global symmetry is[20]

$$W_{\text{ADS}} = C\left(\frac{\Lambda^{b_0}}{\det M}\right)^{1/(N_c-N_f)} \tag{4}$$

This formula is singular when $N_f = N_c$; this is due to new gauge invariant chiral fields, the baryons B and $\overline{B}$, which now have to be included:

$$B = \epsilon_{a_1\cdots a_{N_c}} Q_1^{a_1}\cdots Q_{N_c}^{a_{N_c}}. \tag{5}$$

There is a classical constraint: $\det M = B\overline{B}$. This constraint gets smoothed out by quantum nonperturbative deformations and becomes $\det M - B\overline{B} = \Lambda^{2N_c}$. One of the hypermultiplets can be integrated out and gives back the ADS superpotential. Similarly, for $N_f = N_c + 1$, the superpotential can be written down in terms of the baryons and mesons. As we increase the number of flavors even further, the naive superpotential written in terms of the baryons and mesons no longer has R charge equal to 2. So one needs a better idea. Here Seiberg viewed the baryons as bound states of $\tilde{N}_c = N_f - N_c$ components, i.e. quarks of the dual gauge theory $SU(\tilde{N}_c)$. The proposal is to describe SQCD with $N_f > N_c + 1$ by $SU(\tilde{N}_c)$ SYM coupled to N_f chiral superfields and chiral supermultiplet M^{ij} $(i,j = 1,\cdots,N_f)$, with the superpotential $W = qM\overline{q}$. This proposal has passed various consistency checks, such as 't Hooft anomaly matching and decoupling relations. Further justification comes from considering $N = 2$ SQCD with explicit SUSY breaking terms, as considered by Argyres, Plesser and Seiberg.

Now, with all the developments in string duality, the question is whether we can justify this superpotential from a more fundamental theory, say M–theory. As first investigated by Hanany and Witten, various brane configurations give a very nice and geometrical parametrization of SUSY theories. The idea of D-branes[1] opened up a new and surprisingly simple way to communicate between super symmetric (SUSY) gauge theories and the superstring theories. Many intriguing results about field theories have been derived by investigating gauge theories formulated on the world volume of branes in string theories[2-11] and M-theory[12-16]. In a recent paper [12], Witten provided solutions of $N = 2$ SUSY gauge theories in four dimension[17,18] by reinterpreting configurations of fourbranes and fivebranes in type II superstring as the branes in M theory. Moreover, in a subsequent paper[15], he showed how some of the outstanding problems in particle physics, such as quark confinement and chiral symmetry breaking, can be approached from the M-theory point of view. There, he also suggested a way to calculate the superpotential directly from a brane configuration; thereby he gave a direct evaluation of the tension of the domain wall[19]. A direct comparison with the gauge theory result was given by showing that the minimum of the superpotential of the $N = 1$ $SU(n)$ gauge theory superpotential agrees with the value evaluated with the M-theory method[21]. We have obtained the superpotentials for more general situations with product gauge groups, which corresponds to brane configurations with $n+1$ fivebranes.

3 The Brane Configuration

We begin with brane configurations in type IIA superstring theory. The spacetime coordinates are $(x^0, x^1, \ldots, x^9)$. Our basic brane configuration contains three kinds of branes: NS fivebranes whose world volume occupies $(x^0, \cdots, x^5)$ spacetime coordinates, D fourbranes occupying $(x^0, \cdots, x^3, x^6)$, and D sixbranes occupying $(x^0, \cdots, x^3, x^7, x^8, x^9)$.

The basic brane configuration consists of N_c fourbranes suspended between two parallel fivebranes in Type IIA superstring theory which gives a representation of $N = 2$ SUSY QCD in four dimensions $(x^0, \cdots, x^3)$ with gauge group $SU(N_c)$. Also N_f D sixbranes could be added and this gives N_f hypermultiplets in the fundamental representation of the gauge group $SU(N_c)$.

One of the limitation of type IIA superstring theory is that a fourbrane ending on a fivebrane defies any explicit conformal field theory description. However, by going to M theory, the picture becomes more unified. A type IIA fourbrane becomes an M theory fivebrane that is wrapped over the 11th dimension S^1. Thus D fourbranes and NS fivebranes come from the same basic object in M theory. This $N = 2$ SUSY description can be broken to $N = 1$ by turning on a mass for the adjoint chiral superfield in the $N = 2$ vector multiplet.

Witten's construction[12,15] of the parametric representation of the brane configuration in M theory for $N = 1$ SQCD is as follows. We introduce complex coordinates $v = x^4 + ix^5$, $w = x^7 + ix^8$, $t = \exp(-(x^6 + ix^{10})/R)$. The brane configuration for $SU(n)$ gauge theory is given by the Seiberg-Witten curve[17]

$$t^2 + P_n(v)t + 1 = 0, \tag{6}$$

in $v - t$ space when $w = 0$. Here $P_n(v)$ is a polynomial of the form

$$P_n(v) = v^n + u_2 v^{n-2} + \ldots + u_n, \tag{7}$$

where the u_i are the "order parameters" of the theory.

The brane configuration associated to $N = 1$ SUSY can be obtained by "rotating" the brane configuration that describes $SU(n)$ gauge theory with $N = 2$ SUSY[22]. The rotation is possible only when all the 1-cycles on the corresponding curve are degenerate. This is why the $N = 1$ curve is rational for the two NS fivebrane case[15]. However for more than two NS fivebranes, we can ask whether the curve is generically rational. The generic M-brane configuration corresponds to a curve Σ of genus zero, which has parametric representations. Namely, Σ can be identified by a holomorphic mapping with a Riemann sphere with some points deleted. If λ is the complex coordinate of Σ, the holomorphic functions $v(\lambda), w(\lambda)$, and $t(\lambda)$ on Σ are given by rational functions of λ.

For example, consider the brane configuration with more than two NS fivebranes, where NS fivebranes are located at $\lambda = \lambda_\alpha$, $\alpha = 0, 1, \cdots, n$ at arbitrary angles in $v - w$ space, with k_α D fourbranes suspended between $\alpha - 1^{\text{th}}$ and α^{th} fivebranes. The parametric representation of the curve is:

$$v = \sum_{\alpha=0}^{n} \frac{c_\alpha}{\lambda - \lambda_\alpha}, \quad w = \sum_{\alpha=0}^{n} \frac{s_\alpha}{\lambda - \lambda_\alpha}, \quad t = \prod_{\alpha=0}^{n} (\lambda - \lambda_\alpha)^{-a_\alpha}, \tag{8}$$

for some complex numbers c_α, s_α. We normalize the system by setting $c_0 = 1$, $s_0 = 0$, so that λ_0 is the location of the unrotated NS fivebrane.

4 The Superpotential

Given the brane configurations in M-theory, we calculate the superpotential of the corresponding gauge theory following Witten's idea[15]. Consider a general M–theory compactification on $\mathbf{R}^4 \times X \times \mathbf{R}$ where X is a Calabi-Yau threefold. Suppose in spacetime there are fivebranes of the form $\mathbf{R}^4 \times \Sigma$, Σ being a two-dimensional real surface in X. Choose Σ_0 in the homology class of Σ in $H^2(X, \mathbf{R})$. Then there exists a three manifold B and a map $\Phi_B : B \to X$ such that the boundary of B maps $\Sigma - \Sigma_0$ in X. Let Ω be the holomorphic three form on X. Then as Witten suggested, the superpotential is given by

$$W(\Sigma) - W(\Sigma_0) = \int_B \Phi_B^*(\Omega). \tag{9}$$

This defines $W(\Sigma)$ up to an additive constant. The issue of the Σ_0 independence is a subtle matter and is discussed in the original paper.

The superpotential now becomes

$$W(\Sigma) - W(\tilde{\Sigma}_0) = Rn \int_B \frac{d\tilde{\lambda}}{\tilde{\lambda}} \wedge dv \wedge dw. \tag{10}$$

We can calculate the superpotential for general multi-brane cases:

$$W = -2\pi i R \sum_{\alpha<\beta}^{n} (c_\alpha s_\beta - c_\beta s_\alpha) \left[\frac{(a_\alpha - a_\beta)}{\lambda_{\alpha\beta}^2} + \frac{1}{2}\frac{1}{\lambda_{\alpha\beta}} \sum_{\gamma\neq\alpha,\beta} a_\gamma \left(\frac{1}{\lambda_{\gamma\alpha}} + \frac{1}{\lambda_{\gamma\beta}} \right) \right]$$
$$-2\pi i R \left[(\sum_{\gamma=1}^{n} a_\gamma)(\sum_{\beta=1}^{n} s_\beta) - \frac{1}{2} \sum_{\gamma\neq\beta} \frac{a_\gamma \lambda_\gamma s_\beta}{\lambda_{\gamma\beta}} \right]. \tag{11}$$

Here α, β, γ run from 1 to n. The corresponding calculations in the gauge theory which give this superpotential have not been done; we are looking for field theory counterparts.

The above discussion can easily be generalized to cases with N_f hypermultiplets, corresponding to N_f D sixbranes or semi–infinite D fourbranes. To illustrate this let us consider a situation where N_f semi–infinite D fourbranes are attached to the right hand side of the second NS fivebrane. The parametric form of the curve for such a configuration had already been written down.[14]

$$v = \frac{(\lambda - \lambda_+)(\lambda - \lambda_-)}{\mu\lambda}, \quad w = \lambda, \quad t = \mu^{-N_c} \lambda^{N_c - N_f} (\lambda - \lambda_+)^r (\lambda - \lambda_-)^{N_f - r}, \tag{12}$$

where $\lambda_\pm$ are the two solutions of $v = 0$, and μ is a bare mass for the adjoint chiral multiplet. Employing the same method as above, we get the superpotential

$$W = 2\pi i \frac{R\lambda_+\lambda_-}{\mu}(2N_c - N_f). \tag{13}$$

This precisely agrees with the minimum value of the $N = 1$ gauge theory superpotential ($N_f < N_c$) given as[20,14], up to an overall constant factor;

$$W_{\text{eff}} = (N_c - N_f) \left(\frac{\Lambda_{N=1}^{3N_c - N_f}}{\det M} \right)^{1/(N_c - N_f)} + \frac{1}{2\mu} \left(\mathrm{Tr}(M^2) - \frac{1}{N_c} (\mathrm{Tr} M)^2 \right), \tag{14}$$

where $\Lambda_{N=1}$ is the dynamical scale of $N = 1$ SQCD.This is the first concrete evidence that Witten's proposal for the superpotential is correct.

5 Discussion

In this talk we found a parametric representation of the brane configurations corresponding to $N = 1$ SQCD. Extending Witten's method, we developed a general formalism to calculate superpotentials and compared with known gauge theory results. The comparison with gauge theory result might be interesting when we consider gauge theories for product groups.

Acknowledgements

This work is supported in part by the Ministry of Education(BSRI-97-2442), by KOSEF(961-0201-001-2), and by CTP/SNU through the SRC program of KOSEF.

References

1. J. Polchinski, "TASI Lectures On D-Branes," hep-th/9611050.
2. A. Hanany and E. Witten, Nucl. Phys. B492 (1997) 152.
3. J. de Boer, K. Hori, Y. Oz and Z. Yin, Nucl.Phys. **B502** (1997) 107.
4. S. Elitzur, A. Giveon and D. Kutasov, Phys. Lett. **B400** (1997) 269.
5. S. Elitzur, A. Giveon, D. Kutasov, E. Rabinovici and A. Schwimmer, Nucl. Phys. B505 (1997) 202.
6. N. Evans, et al., Nucl. Phys. **B505** (1997) 251.
7. J. H. Brodie and A. Hanany, Nucl. Phys. **B506** (1997) 157.
8. R. Tartar, hep-th/9704198.
9. A. Marshakov, M. Martellini, and A. Morozov, hep-th/9706050.
10. K. Landsteiner, E. Lopez, and D.A. Lowe, hep-th/9705199.
11. A. Gorsky, S. Gukov, and A. Mironov, hep-th/9707120.
12. E. Witten, Nucl. Phys. **B500** (1997) 3.
13. A. Brandhuber et al., hep-th/9705232.
14. K. Hori, H. Ooguri and Y. Oz, hep-th/9706082.
15. E. Witten, hep-th/9706082.
16. A. Brandhuber et al., Phys. Lett. **B410** (1997) 27.
17. N. Seiberg and E. Witten, Nucl. Phys. **B426** (1994) 19.
18. N. Seiberg and E. Witten, Nucl. Phys. **B431** (1994) 484.
19. G. Dvali, M. Shifman, Phys. Lett. **396B** (1997) 64; A. Kovner, M. Shifman, A. Smilga; B. Chibisov and M. Shifman, hep-th/9706141.
20. I. Affleck, M.Dine and N. Seiberg, Nucl. Phys. **B241** (1984) 493.
21. S. Nam, K.H. Oh, and S. Sin, hep-th/9707247.
22. J. L. F. Barbón, Phys. Lett. **B402** (1997) 59.

IMPLICATIONS OF THE HERA EVENTS FOR R–PARITY BREAKING SUSY SIGNALS AT THE TEVATRON

MONORANJAN GUCHAIT
Department of Physics, Yonsei University, Seoul 120-749, South Korea

I report here on a systematic analysis at the Tevatron of the R–parity violating SUSY signals which correspond to possible explanations of the "anomalous" HERA events.

Both experimental groups at the *ep* collider HERA, H1 and ZEUS, have reported some anomalous high Q^2 events. H1 has seen 12 neutral current events, at $Q^2 > 15{,}000\ GeV^2$ against a Standard Model (SM) prediction of 5 events [1]; while ZEUS has reported 5 events at $Q^2 > 20{,}000\ GeV^2$ against a prediction [2] of 2. The excess of 7 events observed by H1 has clustered around a common e^+q mass of

$$M \simeq 200 \text{ GeV} \tag{1}$$

which is, however, inconsistent with the ZEUS events [3]. These are based on 1994–96 data, corresponding to a combined luminosity of $34pb^{-1}$ for the two experiments. Taken together, there are 10 excess events which corresponds to a cross section

$$\sigma \simeq 0.4 \text{ pb} \tag{2}$$

assuming an 80% detection efficiency for each experiments.

Many mechanisms of new physics were proposed [4,5,6,7] to explain these new events. Among these, the R–parity violating (RPV) SUSY model is the most popular one. The purpose of this work [8] is to study the implications of the RPV SUSY model for the Tevatron.

The Minimal Supersymmetric Standard Model (MSSM) with explicit R-parity breaking[9] contains the following Yukawa terms in the Lagrangian:

$$L = \lambda_{ijk} l_i \tilde{l}_j \bar{e}_k + \lambda'_{ijk} l_i \tilde{q}_j \bar{d}_k + \lambda''_{ijk} \bar{d}_i \tilde{\bar{d}}_j \bar{u}_k, \tag{3}$$

plus analogous terms from the permutation of the tilde, denoting scalar superpartners. We have used the notation of ref. [8]. The terms relevant for the HERA events are

$$\lambda'_{1jk}(l_1 \tilde{q}_j \bar{d}_k + l_1 q_j \tilde{\bar{d}}_k) + h.c. \tag{4}$$

Therefore, for a given quark flux one can easily calculate the cross section for these process at HERA when $\lambda'\sqrt{B}$ is given, where B denotes the squark branching fraction into e^+q. While the size of $\lambda'\sqrt{B}$ is constrained by HERA data as given by Eq.(2), the value of λ' is restricted by upper bounds from some other process[5,8]. It is shown[4,5] that the viable couplings which are allowed by the HERA data as well as by corresponding existing upper limits are λ'_{121} and λ'_{131}, which stand for the scharm and stop squark production from a valence quark, i.e.

$$e^+d \to \tilde{c}_L \to e^+d. \tag{5}$$

$$e^+d \to \tilde{t}_L \to e^+d, \tag{6}$$

where the L subscript stands for left chirality. Therefore, we have concentrated on these two scenarios and called them scharm and stop scenario, corresponding to λ'_{121} and λ'_{131} couplings, respectively.

Since the product $\lambda'\sqrt{B}$ is constrained by the cross section of the HERA events, Eq.(2), it is possible to express B in terms of the MSSM parameters under the assumption that there is only one dominant RPV coupling[8]. The dependence of B on the MSSM parameter space for a given scenario will be discussed below.

It is to be noted that in our analysis we assume a common gaugino and a common sfermion mass at the unification scale, so that the masses of gauginos ($\tilde{g}, \tilde{W}$ and $\tilde{B}$) are related via their gauge couplings at the electroweak unification scale[8]. Under this assumption we have investigated the possibility of finding the stop and scharm scenario at the Tevatron collider.

1) The Stop Scenario at the Tevatron

The dominant mechanisms for stop production at the Tevatron are the leading order QCD processes of quark-antiquark annihilation and gluon-gluon fusion[10]

$$\bar{q}q \to \bar{\tilde{t}}_1\tilde{t}_1 \quad , \quad gg \to \bar{\tilde{t}}_1\tilde{t}_1. \tag{7}$$

The contribution of the RPV couplings is negligible.

The dominant decay mode of stop is the RPV mode

$$\tilde{t}_1 \to e^+d. \tag{8}$$

as long as $M_{\tilde{t}_1} \leq M_{\tilde{W}_1}$, because the decay into neutralinos involves higher order processes due to the large top quark mass. However, the $R-$parity conserving decay mode

$$\tilde{t}_1 \to b\tilde{W}_i \tag{9}$$

will dominate when kinematically allowed. In Fig.1, values of B are shown over the relevant MSSM parameter space. It is clear from this figure that the signal for stop at the Tevatron depends very crucially on the MSSM parameter

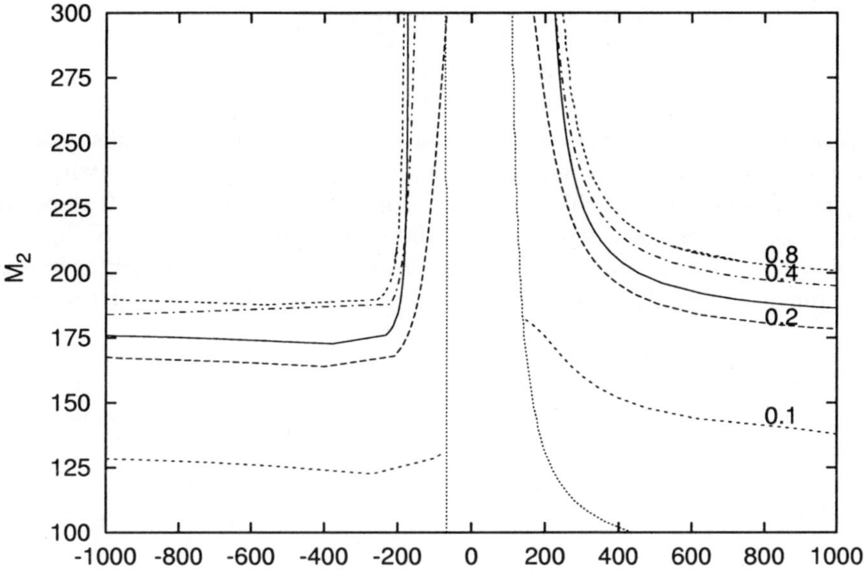

Figure 1: The stop branching fraction (B) is shown as a contour plot in the M_2, μ plane for stop mass $M_{\tilde{t}_1} = 200GeV$, $\tan\beta = 2$ and $\theta_{\tilde{t}} = 0°$. Contours for the lighter chargino mass of 180 GeV (solid line) and 85 GeV (dotted line) are shown.

space. In our analysis, we have considered three types of decay configurations of stop which are discussed below.

(a) Direct leptonic decay: In this case both stop squarks decay in the RPV channel, Eq.(8), resulting in a pair of hard and isolated leptons along with a pair of jets. The corresponding 95% C.L. limits of 3 events as obtained from the 1st generation of Leptoquark search gives a mass bound by CDF M> 210 GeV [11], whereas the corresponding value for the D0 group [12] is M> $225GeV$. Thus CDF and D0 Leptoquark limits would rule out the stop scenario if the direct leptonic decays are dominant, i.e. $M_{\tilde{W}_1} \geq M_{\tilde{t}_1}$. However, the dilepton cross section for $B \leq 0.2$ is so small that it is not possible to probe them in the present data set of CDF and D0. On the other hand, in the main injector (MI) run one will be able to probe the signal down to $B = 0.2$. Nonethless, there is a significant MSSM parameter space where $B \leq 0.2$. This cannot be probed in the MI run, but may be probed at TEV33, where the integrated luminosity will be 20 fb^{-1}.

(b) Mixed Mode: This mode corresponds to direct decay of one stop, eq.(8), while the other undergoes cascade decay as

$$\tilde{t}_1 \to b\tilde{W}_1, \tilde{W}_1 \to q'\bar{q}\tilde{Z}_1, \tilde{Z}_1 \overset{\lambda'_{131}}{\to} \bar{b}\bar{\nu}d + h.c. \tag{10}$$

Thus the final states consists of a very hard $e^{\pm}$, a large number of jets including a pair of b's, and a modest amount of missing p_T carried by the neutrino. The dominant background to this process is from $t\bar{t}$ production. Fig. 2 shows the signal cross section for a stop mass of 200 GeV along with the $t\bar{t}$ background at $\sqrt{s} = 2TeV$ for the following MSSM parameters: $M_2 = 150$ GeV, $\mu = -400$ GeV, $\tan\beta = 2 \Rightarrow M_{\tilde{W}_1} = 158$ GeV, $M_{\tilde{Z}_1} = 77$ GeV. Note that this signal

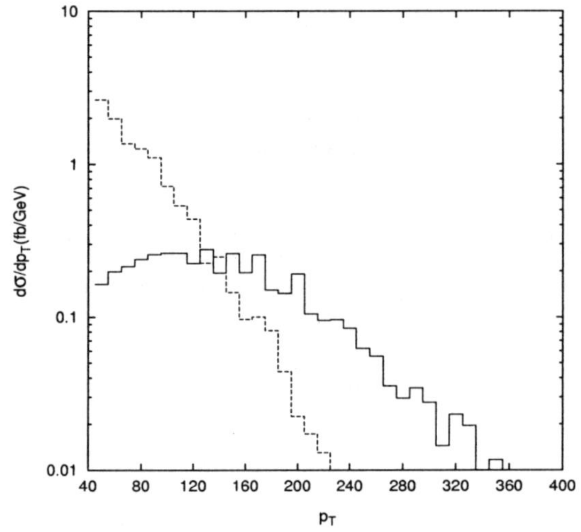

Figure 2: The signal cross section (solid line) for the mixed mode (b) is shown for a stop mass of 200 GeV, along with the $t\bar{t}$ background (dashed line) against the p_T of the electron at 2 TeV. The MSSM parameters are $M_2 = 150$ GeV, $\mu = -400$ GeV, and $\tan\beta = 2$.

rate is quite insensitive to MSSM parameters. From Fig.2 it is clear that the signal can be separated from the background by exploiting the clustering of invariant mass of the electron with the hardest jets at $M \simeq 200GeV$. The size of the signal in Fig.2 is about 50 fb, which means few events are expected for the MI run even for the regions of MSSM parameter space (in Fig.1) where the branching fraction is $B \simeq 7\%$.

(c) Cascade decays: Here both stop squarks decay according to eq.(10), resulting in final states with four b quarks and missing p_T carried by the neutrinos. We need to do triple b tagging to suppress the background from $t\bar{t}$ and $b\bar{b}$ production. We have estimated the signal cross section for two different channels (i) 3 b quarks + missing p_T and (ii) 1 lepton + 3 b + missing p_T. We found no viable signature for detecting the stop pair production for an integrated luminosity of 2 fb^{-1}. But in the MI run it may be possible to probe the entire MSSM parameter space, if the $b-$tagging efficiency is large.

2) The Charm Squark (Scharm!) Scenario at the Tevatron

In this case one expects 8 species of roughly degenerate squarks along with a gluino of comaparable or smaller mass. Only one of them, the left handed charm squark $\tilde{c}_L$, has a R violating decay mode

$$\tilde{c}_L \rightarrow e^+ d. \tag{11}$$

as required to explain the HERA anomaly. However, scharm also has R conserving decay modes:

$$\tilde{c}_L \rightarrow s\tilde{W}_i, s\tilde{Z}_i \tag{12}$$

We have seen that the corresponding branching fraction B for the decay modes eq.(11) is $B \leq 1/20$. Therefore, the direct decay channel eq.(11) will not give any viable SUSY signal. We have instead considered the cascade decays of the squarks and gluinos into the LSP which, gives one electron through its decay

$$\tilde{Z}_1 \stackrel{\lambda'_{121}}{\rightarrow} \bar{c}e^+ d(\bar{s}\bar{\nu}d) + h.c. \tag{13}$$

As a consequence, the pair of LSPs provides a like sign dilepton (LSD) signature in this case due to its Majorana nature. We have considered all possible strong production processes [10], i.e. pair production of gluinos or squarks and associated production, and their subsequent decays according to the masses of the electroweak gauginos [8]. Considering only λ'_{121} as leading RPV coupling, as suggested by the scharm scenario, we have estimated the LSD rates for a given value of charm squark mass (=210 GeV) applying some kinematic cuts to suppress the SM background. We have found that one can expects to see at least half of dozen LSD events with the present Tevatron luminosity of 110 pb^{-1}.

Finally, we conclude with the expectation that the Tevatron groups will analyse their data to probe the RPV SUSY model, and in particular to test the charm squark scenario for the anomalous HERA events.

The work reported here is done in collaboration with D.P.Roy [8].

References

1. H1 Collaboration: C. Adloff et.al, *Z. Phys.* **C74**, 191 (1997).
2. ZEUS Collaboration: J. Breitweg et.al, *Z. Phys.* **C74**, 207 (1997).
3. G. Wolf, hep-ex/9704006; see however M. Drees, hep-ph/9703332
4. D. Choudhury and S. Raychaudhuri, hep-ph/9702392 and hep-ph/9703369.
5. H. Dreiner and P. Morawitz, hep-ph/9703279 (version 2).
6. G. Altarelli, J. Ellis, G.F. Giudice, S. Lola and M.L. Mangano, hep-ph/9703276.
7. .S. Belayev and A.V. Gladyshev, hep-ph/9703251 and hep-ph/9704343; J. Kalinowski, R. Ruckl, H. Spiesberger and P.M. Zerwas, hep-ph/9703288; K.S. Babu, C. Kolda, J. March-Russell and F. Wilczek, hep-ph/9703299; V. Barger, K. Cheung, K. Hagiwara and D. Zeppenfeld, hep-ph/9703311.
8. M. Guchait and D.P. Roy, hep-ph/9707275 (to be published in Physical Review D).
9. H.P. Nilles, *Phys. Rep.* **110**, 1 (1984); H. Haber and G. Kane, *Phys. Rep.* **117**, 75 (1985); S. Weinberg, *Phys. Rev.* **D26**, 287 (1982); N. Sakai and T. Yanagida, *Nucl. Phys.* **B197**, 83 (1982); S. Dimopoulos, S. Raby and F. Wilczek, *Phys. Lett.* **B212**, 133 (1982).
10. G.L. Kane and J.P. Leveille, *Phys. Lett.* **B112**, 227 (1982); P.R. Harrison and C.H. Llewellyn-Smith, *Nucl. Phys.* **B213**, 223 (1983) [Err. *Nucl. Phys.* **B223**, 542 (1983)]; S. Dawson, E. Eichten and C. Quigg, *Phys. Rev.* **D31**, 1581 (1985); E. Reya and D. P. Roy, *Phys. Rev.* **D32** , 645 (1985).
11. CDF Collaboration: C. Grosso-Pilcher, Talk at the Vanderbilt Conference, May 1997.
12. DØ Collaboration: D. Norman, Talk at the Hadron Collider Physics Conference, Stonybrook, June 1997.
13. M. Guchait and D.P. Roy, *Phys. Rev.* **D54**, 3276 (1996).
14. H. Baer, C. Kao and X. Tata, *Phys. Rev.* **D51**, 2180 (1995).

$R-$PARITY VIOLATION IN B DECAYS

JAE SIK LEE

Department of Physics, Korea Advanced Institute of Science and Technology, Taejon 305-701, South Korea

We study purely leptonic and semi–leptonic decays of neutral B mesons in the presence of $R-$parity violation. These B^0 decay modes can constrain $R-$parity violating couplings with one or more heavy generation indices. We find that the effects of λ-type couplings on semi–leptonic B decays are negligible compared with those of λ'-type couplings. The experimental determination of the ratio $\left|\frac{V_{ub}}{V_{cd}}\right|$ can be affected strongly by $R-$parity violation, and $R-$parity violating interactions could decrease the semi–leptonic branching ratio. The upper bound on the ratio of $\tan\beta$ to the charged Higgs boson mass is not affected by $R-$parity violation.

1 Introduction

In supersymmetric extensions of the standard model, there are in general gauge invariant interactions which violate baryon number (B) and lepton number (L). An additional global symmetry is required to prevent the occurrence of these B and L violating interactions. This requirement leads to the consideration of the so called $R-$parity (R_p) [1]. We find that the values of R_p violating couplings involving heavy generation fields are presently not well constrained[2]. This means that we could constrain $R-$parity violating couplings with heavy generation indices from B meson decays [3,4], and $R-$parity violating signals could be found in upcoming experiments on B mesons [5].

In this talk, we derive upper bounds on R_p and lepton flavor violating couplings from purely leptonic and semi–leptonic decays of B-mesons, and discuss physical implications in the minimal supersymmetric standard model (MSSM) with explicit R_p violation ($\not{R}_p$) [3].

In the MSSM, the most general R_p violating superpotential is given by

$$W_{\not{R}_p} = \lambda_{ijk} L_i L_j E_k^c + \lambda'_{ijk} L_i Q_j D_k^c + \lambda''_{ijk} U_i^c D_j^c D_k^c. \tag{1}$$

Here i, j, k are generation indices and we assume that possible bilinear terms $\mu_i L_i H_2$ can be rotated away. L_i and Q_i are $SU(2)$ doublet lepton and quark superfields and E_i^c, U_i^c, D_i^c are singlet superfields. There are 45 different couplings, 36 of which involve lepton flavor violation. We assume that the baryon number violating couplings λ'' vanish in order to avoid too fast proton decay [6], and that the masses of squarks and sleptons are universal with $\tilde{m} = 100$ GeV.

2 Purely leptonic decays

The relevant four fermion interaction Lagrangian is given by

$$\begin{aligned}\mathcal{L}^{\rm eff}_{\not R_p} &= \sum_{n=1}^{3} \frac{2}{m^2_{\tilde{l}_n}} \left[\lambda^*_{njk}\lambda'_{nlm}(\bar{e}_j P_R e_k)(\bar{d}_m P_L d_l) + \text{H.c.}\right] \\ &- \sum_{n,r,s=1}^{3} \frac{1}{2m^2_{\tilde{Q}_n}} V_{nr}V^*_{ns}\lambda'^*_{jrk}\lambda'_{lsm}(\bar{e}_j\gamma^\mu P_L e_l)(\bar{d}_m\gamma_\mu P_R d_k), \end{aligned} \tag{2}$$

where V is the CKM matrix and $P_{R,L} = \frac{1}{2}(1 \pm \gamma_5)$. From this effective Lagrangian we obtain the following decay rates:

$$\begin{aligned}\Gamma(B_q \to e_i^- e_j^+) &= \frac{f^2_{B_q} M^3_{B_q}}{64\pi\tilde{m}^4}\sqrt{1 + (\kappa_i^2 - \kappa_j^2)^2 - 2(\kappa_i^2 + \kappa_j^2)} \\ &\times \left\{(|\mathcal{A}^q_{ij}|^2 + |\mathcal{B}^q_{ij}|^2)(1 - \kappa_i^2 - \kappa_j^2) + |\mathcal{C}^q_{ij}|^2\left[(\kappa_i^2 + \kappa_j^2) - (\kappa_i^2 - \kappa_j^2)^2\right]\right. \\ &+ 4\text{Re}(\mathcal{A}^q_{ij}\mathcal{B}^{q*}_{ij})\kappa_i\kappa_j + 2\kappa_j\text{Re}(\mathcal{A}^q_{ij}\mathcal{C}^{q*}_{ij})(1 + \kappa_i^2 - \kappa_j^2) \\ &\left. + 2\kappa_i\text{Re}(\mathcal{B}^q_{ij}\mathcal{C}^{q*}_{ij})(1 + \kappa_j^2 - \kappa_i^2)\right\}, \end{aligned} \tag{3}$$

where $q = 1(B_d)$ or $= 2(B_s)$, $\kappa_i \equiv m_{e_i}/M_{B_q}$, f_{B_q} is the decay constant, $\tilde{m}$ is the universal soft mass and

$$\begin{aligned}\mathcal{A}^q_{ij} &= 2\sum_{n=1}^{3}\lambda^*_{nij}\lambda'_{nq3}, \quad \mathcal{B}^q_{ij} = 2\sum_{n=1}^{3}\lambda_{nji}\lambda'^*_{n3q}, \\ \mathcal{C}^q_{ij} &= \frac{1}{2}\sum_{n,p,s=1}^{3} K_{np}K^*_{ns}\lambda'^*_{ipq}\lambda'_{js3} = \frac{1}{2}\sum_{n=1}^{3}\lambda'^*_{inq}\lambda'_{jn3}. \end{aligned} \tag{4}$$

From the current measurements by CLEO [7] and CDF [8], neglecting masses of $e^\pm$, $\mu^\pm$ and the contributions of the SM [9], we obtain upper bounds on some R_p violating couplings which are stronger than previously published ones (See Table 1).

3 Semi–leptonic decays

The two relevant pieces of the four fermion interaction Lagrangian are given by [10]

$$\mathcal{L}^{\rm eff}_{\rm MSSM}(b \to q e_l \bar{\nu}_l) = -V_{qb}\frac{4G_F}{\sqrt{2}}\left[(\bar{q}\gamma^\mu P_L b)(\bar{e}_l\gamma_\mu P_L \nu_l) - R_l(\bar{q}P_R b)(\bar{e}_l P_L \nu_l)\right], \tag{5}$$

Table 1: Upper bounds on the magnitudes of products of couplings derived from B^0 decays into two charged leptons.

Decay Mode	Combinations Constrained	Upper bound
$B_d \to e^+e^-$	$\lambda_{121}\lambda'_{213}$, $\lambda_{121}\lambda'_{231}$, $\lambda_{131}\lambda'_{313}$, $\lambda_{131}\lambda'_{331}$	4.6×10^{-5}
$B_d \to e^+\mu^- + e^-\mu^+$	$\lambda_{121}\lambda'_{113}$, $\lambda_{121}\lambda'_{131}$, $\lambda_{121}\lambda'_{231}$, $\lambda_{122}\lambda'_{213}$	4.6×10^{-5}
	$\lambda_{132}\lambda'_{313}$, $\lambda_{132}\lambda'_{331}$, $\lambda_{231}\lambda'_{313}$, $\lambda_{231}\lambda'_{331}$	4.6×10^{-5}
$B_d \to e^+\tau^- + e^-\tau^+$	$\lambda_{123}\lambda'_{213}$, $\lambda_{123}\lambda'_{231}$, $\lambda_{131}\lambda'_{113}$, $\lambda_{131}\lambda'_{131}$	4.9×10^{-4}
	$\lambda_{133}\lambda'_{313}$, $\lambda_{133}\lambda'_{331}$, $\lambda_{231}\lambda'_{213}$, $\lambda_{231}\lambda'_{231}$	4.9×10^{-4}
	$\lambda'_{131}\lambda'_{333}$	5.8×10^{-4}
$B_d \to \mu^+\tau^- + \mu^-\tau^+$	$\lambda_{123}\lambda'_{113}$, $\lambda_{123}\lambda'_{131}$, $\lambda_{132}\lambda'_{113}$, $\lambda_{132}\lambda'_{131}$	6.0×10^{-4}
	$\lambda_{232}\lambda'_{213}$, $\lambda_{232}\lambda'_{231}$, $\lambda_{233}\lambda'_{313}$, $\lambda_{233}\lambda'_{331}$	6.0×10^{-4}
	$\lambda'_{231}\lambda'_{333}$, $\lambda'_{233}\lambda'_{331}$	7.4×10^{-3}
$B_d \to \mu^+\mu^-$	$\lambda_{122}\lambda'_{113}$, $\lambda_{122}\lambda'_{131}$, $\lambda_{232}\lambda'_{313}$, $\lambda_{232}\lambda'_{331}$	2.4×10^{-5}
$B_s \to \mu^+\mu^-$	$\lambda_{122}\lambda'_{123}$, $\lambda_{122}\lambda'_{132}$, $\lambda_{232}\lambda'_{323}$, $\lambda_{232}\lambda'_{332}$	5.5×10^{-5}

where $R_l = r^2 m_{e_l} m_b^Y$ and $r = \frac{\tan\beta}{m_{H^\pm}}$, and

$$\mathcal{L}^{\text{eff}}_{R\!\!\!/_p}\ (b \to q e_l \bar{\nu}_n) = -V_{qb}\frac{4G_F}{\sqrt{2}}\left[\mathcal{A}^q_{ln}(\bar{q}\gamma^\mu P_L b)(\bar{e}_l\gamma_\mu P_L \nu_n) - \mathcal{B}^q_{ln}(\bar{q}P_R b)(\bar{e}_l P_L \nu_n)\right], \tag{6}$$

where

$$\begin{aligned}\mathcal{A}^q_{ln} &= \frac{\sqrt{2}}{4G_F V_{qb}}\sum_{i,j=1}^{3}\frac{1}{2m^2_{d^c_i}}V_{qj}\lambda'_{n3i}\lambda'^*_{lji},\\ \mathcal{B}^q_{ln} &= \frac{\sqrt{2}}{4G_F V_{qb}}\sum_{i,j=1}^{3}\frac{2}{m^2_{\tilde{l}_i}}V_{qj}\lambda_{inl}\lambda'^*_{ij3}.\end{aligned} \tag{7}$$

Note the large numerical factors coming from $\frac{V_{qj}}{V_{qb}}$ for $j = 1, 2$. These factors enhance the effects of $R\!\!\!/_p$ on semi–leptonic decays. We obtain the following decay rates:

$$\begin{aligned}\Gamma^q_{ll} &= \frac{|V_{qb}|^2 G_F^2 m_b^5}{192\pi^3}\left\{|1+\mathcal{A}^q_{ll}|^2\,\Gamma_W + \frac{1}{4}|R_l + \mathcal{B}^q_{ll}|^2\,\Gamma_H\right.\\ &\qquad \left. -2Re\left[(R_l+\mathcal{B}^q_{ll})(1+\mathcal{A}^{q*}_{ll})\right]\frac{m_{e_l}}{m_b}\Gamma_I\right\}, \end{aligned} \tag{8}$$

$$\Gamma^q_{ln}\big|_{l\neq n} = \frac{|V_{qb}|^2 G_F^2 m_b^5}{192\pi^3}\left\{|\mathcal{A}^q_{ln}|^2\,\Gamma_W + \frac{1}{4}|\mathcal{B}^q_{ln}|^2\,\Gamma_H - 2Re\left(\mathcal{B}^q_{ln}\mathcal{A}^{q*}_{ln}\right)\frac{m_{e_l}}{m_b}\Gamma_I\right\},$$

Table 2: Upper bounds on the magnitudes of $R\!\!\!/_p$ couplings from lepton universality in semi–leptonic decays of B-meson.

Processes	Combinations Constrained	Upper Bound
$b \to e\,\bar{\nu}\,X_c$	$\lambda'_{131}\lambda'_{121}$, $\lambda'_{132}\lambda'_{122}$	1.1×10^{-3}
	λ'_{131}, λ'_{132}	1.6×10^{-1}
$b \to \mu\,\bar{\nu}\,X_c$	$\lambda'_{231}\lambda'_{221}$, $\lambda'_{232}\lambda'_{222}$, $\lambda'_{233}\lambda'_{223}$	1.1×10^{-3}
	λ'_{231}, λ'_{232}, λ'_{233}	1.6×10^{-1}

where $\Gamma_{W,H,I}$ denotes contributions from the charged gauge boson, the charged Higgs boson, and interference, respectively. Let us define the total decay rate as the sum of decay rates over the neutrino species, $\Gamma^q_l \equiv \sum_{n=1}^3 \Gamma^q_{ln}$, and the $R\!\!\!/_p$ ratio as the ratio of this total rate to the rate in the SM, $\mathcal{R}^q_l \equiv \frac{\Gamma^q_l(R\!\!\!/_p)}{\Gamma^q_l(SM)}$. We observe that the effects of $R-$parity violation on the semi-leptonic decays of B mesons can be significant within present bounds: $\mathcal{R}^u_l$ can be as large as 10 and $\mathcal{R}^c_l$ 1.7. We also observe that there are no significant contributions from $\mathcal{B} \approx \lambda\lambda'$ taking into accounts the constraints from purely leptonic decays of neutral B mesons.

Experimentally, the ratio of $\left|\frac{V_{ub}}{V_{cb}}\right|$ is determined by [11]

$$\left|\frac{V_{ub}}{V_{cb}}\right|^2 \approx \left[\frac{\Gamma(b \to ul\nu)}{\Gamma(b \to cl\nu)}\right]^{\text{exp.}}_{\text{end point of lepton spectrum}} \tag{9}$$

In the presence of $R\!\!\!/_p$, the left hand side of the above equation should be changed into $\frac{\mathcal{R}^u_{(1,2)}}{\mathcal{R}^c_{(1,2)}}\left|\frac{V_{ub}}{V_{cb}}\right|^2$. $\frac{\mathcal{R}^u_{(1,2)}}{\mathcal{R}^c_{(1,2)}}$ can be as large as 10 within present bounds. This means that the experimental determination of V_{ub} can be affected strongly by $R\!\!\!/_p$. Furthermore, assuming $\mathcal{A}^c_{ij}(i \neq j) = \mathcal{B}^c_{ij} = 0$,

$$0.7 < \mathcal{R}^c_1 < 1.5, \quad 0.7 < \mathcal{R}^c_2 < 1.7, \tag{10}$$

within the present bounds. This means $R\!\!\!/_p$ could decrease the semi-leptonic branching ratio. So, it is possible to explain the discrepancy [12] between the semi-leptonic branching ratios $\mathcal{B}^{\text{exp.}}_{\text{SL}}$ and $\mathcal{B}^{\text{SM}}_{\text{SL}}$ with R_p violation.

From the measurements of lepton universality in $\mathcal{B}_{\text{SL}}$ [13], we derive new, stronger upper bounds on some $R\!\!\!/_p$ couplings (see Table 2).

Turning to the upper bound on $r = \frac{\tan\beta}{m_{H^\pm}}$, assuming only $\mathcal{A}^c_{33}$ are nonzero, we find

$$r \;<\; 0.54 \;\text{GeV}^{-1} \quad \text{for} \quad \mathcal{A}^c_{33} = -0.1,$$

$$r < 0.52 \text{ GeV}^{-1} \quad \text{for} \quad \mathcal{A}^c_{33} = 0,$$
$$r < 0.51 \text{ GeV}^{-1} \quad \text{for} \quad \mathcal{A}^c_{33} = 0.1. \tag{11}$$

So, the effects of $\not{R}_p$ on bounds on r are negligible.

Acknowledgments

I would like to thank the organizers of this workshop. This work was supported in part by the KAIST Basic Science Research Program.

References

1. P. Roy, *Supersymmetric theories with R-parity violation*, talk given at this workshop.
2. See for example, G. Bhattacharyya, Nucl. Phys. Proc. Suppl. **52A**, 83 (1997), G. Bhattacharyya, hep-ph/9709395.
3. J. Jang, J.K. Kim and J.S. Lee, Phys. Rev. **D55**, 7296 (1997), J. Jang, Y.G. Kim and J.S. Lee, Phys. Lett. **408B**, 367 (1997).
4. J. Jang, Y.G. Kim and J.S. Lee, *$B \to X_s l_i^+ l_j^-$ Decays with R-parity Violation*, hep-ph/9711504.
5. T. Nakada, PSI-PR-96-22, hep-ex/9609015. F. Grancagnolo, INFN-AE-90-07. CLEO Collaborations, CLNS-94-1277. J. N. Butler, FERMILAB-PUB-95-363, Nucl. Instrum. Meth. **A368**, 145 (1995), D.E. Kaplan, UW/PT 97-5, hep-ph/9703347.
6. There are interesting bounds on λ''_{112} independent of the lepton number violating couplings in models with a light gravitino or axino, K. Choi, E.J. Chun and J.S. Lee, Phys. Rev. **D55**, 3924 (1997).
7. R. Ammar *et al.*, CLEO Collaboration, Phys. Rev. **D49**, 5701 (1994).
8. CDF Collaboration, Fermilab Conf. 95/201-E(1995).
9. A. Ali, hep-ph/9606324.
10. Y. Grossman and Z. Ligeti, Phys. Lett. **332B**, 373 (1994), J. Kalinowski, Phys. Lett. **245B**, 201 (1990), A.F. Falk, Z. Ligeti, M. Neubert and Y. Nir, Phys. Lett. **326B**, 145 (1994).
11. CLEO Collab., J. Bartlet *et al*, Phys. Rev. Lett. **71**, 4111 (1993), ARGUS collab., Phys. Lett. **255B**, 297 (1991).
12. M. Neubert, CERN-TH/97-19, hep-ph/9702310, J.D. Richman, hep-ex/9701014, and Ref. [9].
13. L3 Collab., M. Acciarri *et al*, Z. Phys. **C71**, 379 (1996).

SMALL INSTANTON CONTRIBUTION TO THE AXION POTENTIAL IN SUPERSYMMETRIC THEORIES

SAEBYOK BAE, KIWOON CHOI, and HYUNGDO KIM

Department of Physics, Korea Advanced Institute of Science and Technology, Taejon 305-701, Korea

In this talk we compute the axion potential induced by small size instantons in various supersymmetric models and examine whether the axion solution is safe or not. It is shown that the resulting high energy axion potential is suppressed by more powers of the soft supersymmetry breaking parameters and/or of the weak scale than what has been anticipated in previous analyses. The axion mechanism turns out to be stable against the effects of small instantons in all models studied here and this appears to be true for a wide class of other supersymmetric models.

1 Introduction

Introducing an axion is an elegant way to solve the strong CP problem. However, there are some classes of theories in which the axion mechanism does not work. If QCD is *not* asymptotically free at some scales above the weak scale M_W, small size QCD instantons may induce a significant high energy axion potential whose minimum does *not* coincide with the minimum of the conventional low energy axion potential. If $V_{\rm HE} \gtrsim 10^{-9} V_{\rm LE}$, the dynamical relaxation mechanism of θ_{QCD} would be spoiled.

The most interesting class of theories in which the effects of small size instantons can be sizable are supersymmetric models. In fact, the small size instanton contribution to the axion potential has been studied before [4,5] for non-supersymmetric models and supersymmetric models based on naive dimensional analysis. We reexamine this problem by computing the axion potential induced by small size QCD instantons in various supersymmetric models. It turns out that the axion solution to the strong CP problem is stable against the effects of small instantons in all models studied in this paper. Our analysis implies that this is also true for a wide class of other supersymmetric models.

Here a parameter which gives the strong CP criterion is introduced. In order for the axion VEV to satisfy the phenomenological bound $\langle a/f_a \rangle \lesssim 10^{-9}$, the high energy axion potential has to be suppressed as $\bar{V}_{\rm HE}/f_\pi^2 m_\pi^2 \lesssim 10^{-9}\gamma^{-1}$, in other words one needs $\delta \equiv \frac{\bar{V}_{\rm HE}}{10^{-9} f_\pi^2 m_\pi^2} \lesssim 1$.

2 Small Instanton Contribution to the Axion Potential

We use spurion field methods to estimate instanton contributions. All soft SUSY breaking parameters are assumed to be of the same order of magnitude as the weak scale. For further details, see [7].

The low energy interactions induced by small QCD instantons can be summarized by

$$\begin{aligned}\mathcal{L}_{\rm eff} &= \int d^2\theta d^2\bar{\theta}\,[e^{-n_1 Y} K_{\rm eff}(\Phi_i, \Phi_i^*; D_\alpha, \bar{D}_\alpha; Z, \bar{Z}^*; \eta) + h.c.] \\ &+ [\int d^2\theta e^{-n_2 Y} W_{\rm eff}(\Phi_i, Z) + h.c.], \end{aligned} \tag{1}$$

where n_1 and n_2 are positive integers corresponding to the instanton winding number and D_α denotes the supercovariant derivative on the superspace. The axion potential can be derived from the above by replacing the spurion Y by $Y + ia/f_a$ and integrating out the Grassmann coordinates θ and $\bar{\theta}$. Let $m_{\rm soft}$ collectively denote the soft mass parameters. Usually any nonzero VEV of the auxiliary F-component of the observable sector field Φ_i is due to the presence of soft SUSY breaking terms and thus it is suppressed by some powers (typically a single power) of $m_{\rm soft}$. It is then obvious that the axion potential arising from $K_{\rm eff}$ is suppressed *at least* by two powers of $\rho m_{\rm soft}$, while the axion potential from $W_{\rm eff}$ is suppressed *at least* by a single power of $\rho m_{\rm soft}$. (Here ρ denotes the size of instantons generating the above effective interactions.)

The low energy effective superpotential is severely constrained by selection rules and the requirement of sensible limiting behavior. *None* of the models studied in this paper allows an effective superpotential satisfying these requirement together with holomorphy. We then identify the terms in the expansion of $K_{\rm eff}$ which would give rise to the leading contribution to the axion potential. Finally the loop suppression factor is determined through the instanton diagram. Among the various models we will restrict ourselves to the NMSSM with 3 generations. All other models and detailed results are in [7]. The next to minimal supersymmetric standard model (NMSSM) has an additional gauge singlet superfield S compared to the MSSM. The NMSSM has global symmetries under which spurions also transform. The small QCD instanton corrections from the Kähler potential and superpotential are given in (1), and the most dominant terms arise when $n_1 = n_2 = 1$ (single instanton).

Let's consider the superpotential part first. The small instanton induced superpotential takes the form:

$$W_{\rm eff} \propto \rho^{-3}{\rm det}(Z_u Z_d) {Z_1}^{N_1} (\rho^2 H_u H_d)^{N_2} (\rho S)^{N_3} {Z_2}^{N_4}. \tag{2}$$

Holomorphy and sensible limiting behavior of coupling constants restrict the N_i ($i = 1, ..., 4$) to be non–negative integers. Global $U(1)_R$ invariance gives another constraint, $N_1 + N_4 + 2 = 0$, which cannot be satisfied for nonnegative integers N_i. This proves that there is no contribution from the superpotential W_{eff}. This is also the case for other models considered here.

There are thus at least two powers of (ρm_{soft}) SUSY suppressions, because the entire correction to the axion potential is obtained from the Kähler potential. The small instanton induced Kähler potential in (1) is given by

$$K_{\text{eff}} \simeq \rho^{-2}(\det Z_u Z_d) {Z_1}^{N_1} (\rho^2 H_u H_d)^{N_2} (\rho S)^{N_3} {Z_2}^{N_4} (\rho D^2)^{N_5} F(\eta). \quad (3)$$

Negative powers of some fields or spurions are expressed through positive powers of their hermitian conjugates in our notation. For example, ${Z_1}^{-1}$ simply means ${Z_1}^*$.

The magnitude of $\bar{V}_{\text{HE}}$ crucially depends on N_2, N_3 and N_5. To see this more clearly, if we replace $\langle H_u \rangle = \langle H_d \rangle = \langle S \rangle = m_{\text{W}}$, the instanton induced axion potential is

$$\bar{V}_{\text{HE}} \simeq e^{-\frac{2\pi}{\alpha(\rho)}} \rho^{-4} (\rho m_{\text{soft}})^{2+|N_5|} (\rho m_{\text{W}})^{2|N_2|+|N_3|} (\det Z_u Z_d) {Z_1}^{N_1} {Z_2}^{N_4}. \quad (4)$$

The instantons are very small, so that higher powers of (ρm_{W}) and (ρm_{soft}) are more strongly suppressed. The solution with the smallest number of $2|N_2| + |N_3| + |N_5|$ gives the dominant contribution, which makes it is easy to find the leading term.

Let's consider the case of three generations. The small instanton induced axion potential is

$$\bar{V}_{\text{NMSSM}}(N_g = 3) \simeq \left(\frac{1}{16\pi^2}\right)^7 e^{-\frac{2\pi}{\alpha(\rho)}} \rho^{-4} (\rho m_{\text{soft}})^3 (\det \lambda_u \lambda_d) {\lambda_1^*}^3 \lambda_2. \quad (5)$$

We can easily extract the loop factor 7 from Feynman diagrams. All other models are analysed in the same way stated above.

We can see the dependence of the potential on the size of instantons ρ from the running behavior of the strong coupling constant α. Final results are given in the table. $\delta = 1$ is the strong CP criterion and $\delta \lesssim 1$ is the requirement for maintaining the axion solution. All the models except one (an E_6 model with intermediate scale, $N_g = 4$) satisfy the strong CP criterion.

4. Conclusions

We studied the stability of the axion solution of the strong CP problem against small size instanton corrections in several supersymmetric models and showed

Model		δ(scale)
MSSM	$N_g = 3$	$10^{-34}(m_{\text{soft}})$
	$N_g = 4$	$10^{-34}(m_{\text{soft}})$
NMSSM	$N_g = 3$	$4 \times 10^{-39}(m_{\text{soft}})$
	$N_g = 4$	$2.5 \times 10^{-41}(m_{\text{soft}}$ or $M_{\text{GUT}})$
	$N_g = 5$	$10^{-19}(M_{\text{GUT}})$
E_6(no M_I)	$N_g = 3$	$2.5 \times 10^{-15}(M_{\text{GUT}})$
	$N_g = 4$	10^{-5} (10^{13} GeV)
$E_6(M_I)$	$N_g = 3$	$2.5 \times 10^{-23}(m_{\text{soft}})$
	$N_g = 4$	$2.5 \times 10^{7}(M_{\text{GUT}})$

Table 1: Strong CP criteria δ

that the axion solution is valid in all these cases. This result strongly suggests that the axion solves the strong CP problem for a wide class of supersymmetric models.

We confirmed that axion is an elegant way to solve the strong CP problem in various supersymmetric models like the MSSM, the NMSSM, E_6 models without intermediate scale, and E_6 models with intermediate scale.

Acknowledgements

HK would like to thank C. Munoz for helpful discussions. This work is supported in part by KOSEF Grant 951-0207-002-2 (KC, HK).

References

1. R. D. Peccei and H. R. Quinn, *Phys. Rev. Lett.* **38**, 77 (1440); *Phys. Rev.* D **16**, 77 (1791).
2. J. E. Kim, *Phys. Rev. Lett.* **43**, 103 (1979).
3. M. Dine, W. Fischler and M. Srednicki, *Phys. Lett.* B **104**, 199 (1981).
4. M. Dine and N. Seiberg, *Nucl. Phys.* B **273**, 109 (1986).
5. J. Flynn and L. Randall, *Nucl. Phys.* B **293**, 731 (1987).
6. K. Choi, H. B. Kim, and J. E. Kim, *Nucl. Phys.* B **490**, 349 (1997).
7. S. Bae, K. Choi, and H. Kim, in preparation.

Plenary Session Program

Friday Morning Session

Time	Speaker	Title
		Chair: J.B. Choi
09:00 – 09:10	M. Kobayashi, B.K. Chung	Opening addresses
		Chair: H.Y. Cheng
09:10 – 09:40	T. Mori	From LEP1 through LEP2 beyond Standard Model
09:40 – 10:10	S. Matsumoto	Status of the Standard Model
10:10 – 10:40	W.G. Li	Results from BEPC
10:40 – 11:00		*Coffee Break*
		Chair: K.S. Kang
11:00 – 11:30	Y.K. Kim	Tevatron results
11:30 – 12:00	I.H. Park	HERA results
12:00 – 12:30	P. Roy	Supersymmetric theories with R-parity violation
12:30 – 12:40		Photo session

Saturday Morning Session

Time	Speaker	Title
		Chair: H.S. Song
09:00 – 09:30	A. Ali	Flavour Changing Neutral Current processes and CKM phenomenology
09:30 – 10:00	Y.J. Kwon	CLEO results
10:00 – 10:30	H.Y. Cheng	Charmless B decays
10:30 – 11:00		*Coffee Break*
		Chair: B.K. Chung
11:00 – 11:30	T. Onogi	Standard Model parameters from Lattice QCD
11:30 – 12:00	C.S. Kim	Recent theoretical progress on B and CP-violation physics
12:00 – 12:30	M. Hazumi	Physics at B factories

Sunday Morning Session

Time	Speaker	Title
		Chair: J.S. Kang
09:00 – 09:30	H. L. Yu	Perturbative QCD analysis of heavy hadron decays
09:30 – 10:00	P. Ko	Recent developments in heavy Quarkonium physics
10:00 – 10:30	K. Hikasa	Physics at the Energy Frontier
10:30 – 11:00		*Coffee Break*
		Chair: Probir Roy
11:00 – 11:30	X. Tata	Supersymmetry phenomenology at colliders and supercolliders
11:30 – 12:00	T. Han	Electroweak Symmetry Breaking Sector at future colliders
12:00 – 12:30	H.T. Wong	Overview of neutrino physics in Taiwan

Sunday Afternoon Session

Time	Speaker	Title
		Chair: W. Namgung
14:00 – 14:45	M. Nakahata	Results from Super-Kamiokande
14:45 – 15:15	S. Pakvasa	Neutrinos
15:15 – 15:45	Y. Ho	Recent progress in astro-particle physics
15:45 – 16:15	S.J. Rey	Phenomenological impact of recent developments in string theory
16:15 – 16:45		*Coffee Break*
		Chair: M. Kobayashi
16:45 – 17:15	M. Peskin	SUSY breaking via Horava-Witten
17:15 – 17:45	K. Choi	Axions and the strong CP problem
17:45 – 18:15	Z. Fodor	Electroweak phase transition
		Chair: J.B. Choi
18:15 – 18:25	K. Hagiwara	Closing remarks

Parallel Sessions Program

Friday Afternoon, Room A: B/Flavor Physics, QCD

Time	Speaker	Title
		Chair: E. Takasugi
14:00 – 14:25	Z.Z. Xing	Leptonic CP violation at B factories: ϵ_B and new physics
14:25 – 14:50	Y.Y. Keum	Probing penguin contributions with the B_d/B_s lifetime ratio
14:50 – 15:10	Y.G. Kim	The parton model vs. the ACCMM model
15:10 – 15:35	M. Ahmady	Non-spectator contributions to $B \to X_s \eta'$ decays
15:35 – 16:00	L.T. Handoko	Determination of CKM matrix from semi–leptonic B decays
16:00 – 16:30		*Coffee Break*
		Chair: W.G. Li
16:30 – 16:50	S.K. Kang	New class of quark mass matrix and flavor mixing matrix
16:50 – 17:20	M. Kuriki	Evidence for the decay $K^+ \to \pi^+ \nu \bar{\nu}$
17:20 – 17:45	T.W. Yeh	$B \to \rho l \nu$ decays: A varying external field approach to QCD sum rules
17:45 – 18:10	C. Liu	Excited heavy mesons from QCD sum rules
18:10 – 18:35	B. Tseng	Nonfactorizable soft gluons in nonleptonic heavy meson decays
18:35 – 19:00	D.K. Hong	Chiral symmetry breaking under an external field

Friday Afternoon, Room B: Collider physics, NRQCD

Time	Speaker	Title
		Chair: M. Drees
14:00 – 14:25	E. Won	Top quark production at D0
14:25 – 14:45	Jake Lee	T–odd $t\bar{t}g$ coupling at the LHC
14:45 – 15:10	A. Bodek	Physics with W and Z bosons and high mass Drell–Yan pairs at CDF
15:10 – 15:35	A. Miyamoto	Status of Linear Collider R&D in Japan
15:35 – 16:05	M. Nojiri	Light sparticle production as probe of the squark mass scale
16:05–16:30		*Coffee Break*
		Chair: M. Nojiri
16:30 – 16:55	S.Y. Choi	Chargino production and decays at LEP2
16:55 – 17:20	R. Szalapski	Searching for new physics via $e^+e^- \to W^+W^-$
17:20 – 17:40	K.Y. Lee	New electroweak tests in and beyond the SM
17:40 – 18:00	S. Kanemura	Enhancement of loop-induced H^+WZ vertex in 2 Higgs doublet model
18:00 – 18:20	S. Baek	Polarized J/ψ production at CLEO
18:20 – 18:40	Jungil Lee	J/ψ production in Z decays at LEP
18:40 – 19:05	K. Sridhar	Aspects of hadronic J/ψ production

Saturday Afternoon, Room A: Lepton physics/Cosmology

Time	Speaker	Title
		Chair: D. Son
14:00 – 14:30	M. Sakuda	Status of the long baseline neutrino oscillation experiment KEK-PS E362
14:30 – 14:50	M. Koike	CP violation in long baseline neutrino oscillation experiments
14:50 – 15:15	J. Sato	CP and T violation effects in long baseline neutrino oscillation experiments
15:15 – 15:40	H.J. Kim	Evidence for neutrino oscillation from the LSND experiment
15:40 – 16:00	J.H. Song	Pulsar velocities from 3-neutrino oscillations in non-adiabatic processes
16:00–16:30		*Coffee Break*
		Chair: S.K. Oh
16:30 – 17:00	K. Kang	Neutrino mass matrix and oscillation
17:00 – 17:20	M.S. Baek	CP violation in τ pair production at e^+e^- colliders
17:20 – 17:50	E. Takasugi	Gravitational waves from a binary

Saturday Afternoon, Room B: SUSY Models/String

Time	Speaker	Title
		Chair: J.H. Kim
14:00 – 14:30	C. Muñoz	String phenomenology
14:30 – 14:55	N. Haba	Model building using SUSY QCD
14:55 – 15:15	E.J. Chun	Gauge unification and dynamical SUSY breaking
15:15 – 15:40	M. Yamaguchi	Supersymmetry breaking in M theory
15:40 – 16:00	H.B. Kim	Cosmology of moduli lighter than the weak scale
16:00–16:30		*Coffee Break*
		Chair: H.L. Yu
16:30 – 16:55	S. Nam	M–Theory and dynamics of supersymmetric gauge theories
16:55 – 17:15	M. Guchait	Implications of the HERA events for the R–parity breaking SUSY signal at the Tevatron
17:15 – 17:35	J.S. Lee	R–parity violation in B decays
17:35 – 17:55	H. Kim	Small size instanton contributions to the Axion potential in supersymmetric theories

List of Participants

Ahmady, Mohammad (RIKEN, Tokyo) ahmady@riken.go.jp
Ali, Ahmed (DESY, Hamburg) ali@x4u2.desy.de
Bae, Mi-young (Sung Kyun Kwan U., Seoul) mybae@hep.skku.ac.kr
Baek, Min-Seok (Yonsei Univ., Seoul) msbaek@theory.yonsei.ac.kr
Baek, Seungwon (Seoul Natl. Univ.) swbaek@phya.snu.ac.kr
Bodek, Arie (Univ. of Rochester) bodek@urhep.pas.rochester.edu
Chay, Junegone (Korea Univ., Seoul) chay@kupt.korea.ac.kr
Cheng, Hai-Yang (Academia Sinica, Taipei) phcheng@ccvax.sinica.edu.tw
Choi, Jong B. (Chonbuk Natl. Univ., Chonju) jbchoi@moak.chonbuk.ac.kr
Choi, Kiwoon (KAIST, Taejon) kchoi@chep6.kaist.ac.kr
Choi, Seong-Youl (Yonsei Univ., Seoul) sychoi@theory.yonsei.ac.kr
Choi, So Myeong (Yonsei Univ., Seoul) choi@phya.yonsei.ac.kr
Choi, Soo-Kyung (Gyeongsang Univ., Kyungnam) choi@hep1.gsnu.ac.kr
Choi, Young-Il (Sung Kyun Kwan U., Seoul) yichoi@yurim.skku.ac.kr
Chun, Eung Jin (KIAS, Seoul) ejchun@supy.kaist.ac.kr
Drees, Manuel (APCTP, Seoul) dress@apctp.kaist.ac.kr
Fodor, Zoltan (KEK, Tsukuba) fodor@thwgs1.cern.ch
Guchait, Monoranjan (Yonsei Univ., Seoul) guchait@theory.tifr.res.in
Haba, Naoyuki (mie Universty) haba@phen.mie-u.ac.jp
Hagiwara, Kaoru (KEK, Tsukuba) kaoru.hagiwara@kek.jp
Han, Tao (UC Davis) than@ucdhep.ucdavis.edu
Handoko, Laksana Tri (hiroshima U.) handoko@theo.phys.sci.hiroshima-u.ac.jp
Hazumi, Masahashi (Osaka U.) hazumi@champ.hep.sci.osaka-u.ac.jp
Hikasa, Ken-ichi (Tohoku Univ., sendai) hikasa@tuhep.phys.tohoku.ac.jp
Ho, Yunsil (Columbia Univ., New York) ho@nevis.columbia.edu
Hong, Deog Ki (Pusan Natl. Univ.) dkhong@hyowon.cc.pusan.ac.kr
Hwang, Dae Sung (Sejong Univ., Seoul) dshwang@phy.sejong.ac.kr
Hwang, Soung Soon (Yonsei Univ., Seoul) sshwang@theory.yonsei.ac.kr
Jang, Dong-wook (Yonsei Univ., Seoul) jang@hep1.yonsei.ac.kr
Jang, Hyun Ki (Seoul natl. Univ.) top@hep1.snu.ac.kr
Jeong, Kyoungkeun (Yonsei Univ., Seoul) kkjeong@theory.yonsei.ac.kr
Kanemura, Shinya (KEK, Tsukuba) kane@theory.kek.jp
Kang, Kyungsik (brown Univ., Providence) kang@mail.het.brown.edu
Kang, Sin Kyu (KIAS, seoul) skkang@chep6.kaist.ac.kr
Keum, Yong-Yeon (APCTP, Seoul) keum@apctp.kaist.ac.kr
Kim, Chang Hwan (Seoul Natl. Univ.) chkim@hep1.snu.ac.kr

Kim, Choong Sun (Yonsei Univ., Seoul)
Kim, Donguk (Sung Kyun Kwan U., seoul)
Kim, Hang Bae (KAIST, Taejon)
Kim, Hee-Jong (Yonsei Univ., Seoul) heejong@hep1.yonsei.ac.kr
Kim, Hong Joo (Seoul natl. Univ.) hjkim@hep1.snu.ac.kr
Kim, Hyungdo (KAIST, Taejon) hdkim@supy.kaist.ac.kr
Kim, Inbo (Sogang Univ., seoul) ibkim@phycmt1.sogang.ac.kr
Kim, Seyong (KIAS, Seoul) skim@ctp.snu.ac.kr
Kim, Yeong Gyun (KAIST, Taejon) ygkim@chep6.kaist.ac.kr
Kim, Yoonbai (Sung Kyun Kwan U., Seoul) yoonbai@cosmos.skku.ac.kr
Kim, Young-Kee (UC Berkeley) ykkim@fnald.fnal.gov
Ko, Pyungwon (KAIST, Taejon) pko@chep6.kaist.ac.kr
Kobayashi, Makoto (KEK, Tsukuba) kobayath@theory.kek.jp
Koike, Masao (KEK, Tsukuba) Koike@icrr.u-tokyo.ac.jp
Kou, Emi (Ochanomizu Univ., Tokyo) kou@fs.cc.ocha.ac.kr
Kuriki,Masao (KEK, Tsukuba) mkuriki@insuty.tanashi.kek.jp
Kwon, Youngjoon(Yonsei Univ., Seoul) kwon@phya.yonsei.ac.kr
Lee, Haeshim (Chungnam Natl. Univ.) hslee@nsphys.chungnam.ac.kr
Lee, Jae Sik (KAIST, Taejon) jslee@chep6.kaist.ac.kr
Lee, Jake (Yonsei Univ.,Seoul) jslee@theory.yonsei.ac.kr
Lee, Kang Young (KAIST, Taejon) kylee@chep5.kaist.ac.kr
Li,Weiguo (IHEP, Beijing) liwg@sun.ihep.ac.cn
Liu, Chun (Seoul Nat'l. Univ.) liuc@ctp.snu.ac.kr
Matsumoto, Seiji (CERN, Geneva) seiji.matsumoto@cern.ch
Miyamoto, Akiya (KEK, Tsukuba) akiya.miyamoto@kek.jp
Mori, Toshinori (ICEPP, Tokyo) mori@icepp.s.u-tokyo.ac.jp
Muñoz, Carlos (KAIST and Madrid Univ.) cmunoz@chep6.kaist.ac.kr
Myung, Seong Sook (korea Univ., Seoul) myung@hep.korea.ac.kr
Nakahata, Masayuki (kamioka Observ.) nakahata@icrr.u-tokyo.ac.jp
Nam, Jiwoo (sung Kyun Kwan U., Seoul) jwnam@hep.skku.ac.kr
Nam, Soonkeon (kyunghee Univ., Seoul) nam@nms.kyunghee.ac.kr
Nojiri, Mihoko (YITP, Kyoto) nojiri@yukawa.kyoto-u.ac.jp
Onogi, Tetsuya(Hiroshima U.) onogi@theo.phys.sci.hisroshima-u.ac.jp
Pahk, Sung-Min(Yonsei Univ., Seoul)smpahk@npl1.yonsei.ac.kr
Pakvasa, Sandip(U.Hawaii, Honolulu)pakvasa@uhheph.phys.hawaii.edu
Park, Il H.(Kyungpook Natl., Taegu)ipark@knuhep.kyungpook.ac.kr
Park, Kee-Sung(Seoul Natl.Univ.)kspark@hep1.snu.ac.kr)
Peskin, Michael(SLAC, Stanford) mpeskin@slac.stanford.edu
Rey, Soo-Jon (Seoul Natl. Univ.) sjrey@phya.snu.ac.kr
Ro, Sangryul (Sung Kyun Kwan U., Seoul) sro@hep.skku.ac.kr

Roy, Probir (TIFR, Mumbai) probir@theory.tifr.res.in
Sakuda, Makoto (KEK, Tsukuba) sakuda@kekvax.kek.jp
Sato, Joe (Tokyo University) joe@hep-th.phys.s.u-tokyo.ac.jp
Song, Hee Sung (Seoul Natl. Univ.) hssong@ctp.snu.ac.kr
Song, JeongHyeon (Seoul Natl. Univ.) jhsong@ctp.snu.ac.kr
Sridhar, K. (TIFR, Mumbai) sridhar@theory.tifr.res.in
Sul, BuKyung (Seoul Natl. Univ.) sul@zoo.snu.ac.kr
Szalapski, Rob (KEK, Tsukuba) robs@theory.kek.jp
Takasugi, Eiichi (Osaka U.) takasugi@phys.wani.osaka-u.ac.jp
Tata, Xerxes (U. Hawaii, Honolulu) tata@uhheph.phys.hawaii.edu
Tetsuya, Takaishi (Hiroshima University) tt-taka@hue.ac.jp
Tseng, Benjamin (Cheng-Kung U., Tainan) btseng@ibm65.phys.ncku.edu.tw
Won, Eunil (Seoul Natl. Univ.) eiwon@hep1.snu.ac.kr
Wong, henry Tsz-King (Academia Sinica, Taipei) htwong@phys.sinica.edu.tw
Xing, Zhi-Zhong (Nagoya U.) xing@eken.phys.nagoya-u.ac.jp
Yamaguchi, Masahiro (Tohoku U., Sendai) yama@tuhep.phys.tohoku.ac.jp
Yee, Sooman (Soonchunhyang U., Chungnam) smyyee@asan.sch.ac.kr
Yeh, Tsung-Wen (Academia Sinica, Taipei) kcyang@phys.sinica.edu.tw
Yu, Chae Hyun (Seoul Natl. Univ.) freman@phya.snu.ac.kr
Yu, Hoi-Lan (Academia Sinica, Taipei) phhlyu@ccvax. Sinica.edu.tw